GEMOLOGY

GEMOLOGY

CORNELIUS S. HURLBUT, JR., Ph.D.

PROFESSOR EMERITUS OF MINERALOGY
HARVARD UNIVERSITY

GEORGE S. SWITZER, Ph.D.

CURATOR EMERITUS OF MINERALOGY
SMITHSONIAN INSTITUTION

A WILEY-INTERSCIENCE PUBLICATION

JOHN WILEY & SONS, New York • Chichester • Brisbane • Toronto • Singapore

Library of Congress Cataloging in Publication Data

Hurlbut, Cornelius Searle, 1906–
Gemology.

"A Wiley-Interscience publication."
Includes bibliographical references and indexes.
1. Precious stones. I. Switzer, George S., joint author. II. Title.

QE392.H87 553'.8 78-13262
ISBN 0-471-42224-X

Printed in the United States of America

10 9 8 7 6 5

A Gemstone

A gemstone is a lovely thing.
It gleams and sparkles in a ring
And makes a pin a wondrous sight
By playing magic with the light.

It makes one very proud to wear
A flashing diamond solitaire,
Or garnet with a subtle hue,
Or emerald green, or sapphire blue.

But would it not mean more to you
To learn why it is green or blue
And make you very pleased to know
The reason why it sparkles so?

I think it would mean even more
To know its history and lore
And was it part of nature's plan
Or from a furnace, made by man.

C.S.H.

PREFACE

With the rise in living standards during the present century, gems, once restricted to the wealthy, have come within the reach of many. The resulting increased interest in gems and jewelry has paralleled, if not created, a rapid growth in the hobbies of gem cutting, jewelry making, and mineral collecting. At the same time people in the jewelry industry and the gem trade have recognized the need for greater technical knowledge of gems. This increased awareness of gems at all levels has led to the publication of many books on gems and related subjects. Some of these are comprehensive reference works; others are highly specialized, concerned with a narrow field such as description of a great collection, a single gem material, gem cutting, or jewelry making. In addition many "coffee table" books have appeared with little text and featuring large numbers of color photographs of minerals, gems, and jewelry.

Despite this flood of material, no book has appeared with a textbook approach, that is, a concise yet comprehensive and systematic treatment. It is to fill this gap that we have prepared this volume. Some of the material is by nature technical, but an attempt has been made to present it in a manner understandable to anyone with a modest scientific background. As a result, we believe that the book will be useful not only to the dealer, lapidary, and jeweler but also to the hobbyist who, through increased knowledge, will acquire a greater appreciation of gems. Furthermore, it is hoped the book will serve as an up-to-date textbook in gemology for college courses, adult education classes, and correspondence courses.

To understand all but the most superficial descriptions of gems, one should have a knowledge of the properties minerals possess that make them gems and by which they are characterized. Thus the order of material as presented in this book is from the general to the specific. The second chapter, dealing with the genesis and occurrence of gem minerals, is followed by chapters on the chemical, crystallographic, physical, and optical properties of gems and the methods and instruments used in their determination. Brief chapters deal with gem inclusions and imperfections, synthesis, imitation, and cutting and polishing. Chapter 13, the main portion of the latter part of the book, is devoted to descriptions of about 100 individual gems. The list could be considerably expanded to include many minerals that, as curiosities, have been fashioned into gems but that have only a few such representatives. We feel that their inclusion would considerably lengthen the book while adding little basic information.

Chapter 14 comprises two determinative tables. In the first table gems are listed in order of increasing refractive index; in the second they are arranged in order of increasing specific gravity and cross-referenced to the first table, where more data are given. There are both a subject and a gem index to facilitate obtaining commonly sought information; the name in the gem index is followed by the chemical composition, crystal system, and optical and physical properties.

In writing this book we have drawn freely on the advice and counsel of numerous people. We owe particular thanks to Pete J. Dunn of the Smithsonian Institution for his careful and critical reading of the complete manuscript, and to Kurt Nassau of Bell Laboratories for reading the chapter on gem synthesis and supplying photographs of synthetic gems. The helpful cooperation of many individuals and corporations in furnishing illustrative material is gratefully acknowledged. Special thanks must go to A. Ruppenthal of Idar-Oberstein, West Germany, for allowing us to reproduce their superb photographs in six of the color plates. Above all we are indebted to the Gemological Institute of America and particularly to its President, Richard T. Liddicoat, for encouragement and help in all phases of preparation of this book.

Cornelius S. Hurlbut, Jr.
George S. Switzer

Cambridge, Massachusetts
Washington, D.C.
September 1978

CONTENTS

GEMOLOGY

1 INTRODUCTION

Because gemstones are traditionally minerals, their history is intimately and inextricably associated with the history of minerals. Early man undoubtedly discovered by accident that quartz and flint could be broken to produce sharp cutting edges that could be used as tools or weapons. It was through the use of minerals that he began his long journey to civilization. Throughout the following evolutionary period, perhaps as much as a million years, the art of shaping the raw materials into specialized tools became more refined. In searching for suitable materials, the toolmaker no doubt came upon pleasingly colored stream pebbles or flashing crystals embedded in a rocky ledge. Although of no utilitarian value, these first gemstones were recovered and retained as things of beauty.

Early Gems

As the Stone Age gave way to the Bronze Age, other minerals were sought from which metals could be derived, and at the same time interest in gem minerals increased. Paintings in tombs constructed 5000 years ago in ancient Egypt show the advance in the mineralogical arts. They depict craftsmen smelting ores for metals, weighing gem materials, and fashioning gems from lapis lazuli and malachite. Far more compelling evidence is given by Egyptian necklaces, called broad bead collars, dating from the IV Dynasty. The necklace shown in Figure 1.1 is from the VI Dynasty, 2345–2181 B.C. The cylindrical beads are mostly blue faīence but some are turquoise. In the single strand at the top are rounded beads of gold, lapis lazuli, carnelian, and turquoise.

From the written record as well as from archaeological finds, we learn that during the ancient Greek and Roman civilizations many minerals were used as gems and the art of carving them reached a high degree of proficiency.

The early gems were valued and used much as they are today for personal adornment. But of equal or greater importance to the wearer were their presumed magical powers. To each mineral was attributed a certain virtue and thus gemstones were worn as amulets or talismans. A few of the powers as-

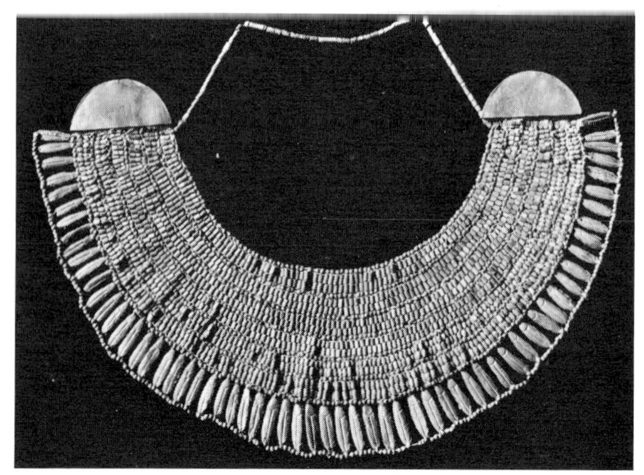

FIG. 1.1. Egyptian necklace from tomb of Impy, VI Dynasty Old Kingdom. Courtesy of Museum of Fine Arts, Boston, Massachusetts.

cribed to them were to make one invulnerable, to prevent disease, to render one invisible, to produce sleep, to endow one with wisdom, and to prevent drunkenness. These superstitions not only persisted well into the Christian era but were added to and elaborated on. Remnants of these beliefs linger today and are found in lucky charms, birthstones, and religious symbols.

What Is a Mineral?

Since most gem materials are minerals, it is important at the beginning of a study of gemology to understand the term *mineral*. It is impossible to give a brief definition that is acceptable to all mineralogists. Yet with minor qualifications most of them agree with the following: *a mineral is a naturally occurring, crystallized chemical element or compound having either a definite chemical composition or range in composition, and usually formed as a product of inorganic processes.* Let us consider each of the parts of this definition and point out the exceptions or qualifications.

Naturally occurring implies that minerals are products of nature. However, many compounds that are found as minerals have been purposely synthesized by man and have identical properties. Strictly these are not minerals but may be called synthetic or man-made minerals to distinguish them from natural products.

A *crystalline chemical element or compound* implies that minerals are solids composed of atoms or ions arranged in a regular three-dimensional pattern. Such solids have compositions expressible by chemical formulas. Some minerals have definite compositions; for example, quartz (rock crystal and amethyst) is silicon dioxide, SiO_2, and corundum (ruby and sapphire) is aluminum oxide, Al_2O_3. Others have a range in composition in which atoms of one element substitute for atoms of another element in the crystal structure. For example, the mineral group olivine has as one member pure magnesium silicate, Mg_2SiO_4; but iron usually substitutes for some magnesium as in the gem variety peridot, and the formula is written $(Mg,Fe)_2SiO_4$.

The restriction that minerals be crystalline excludes liquids and noncrystalline solids, although they are like minerals in chemistry and occurrence. In a rigorous classification such substances are called *mineraloids*. Although opal is essentially noncrystalline and thus falls in this group, it is treated in the text as a mineral.

There are a few natural noncrystalline solids called *metamict minerals* that are classed as mineraloids. When originally formed these were crystalline, but the crystal structure has been destroyed by radiation from radioactive elements. The only important metamict mineral used as a gem is one type of zircon.

Usually formed as a product of inorganic processes implies that substances formed under biogenic processes are normally excluded. A few natural substances can originate through both organic and inorganic agencies with the two types having identical properties. Such organically produced materials are here considered minerals. The two most important gem materials containing a mineral having this dual origin are pearl and mother-of-pearl.

What Is a Gemstone?

The term *gemstone* is applied to those materials, usually after they are cut and polished, that are sufficiently attractive to be used for personal adornment. The modifiers *precious* and *semiprecious* have been used to distinguish the more valuable from the less valuable gems. Consistently diamond, emerald, ruby, and sapphire have been classed as precious stones, but at times opal, pearl, and alexandrite also have been considered precious. Since there is no rigid set of criteria that separate gems of great value from those of less value, the term semiprecious should be abandoned and all gems referred to as precious.

The various cuts of gemstones are discussed in Chapter 12. Because reference is made to them in earlier chapters it is desirable that the reader be aware at the outset of the different cuts and the nomenclature used to describe them. There are two basic types, the cabochon cut and the faceted cut. There are various cabochon cuts but the most common has a smooth domed top and a flat base (Fig. 1.2).

Faceted cut gemstones are bounded by smooth plane surfaces (facets) to which different names are given depending on their positions on the stone. Although faceted gems have different shapes, the brilliant cut is most common for diamonds (Fig. 1.3). The top of the stone is called the *bezel, crown,* or *top* (Fig. 1.3*a*); the lower part of the stone is called the *pavilion, base,* or *back* (Fig. 1.3*b*). The top facet (usually the largest) is the *table, T;* the small facet at the bottom of the stone parallel to the table (which may not be present) is the *culet, C.* The edge between the bezel and pavilion is the *girdle,* G (Fig. 1.3*c*).

Of the approximately 2500 known minerals about 70 possess one or more of the qualifications that place them in the special category of gemstones. Most members of this relatively small group are rare, and only about 20 are commonly encountered as gems.

The one attribute common to all gems is *beauty*. That beauty is in the eye of the beholder is as true with gems as with many other things. To one, beauty may be in the color; to others it may be in luster, transparency, or brilliancy and fire resulting from skillful cutting by the lapidary. The beauty of most gems results from a combination of two or more of these properties, but in some nontransparent stones, such as turquoise and malachite, it rests in color alone. All the factors contributing to beauty are found in a few precious stones. But the colored diamond stands preeminent, for it not only possesses them all but to a high degree.

FIG. 1.2. Cross section of cabochon cut gemstone with domed top and flat base.

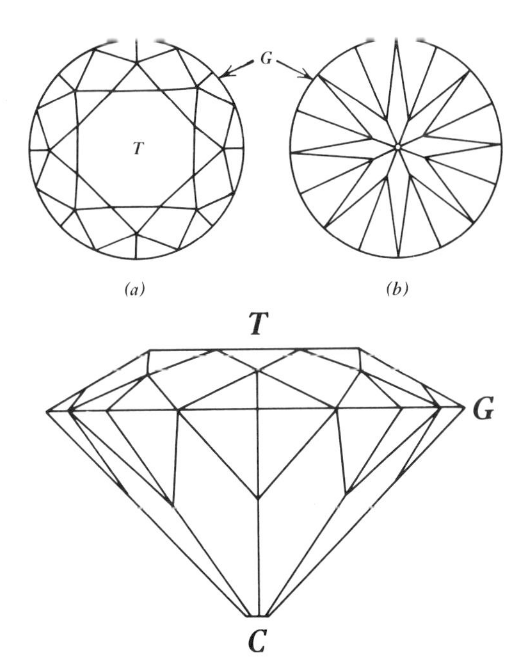

FIG. 1.3. A faceted gemstone (brilliant cut) with 58 facets. (*a*) The bezel showing the top facets. (*b*) The pavilion showing the bottom facets. (*c*) Front view showing the table (*T*), culet (*C*), and girdle (*G*).

Many museums exhibit faceted stones cut from minerals that have the gemstone qualification of beauty but are rarely seen in jewelry. A major reason for this is that these minerals are so soft that if worn in a ring or even in a brooch, they would soon become scratched and dull and lose their attractiveness. *Durability* is thus the second requisite of a gem mineral and depends on two physical properties, hardness and toughness. The resistance a mineral offers to scratching or abrasion is known as its hardness. In general, the harder the mineral, the more desirable it is as a gemstone. Diamond, the hardest of substances, can be scratched only by another diamond. Near the other end of the hardness spectrum is gypsum, which, although it too can be fashioned into attractive stones, is so soft it can be scratched by a fingernail. The high hardness of diamond, ruby, sapphire, and emerald is a major factor in making them the most prized of gems. However, some hard minerals are brittle and can be chipped or fractured if struck a sharp blow, whereas other minerals of less hardness may be tough and resist fracturing. The latter case is particularly true of nephrite jade, which is composed of a felted aggregate of fibrous crystals. The blow that would shatter a diamond would have little effect on the far softer jade. Thus jade owes its durability to toughness rather than hardness.

Durability is not necessarily an overriding factor in determining the worth of a gemstone. Such is the case with opal, which, although relatively soft, is held by many in as high esteem as the harder stones mentioned above. This is because of its incomparable beauty resulting from an internal display of colors.

Quartz, one of the most abundant minerals on the earth's surface, is ever present as tiny particles in the air we breathe and in the dust that settles on our clothing. Therefore, ideally a gem should be harder than quartz, a qualification met by only 10 or 12 gemstones. This does not mean that others are not highly desirable but they must be worn with more care, particularly when set in a ring. However, in other forms of jewelry such as a brooch or pin, gemstones are subjected to little abrasion and gems mounted in this way, even lacking the hardness of quartz, can be worn for many years and retain their brilliance.

Beauty and durability are the two inherent properties of a mineral on which its use as a gem depends. Compared to them other qualifications are superficial and depend mostly on the whims of man. Since it is human nature to desire something rare, *rarity* is another criterion of a valued gem. If there is only a limited quantity of a gem mineral reaching the market, the price is high and because of this the gem is considered desirable by the few who can afford it. With the discovery of new sources, resulting in a great increase in supply, the price falls. Although its beauty and durability are the same, it becomes less desirable, for now it is within the reach of many. An interesting example is amethyst, the lovely purple variety of quartz. For hundreds of years the supply, mostly from European sources, was small and it remained a highly prized gem. But with the discovery of large deposits of high-quality amethyst in South America its value dropped and it has never regained its esteemed position.

The demand for and hence the value of certain gems fluctuate with the *fashion* of the times. For example, a dark red variety of garnet found in Bohemia was extremely popular during the nineteenth century and is seen in many pieces of jewelry dating from that time. Today the same garnets are available but have fallen from favor and are little used. An opposite trend has developed in recent years for turquoise. An upsurge of interest in turquoise-set Navaho Indian jewelry has resulted in an enormous increase in price.

Portability is another factor that should be added to the desirable properties of gems. This is particularly appealing for people living in politically unstable countries from which they may be forced to flee on short notice. For gems, more than any other commodity, offer high value combined with small volume and weight and maintain their value in times of depressed currency and political upheaval.

Although quite unrelated to the intrinsic properties of a gem, rarity, fashion, and portability play a large part in determining its value.

Synthetic Minerals and Gemstones

Because of the high value placed on gemstones, man has long been intrigued with the thought of duplicating natural stones in the laboratory. The countless early efforts of synthesis either met with complete failure or produced tiny crystals, too small for cutting. However, at the beginning of the twentieth century a process was developed by which corundum (ruby and sapphire) and spinel were grown large enough to be cut into faceted stones. With advancing technology following World War II other gem minerals have been manufactured. These include emerald, rutile, quartz of several colors, and most recently, chrysoberyl and opal. Although several experimenters had claimed to have made diamonds, the first authenticated diamond crystals were produced in 1955. They were tiny but with improved techniques larger and larger crystals have been made. In all probability the future will see synthetic diamonds competing with the natural in the jeweler's window.

These man-made gemstones are not "imitations" or "fakes" but have the same chemical compositions and physical properties as the natural minerals. One of the major tasks of the gemologist, and frequently a difficult one, is to determine whether a given gem is natural or synthetic. Such determinations may be very important, for the rare natural gem may have a value several hundred fold greater than its abundant synthetic rival.

By the techniques of mineral synthesis, several compounds have been manufactured that are not found as minerals that have some of the qualifications of gemstones. These materials as well as glass imitations are common, and the gemologist must be familiar with their properties to distinguish them from the natural gemstones they may resemble.

Gemology and Mineralogy

Gemology is a relatively new science that is so deeply rooted in mineralogy it is impossible to draw a line between them. The gemologist can be considered a mineralogist specializing in the gem minerals. In the broad sense these include, in addition to the minerals cut into gemstones, the natural ornamental materials used for carvings and decorative purposes.

Mineralogy is not a fundamental science but a synthesis of chemistry, physics, and geology. As a specialized mineralogist, the gemologist must draw on these sciences for a complete understanding of his subject. The physics and chemistry of minerals are intimately related, for the complex physical and chemical properties that characterize a mineral depend on the chemical composition and the arrangement of the atoms in the crystal structure. From geology comes an understanding of the physicochemical conditions under which minerals form. Furthermore, only with a knowledge of geology can there be an intelligent approach to prospecting for gems or an appreciation of the problems of extracting them from the enclosing rock.

Although similar equipment and similar methods of study are used by both mineralogist and gemologist, there is a fundamental difference in approach. The mineralogist can scratch, powder, or dissolve in acid the objects of his study with impunity. The gemologist presented with a valuable faceted stone is restricted to nondestructive tests and measurements. To gather sufficient data to identify such a gemstone, the gemologist has developed special instruments and techniques. In the text these devices and methods are emphasized.

2 ORIGIN AND OCCURRENCE OF GEMSTONES*

Although gemology is a relatively new science, gems have been known for a long time. Ever since the first chance discoveries by early man, gems have been eagerly sought and highly valued. Gemstones are traditionally minerals; that is, they are elements or inorganic compounds formed as products of nature. With advancing technology, man has been able to manufacture substances that are used as gems, some identical, or nearly identical, to minerals and others that have no natural counterparts. Therefore, a complete discussion of the origin of gemstones must include the methods used by man as well as the earth processes that have been working throughout geologic time. This chapter gives a brief account of the origin and occurrence of gemstones in nature. Man-made gems are considered in Chapter 10.

To discuss the origin of gemstones is to discuss the origin of minerals, for all the major processes by which minerals form have somewhere given birth to a gem. Furthermore, many minerals originate in a variety of ways, and it is the exceptional mineral for which a single explanation can account for all occurrences. So it is with gems. But in the limited space available, only the principal processes responsible for gem formation are considered.

Only six of the approximately 2500 minerals are universally abundant. These major rock-forming minerals or mineral groups that make up 95% of the earth's crust are listed in Table 2.1. With the exception of quartz, the names in the table are of mineral groups, each of which has several members that differ chemically from each other. But all these common and abundant minerals, and quartz as well, have varieties which, because of unusual conditions of formation, occur in gem quality material. An exception is mica. This soft, platy mineral lacks gem qualifications, but even it is sometimes fashioned into ornamental objects.

The materials available for mineral formation are the 92 elements found on earth of which only eight make up nearly 99% of the rocks. About half

* Much of the material in this chapter has been drawn from an article, *Gemstone Genesis*, by C. S. Hurlbut, Jr. that appeared in Jewelers' Circular-Keystone, October 1972/Part II and is reprinted with permission of the publisher.

Table 2.1
MAJOR ROCK-FORMING MINERALS

Mineral	% of Earth's Crust	Principal Elements
Feldspar	60	Na, K, Ca, Al, Si, O
Quartz	13	Si, O
Amphibole and pyroxene	17	Mg, Fe, Ca, Na, Al, Ti, Mn, Si, O
Mica	4	K, Mg, Fe, Al, Si, O
Olivine	1	Mg, Fe, Si, O

of the gem minerals are also composed wholly of these same eight elements. If we expand the list to the 20 most abundant elements, all but six are included that are major constituents of minerals normally considered as gems. These 20 elements are listed in Table 2.2 in order of decreasing abundance in the earth's crust. Also listed are a few of the rarer elements either present in gem minerals or otherwise of interest to the gemologist.

When one of the rarer elements is present in more than trace amounts, it indicates that some natural process has brought about a concentration. Such concentrations do occur, resulting in rare and unusual minerals, and their explanation forms a fascinating chapter in the science of geochemistry. Gem varieties of some of these minerals, although rare, may in places occur in appreciable amounts. Thus in considering the origin of gem minerals we must account for both the unusual conditions under which the common elements are brought together and the significant concentrations of minor elements.

Gems and Common Rocks

All rocks of the earth's crust belong to one of three major types: igneous, sedimentary, and metamorphic. Igneous rocks form by the crystallization of minerals from a hot molten mass known as magma. Sedimentary rocks are composed of materials resulting from the mechanical or chemical breakdown of preexisting rocks. Deposited in layers, the fragmented mineral particles and chemically precipi-

Table 2.2
ABUNDANCE OF ELEMENTS IN THE EARTH'S CRUST

The 20 Most Abundant Elements			Some Less Abundant Elements		
Element	Symbol	Wt %	Element	Symbol	Wt %
Oxygen	O	46.60	Nickel	Ni	0.008
Silicon	Si	27.72	Zinc	Zn	0.0065
Aluminum	Al	8.13	Copper	Cu	0.0045
Iron	Fe	5.00	Lithium	Li	0.0030
Calcium	Ca	3.63	Cobalt	Co	0.0023
Sodium	Na	2.83	Lead	Pb	0.0015
Potassium	K	2.59	Tin	Sn	0.0003
Magnesium	Mg	2.09	Boron	B	0.0003
Titanium	Ti	0.44	Beryllium	Be	0.0002
Hydrogen	H	0.14	Uranium	U	0.0002
Phosphorus	P	0.10	Silver	Ag	0.00001
Manganese	Mn	0.09	Platinum	Pt	0.0000005
Fluorine	F	0.07	Gold	Au	0.0000005
Sulfur	S	0.052			
Strontium	Sr	0.045			
Barium	Ba	0.040			
Carbon	C	0.032			
Chlorine	Cl	0.020			
Chromium	Cr	0.020			
Zirconium	Zr	0.016			

tated substances become *rock* when firmly cemented or compacted. Metamorphic rocks were at one time igneous or sedimentary rocks but have changed in texture and/or mineral composition. The changes are usually brought about by high temperatures and high pressures resulting either from burial deep within the earth's crust or from the intrusion of an igneous magma.

Whether igneous, sedimentary, or metamorphic every rock, with a few exceptions such as recent lava flows, either formed, or at one time was buried, beneath a great thickness of overlying rock. The agents of weathering and erosion working throughout geologic time have stripped thousands of feet of rock from the continental areas of the earth to expose the rocks we now find at the surface.

Igneous Rocks

It is now generally believed that the earth had its beginning about 4½ billion years ago as a hot, fiery sphere of molten material. As heat gradually escaped into space, minerals crystallized and eventually the surface of the primitive earth was composed of an interlocking mass of crystals forming a thin layer of igneous rock. In the beginning, then, all rocks were igneous and the other rock types we now recognize were derived from them.

The name igneous (from the Latin for "fire") is applied to any rock that has formed by a cooling magma. If such a molten mass is deep within the earth's crust, it cools very slowly and the atoms of which it is composed have ample time to migrate through the melt to form relatively large crystals resulting in a coarse-grained *intrusive* rock. If, however, a magma is poured onto the earth's surface as a lava flow, it cools quickly, many tiny crystals form, and a fine-grained *extrusive* or *volcanic* rock results. Very rapid cooling of a lava permits no time for crystallization and a glass is formed. Obsidian is a black volcanic glass used as a gem material.

Igneous rocks vary in mineral composition from place to place, reflecting the chemical compositions of the magmas from which they crystallized; to them have been given a host of names. To add to the complexity of the nomenclature, a fine-grained extrusive rock is given a name different from its coarse-grained analogue. Most igneous rocks have as major constituents one or more of the minerals quartz, feldspar, nepheline, mica, amphibole, pyroxene, and olivine. It is by the relative amounts of certain of these rock-forming minerals and by the presence or absence of others that igneous rocks are classified. The commonest intrusive is granite, a light-colored, coarse-grained rock, composed essentially of feldspar, quartz, and mica. Basalt, the most abundant extrusive rock, is fine-grained and black and made up essentially of feldspar and pyroxene (with or without olivine). Although granite and basalt are the most important igneous rocks, some gemstones are genetically related to other types that will be mentioned in the occurrences of specific gem minerals.

Extrusive Rocks

There are some lava flows that contain a few large crystals disseminated through an otherwise fine-textured rock. It is believed that these crystals formed while the magma was cooling slowly at great depth, and had it remained there, a coarse-grained rock would have resulted. However, because of earth movements the still largely molten material was forced quickly upward to erupt on the earth's surface. The already formed crystals then became frozen in a fine matrix. When the common mineral olivine is found in this way, it may be plucked from the enclosing lava. Such is the occurrence of peridot crystals in the Navaho Indian country of New Mexico and Arizona. In a few places early formed crystals have been liberated from the rock by a gentle weathering process to lie loose on the surface. In Kenya, moonstone (as crystals of feldspar) may be picked up from the surface of a weathering lava flow.

In general we cannot look to the primary minerals of extrusive rocks as a significant source of gems. However, when a lava is rapidly brought to the earth's surface, the reduced pressure causes the contained gases to expand, forming cavities. Later, solutions moving through the rock may deposit minerals in the open spaces. Such mineral-lined cavities are called *geodes*. Chalcedony, fine-grained quartz, is the commonest geode-forming mineral. Deposited in concentric layers as agate it may completely fill or only partially fill the cavity. If only partially filled, crystals of quartz may project into the hollow center (Fig. 2.1). Although the quartz is usually colorless, some may be amethystine. Some of the finest-quality amethyst crystals occur in this way. Other crystals, but rarely of gem minerals, may also be found in the central cavity. Geodes are not confined to volcanic rocks but also are found in cavities in some sedimentary rocks.

Intrusive Rocks

In contrast to extrusive rocks, the processes that give rise to the coarse-grained igneous rocks play a major role in the formation of several important gemstones. In intrusive rocks, in addition to the rock-forming minerals of which they are largely composed, are other minerals in minor amounts. These *accessory minerals,* which play no part in rock classification, may crystallize from the magma to form large enough and perfect enough crystals to make gems. Some zircons, sapphires, and garnets have originated in this way. But most commonly these minerals are recovered from accumulations in sands and gravels after weathering has released them from the rocks in which they formed.

However, the rock itself may be mined if the accessory mineral is of sufficient value. For example, sapphire is recovered on a large scale from an igneous rock at Yogo Gulch, Montana. But diamond is the outstanding example. Pipes of kimberlite (a rock composed largely of altered olivine), in which diamond is found as an accessory mineral, formed millions of years ago at high pressures and high temperatures deep within the earth's crust. Before the kimberlite rock completely crystallized, it forced its way toward the surface carrying with it the diamonds.

FIG. 2.1. Geode (7 in. long) weathered from a lava flow in southern Brazil. The outer portion is layered chalcedony (agate) from which amethyst crystals project into the central cavity.

Pegmatites

The small geologic bodies known as pegmatites are the home of a large number of gem minerals. Pegmatites are a special type of igneous rock composed mostly of very large crystals of rock-forming minerals and containing a high percentage of rare and unusual elements. Pegmatites are genetically related to large masses of igneous rocks, usually granite, and are considered to represent the end phase of the crystallization of a magma. As minerals slowly separate from the molten rock material, water and other volatile constituents such as boron and fluorine become concentrated in the still liquid portion. This highly mobile fluid, from which pegmatites form, is expelled from the magma chamber into the already crystallized granite and into the surrounding rocks. With it go all the rare elements whose atoms were either too large or too small to be incorporated into the crystal structures of the rock-forming minerals, as well as the abundant elements that form the common minerals. Eventual crystallization of this mobile magma results in large, sometimes gigantic, crystals of quartz, feldspar, and mica. Of greater interest to us are the gem minerals, whose presence in the pegmatites depends on the concentration of rare elements.

When the rare minerals of pegmatites are embedded firmly in a rock matrix, they are commonly fractured. But the same minerals growing as free crystals into a cavity may be clear and flawless. Such "pockets" in pegmatites have been appropriately called *nature's jewel box,* for in this type of occurrence are found a greater variety of gem minerals than in any other (see Fig. 2.2).

The following are important "jewels" found in pegmatite pockets; the rare elements to which they owe their presence are given in parentheses: beryl (beryllium) in clear transparent crystals as aquamarine, morganite, and golden beryl; tourmaline (boron, lithium) in various shades of green, red, yellow, and blue; topaz (fluorine) as clear crystals or colored yellow, pink, blue, or green; spodumene (lithium) in colorless and yellow crystals or as the gems pink kunzite and green hiddenite; chrysoberyl (beryllium) in yellow-green crystals; and apatite (phosphorus) in shades of blue, yellow, and purple. Less common gem minerals are beryllonite (beryllium); brazilianite and amblygonite (phosphorus); zircon (zirconium); cassiterite (tin); and euclase (beryllium).

Growing into pegmatite pockets and frequently associated with the gem minerals listed above may be gem quality crystals of the abundant rock-forming minerals. Quartz is common as rock crystal, citrine, and smoky quartz. Rose quartz is typical of pegmatites and in many localities occurs in large formless masses. Crystals of rose quartz are rare but in a few places, notably in Brazil, they have been found in pegmatite pockets with other gem minerals. Feldspar, the most abundant mineral on earth, may be found in a variety of gemmy crystals lining cavities. There are yellow orthoclase, green microcline (amazonstone), and albite both as opalescent moonstone and as the delicately iridescent peristerite. One form of feldspar has orange or red reflections from spangles of inclusions and is known as sunstone or aventurine feldspar.

FIG. 2.2. Removing gem tourmaline crystals from a pegmatite cavity at Newry, Maine. Although Newry is an old mineral locality, its reopening in 1972 is reported to have resulted in the largest find of high-quality gem tourmaline on record.

Sedimentary Rocks

Any rock—igneous, metamorphic, or sedimentary—exposed at the earth's surface is subjected to the processes of weathering, both mechanical and chemical. Mechanical weathering is a fragmentation in which the small broken pieces have essentially the same mineral composition as the original rock. Chemical weathering, however, can bring about profound changes. Rainwater with dissolved carbon dioxide and the weak acids of the soil reacts with some minerals to form new compounds. Thus feldspar, abundant in most igneous and metamorphic rocks, may be converted largely to clay and in the process certain soluble salts are formed. However, quartz and most of the gem minerals are inert to this chemical attack, and weathering merely liberates them from the parent rock. These products of weathering are the materials available for the formation of sedimentary rocks.

The weathered material may be removed by wind or glacier to build up deposits on land, but usually it is carried by streams and ocean currents to eventually settle out of the water in parallel beds or layers. When these sediments are firmly cemented or compacted they become sedimentary rocks. Sedimentary rocks can be divided into two broad groups, mechanical and chemical.

The mechanical sedimentary rocks are those composed of material that was transported as discrete particles to the place of deposition. Fine material, such as clay, can be carried far from its source to settle as mud. When compacted, it becomes a *shale.* When firmly cemented, sand becomes *sandstone* and coarse, rounded pebbles become *conglomerate.*

Chemical sedimentary rocks are precipitated on the floor of a body of water, usually the sea, by chemical reactions, evaporation, or organisms. Beds of rock phosphate, gypsum, and rock salt are chemical sedimentary rocks. But most important is limestone, composed essentially of the mineral calcite. Although in part a chemical precipitate, limestone is formed chiefly by the accumulation of calcareous shells of marine organisms.

Placer* Minerals

In the weathering of rocks, minerals that are resistant to chemical attack are liberated essentially intact. Many gem minerals thus freed from the parent rock are washed into streams to become part of the alluvium, the sand and gravel of the stream bed. Rolled and tumbled about, they are rounded and waterworn as they slowly move down stream. Most gemstones have a density greater than that of quartz, the principal mineral of most sand and gravel. Because of their high density, the gemstones work their way to the bottom of the stream bed and lodge behind obstructions in the bedrock. In many places in the world gems are recovered from such accumulations known as *placer* deposits (Fig. 2.3).

To accumulate in placers a mineral must be chemically inert and mechanically strong and have a relatively high density and high hardness. These attributes belong to the gem minerals diamond, zircon, corundum, garnet, tourmaline, chrysoberyl, spinel, topaz, and andalusite, and to a number of lesser gems. Gold and platinum also have all the necessary properties and are frequently found in placers.

If ancient sands and gravels have been consolidated into sandstone and conglomerate, any gem minerals they contain are firmly cemented in the hard rock. In the eighteenth century, in Brazil, the discovery of diamond crystals in coarse sandstone and conglomerate (Fig. 2.4) gave rise to the belief that the home of diamonds was in sedimentary rocks. We now know these rocks are merely foster homes and the diamonds had their beginning in an igneous rock.

FIG. 2.3. Washing gem gravels in Sri Lanka (Ceylon).

* Placer is pronounced plas'er; rhymes with passer.

Metamorphic Rocks

Metamorphic rocks form from igneous or sedimentary rocks by their recrystallization in response to changes of pressure and temperature. The process takes place essentially in the solid state but solutions aid in the chemical reactions that result in new minerals and new textures. There are two types of metamorphism, *regional* and *contact.*

Regional Metamorphism

This type of metamorphism affects large areas and brings about changes in great masses of rocks. If an igneous or sedimentary rock becomes deeply buried, the adjustment to the higher temperatures and higher pressures may completely alter the original features. Mineral particles may be broken and flattened and others recrystallized to form in parallel layers. The most striking feature of metamorphic rocks is the banded or laminated appearance resulting from a parallel arrangement of mineral grains. The principal types of regionally metamorphosed rocks are gneisses and schists. In gneisses coarse bands of light-colored minerals alternate with bands of dark minerals. Schists are finely foliated and can be broken easily into slablike fragments owing to the parallel orientation of the platy minerals such as mica.

The metamorphic rock that forms depends partly on the chemical composition of the original rock and partly on the pressure–temperature con-

FIG. 2.4. Diamond crystal firmly cemented in conglomerate, Brazil. Because of its great hardness, the diamond (center) has retained smooth crystal faces although subjected to the same stream action that has rounded the quartz pebbles.

ditions under which recrystallization took place. But since there is little change in the bulk chemical composition during metamorphism, the original rock is frequently the controlling factor. This is particularly true if the original rock is composed of a single mineral. Sandstone, essentially an aggregate of quartz grains, is metamorphosed to *quartzite,* made up of little else than interlocking quartz grains. A pure limestone is composed of calcium carbonate, calcite, and its metamorphism yields a coarse-grained marble composed of the same mineral. If, however, aluminum is present in mineral impurities in the limestone, ruby and sapphire (aluminum oxide) and spinel (magnesium aluminum oxide) may form during recrystallization. Such is the origin of these superlative gems in marble in Burma. A similar origin is ascribed to tanzanite, the gem zoisite from Tanzania.

The effect of temperature and pressure is more pronounced in the rocks that form by metamorphism of shale than of most other sedimentary rocks. As the intensity of metamorphism increases, old minerals disappear as they react to form new minerals that are in equilibrium with the changed conditions. Slate first develops as the lowest grade metamorphic rock. Increase in intensity of heat and pressure results in an increase in grain size and a transition from slate to mica schists of higher and higher grade. From these schists may come gem crystals of garnet, staurolite, andalusite, kyanite, and chrysoberyl (see Fig. 2.5). Cordierite (iolite) is also a metamorphic mineral found in both schists and gneisses.

Contact Metamorphism

When a rock, particularly a limestone, is invaded by an igneous magma, changes may take place at the contact. This contact metamorphism, brought about by heat from the magma, causes a recrystal-

FIG. 2.5. Crystals of kyanite (light) and staurolite (dark) in mica schist, St. Gotthard, Switzerland.

lization of the country rock. If solutions emanating from the magma introduce additional elements, the rocks are said to have undergone *metasomatism*. Both heat and solutions from the magma result in the formation of new minerals. Among the common contact metamorphic minerals that are sometimes found in crystals of gem quality are garnet, diopside, epidote, and idocrase. Less common are ruby, sapphire, spinel, axinite, and lazurite. The mineral mixture lapis lazuli is also a product of contact metamorphism.

Gems from Solutions

In contrast to igneous minerals which crystallize from a melt and metamorphic minerals which form in an essentially solid rock, some minerals crystallize from fluid, water-rich solutions. These may be extremely hot solutions, moving upward through the earth's crust, or relatively cool, surface waters seeping downward. When the cracks and fissures through which the fluids move become filled with mineral matter, they are called veins.

Vein Minerals

The origin of rising *hydrothermal solutions* is believed to be similar to the fluid magmas from which pegmatites form. That is, they originated in a crystallizing magma and carry with them many of the rare elements. A succession of minerals may be deposited from these solutions as they move upward into cooler and cooler environments. Ores of many of the metals including gold, silver, tin, lead, zinc, and copper are believed to have originated in this way. The same process accounts for the formation of some gem minerals which in places are associated with ores of the metals. Beautifully banded rhodochrosite comes from a silver mine in Argentina and rhodonite from the famous Broken Hill mine in Australia. Fluorite and calcite are common associates of ore minerals, but elsewhere each may crystallize by itself in veins or cavities. Commonest of the gem minerals found in veins is quartz as colorless crystals and more rarely as amethyst (Fig. 2.6).

An interesting origin has been attributed to a group of minerals, some of them gems, found in veins and cavities in the Swiss Alps. It is suggested that the minerals formed from elements leached from the adjacent rock by hot rising solutions. The minerals of these cavities, all beautifully crystallized, include smoky quartz, adularia feldspar, fluorite, axinite, sphene, anatase, and brookite.

Secondary Minerals

Secondary minerals are those that form from material derived by alteration or solution of preexisting minerals. The process takes place near the surface

FIG. 2.6. Veins of amethyst in an altered gneiss, Zambia.

and usually is brought about by rainwater slowly seeping downward with dissolved carbon dioxide, oxygen, and acids of the soil. The solutions, although dilute, can make profound changes in the rocks through which they pass. Which secondary minerals form of course depends on the elements present in the *primary* minerals, and only a few have the properties necessary for gems. If the solutions encounter an ore vein containing zinc and copper minerals, the minerals go into solution and the elements composing them are carried downward, perhaps to leave the region in the underground drainage. If, however, limestone forms the vein wall, as is frequently the case, the metals in solution react with it forming secondary carbonate minerals: zinc to form smithsonite and copper to form azurite and malachite (Fig. 2.7). If silica is present, the reacting solutions may form the secondary copper minerals dioptase and chrysocolla.

FIG. 2.7. Malachite cut and polished. The banding is in various shades of green.

Turquoise and variscite are secondary minerals formed at or near the earth's surface by downward moving solutions. They both contain phosphorus and are thus found in areas where this comparatively rare element can be derived from the overlying rocks.

Opal is the most valued of the gems formed under the near-surface conditions of low pressure and low temperature. It is frequently difficult to determine whether the opal at a given locality was deposited from rising hydrothermal solutions as they approach the surface or from circulating ground water. In either case the common elements silicon and oxygen, of which opal is chiefly composed, are abundantly available. Common opal, as the name implies, is far more plentiful than precious opal but both types form in a similar way. They are found in veins close to the surface and as the petrifying material in petrified wood present in recent volcanic ash and sedimentary rocks.

. . .

We have seen that there is no common origin of gem minerals. Some were formed in the remote geologic past deep within the earth at exceedingly high temperatures and high pressures. Others have formed and are still forming today at or near the earth's surface under essentially atmospheric conditions. Man, in his efforts to synthesize gems, has first to determine and then to duplicate the natural conditions of formation. Although he has been highly successful with many, a few still remain problems for the future. Perhaps this is because time and experience are on nature's side. It must be remembered that nature has been making minerals and their gem relatives since the world began more than 4 billion years ago.

3 CRYSTAL CHEMISTRY

Introduction

It was recognized long ago that a relation exists between the chemistry of minerals and their crystallography. However, an understanding of this relationship had to wait until crystal structures were determined by means of X-ray diffraction. With it came the realization of the interdependency of crystallography and chemistry and the impossibility of adequately discussing one without drawing on the other. As a result, a new science came into being: *crystal chemistry*, a science that related chemical composition to internal structure and physical properties in crystalline solids.

Gemology as a science rests, as does mineralogy, on the fundamental factors of chemistry and crystal structure. For from their interrelation stem the complex optical, mechanical, and chemical properties that characterize each mineral, Thus a knowledge, even though elementary, of these rather technical fields will give the gemologist a greater understanding of the objects which he studies.

Crystal vs. Crystalline

Our definition of *mineral* in Chapter 1 states, in part, that it is ". . .a crystalline chemical element or compound. . . ." *Crystalline* implies a regular internal arrangement of the atoms or ions of which it is composed. In this book we use the word *crystal* in reference to a crystalline solid bounded by plane surfaces arranged in geometric forms that are the outward expression of the ordered internal structure. With a few exceptions minerals are crystalline, but it is the unusual specimen that is found in well-formed crystals with smooth shining faces. The geometry of crystals with their symmetrical arrangement of faces is discussed in Chapter 4, *Crystallography*. But first we consider the reasons for the internal order reflected in the external form.

Chemical Composition

A complete description of any mineral must include its chemical composition. This is usually expressed as a formula derived from a quantitative chemical analysis giving the percentages of the elements present. Although for this information we must rely on the chemist with his specialized skills, qualitative chemical tests, that is, the determination of the presence (but not the amount) of an element, can be easily made in most mineralogical laboratories. However, traditional chemical tests destroy the material under investigation and therefore can not be used by the gemologist studying a cut gemstone. But in recent years highly sophisticated equipment, such as the electron microprobe, permit a quantitative chemical analysis on an extremely small area of a polished surface (less than 0.01 mm in diameter). Although such analyses are rarely made on gems, the gemologist can detect the presence of certain elements with a spectroscope (see Chapter 8).

Classification of Minerals

In early chemical classifications, minerals having a common metal were placed together. There were thus separate mineral groups of iron, copper, zinc, calcium, etc. It became apparent, however, that except for the metal, the minerals in each group had little in common. As a result, another more meaningful classification took form about the middle of the nineteenth century and is generally accepted today. This, also based on chemical composition, arranges the minerals in groups according to the dominant anion (nonmetal) or anionic group (see *Ions and Chemical Bonding* below). It has proved to be a valid method of classification, for in it are brought together minerals that have not only like chemical properties but also similar origins and occurrence and similar physical properties.

This classification, with classes lacking any significant gem mineral representatives omitted, is as follows:

Chemical Class	Gem Representative
Native elements	Diamond
Sulfides	Sphalerite
Oxides	Corundum
Carbonates	Calcite
Phosphates	Apatite
Sulfates	Gypsum
Borates	Rhodizite
Silicates	Zircon

On these classes, the silicates are by far the most important, for this group includes about 60% of the gem minerals. The oxides are next in importance.

Although this arrangement had much to recommend it over earlier classifications, it left many puzzling problems. The properties of many minerals deviate greatly from those expected on a purely chemical basis. Also it leaves unexplained why two minerals, such as diamond and graphite, with identical composition should have vastly different properties. We now know this difference depends on their internal structures.

The study of crystal structures which began in 1912 has answered many of the troublesome questions arising from the classification of minerals on chemical composition alone. Whereas a chemical analysis gives the relative amounts of the elements present, an X-ray structure determination yields information about the internal geometric arrangement of the atoms or ions of the elements and the nature and magnitude of the electrical forces holding them together.

Polymorphism

X-ray structure investigations have also revealed the reasons why a specific chemical element or compound occurs in two or more structural forms. Consider diamond and graphite. The crystal structure of these two forms of carbon shows that diamond is harder and denser than graphite because the carbon atoms are more closely packed and held together by stronger bonds. Similarly calcium carbonate ($CaCO_3$) is found as two different minerals, calcite and aragonite, with differences in their internal structures. Although the gem minerals kyanite, andalusite, and sillimanite share the same chemical composition (Al_2SiO_5), their structural differences give each its own distinctive physical properties.

The phenomenon of a chemical element or compound existing in two distinct physical forms is called *dimorphism*, in three forms *trimorphism*. However, more generally used is the term *polymorphism* for an element or compound that exist in two or more structural modifications.

Crystal structure not only accounts for polymorphic forms but also resolves other apparent inconsistencies. For example, it explains the disturbing fact that ruby, composed of the light elements aluminum and oxygen, has nearly as high a density as sphalerite, made up of the heavier elements zinc and sulfur.

Atoms, Ions, and the Periodic Table

As a basis for understanding how the physical properties of minerals, and of synthetic gem materials as well, relate to chemistry and structure, we briefly discuss the structure of atoms and the bonding forces that lock them together in crystals.

The units of which all matter is composed are atoms. Although atoms themselves are made up of smaller particles, they are the smallest subdivisions that retain the characteristics of the elements. The model of the atom most generally accepted was proposed by the Danish physicist Niels Bohr in 1912. According to his theory, the atom resembles a minute solar system with a nucleus at the center circled by electrons at different distances or energy levels (Fig. 3.1). Except for the hydrogen atom, the nucleus is composed of protons and neutrons. The hydrogen nucleus is a single proton. Protons and neutrons have essentially the same mass but the proton carries a positive electric charge, whereas the neutron is electrically neutral. Each of the electrons that orbits the nucleus, as planets orbit the sun, carries a negative electric charge. Because the atom as a whole is electrically neutral, the number of electrons equals the number of protons. The mass of the atom is concentrated in the nucleus, for a proton is 1857 times heavier than an electron. Although both proton and electron are exceedingly small, the rapidly orbiting electrons give the atoms relatively large effective diameters, up to 100,000 times the diameter of the nucleus. Yet a single atom is far too small to be seen with the highest magnification of the electron microscope. Nevertheless, diameters of atoms have been determined and are expressed in angstrom units (Å). An angstrom is one ten-millionth of a millimeter. For example, the diameter of an atom of gold is 2.88 Å but is usually given as the radius, 1.44 Å.

There are 92 natural elements whose atoms differ from each other in the number of protons, from 1 in hydrogen to 92 in uranium. The number of protons, which equals the number of electrons, is called the *atomic number*. In the periodic table (Table 3.1) the elements are arranged by increasing atomic number. The *atomic weight* of an element is a number expressing its relative weight in terms of oxygen equal to 16.0 (see Table 3.2). In general atomic weight increases, though not uniformly, with increase in atomic number.

The paths of the fast-moving electrons lie in energy levels or orbital shells concentric to the nucleus. The shells are referred to by numbers 1 to 7 or by the letters *K, L, M, N, O, P,* and *Q*, with *K* equal to 1, closest to the nucleus, and with *Q* equal

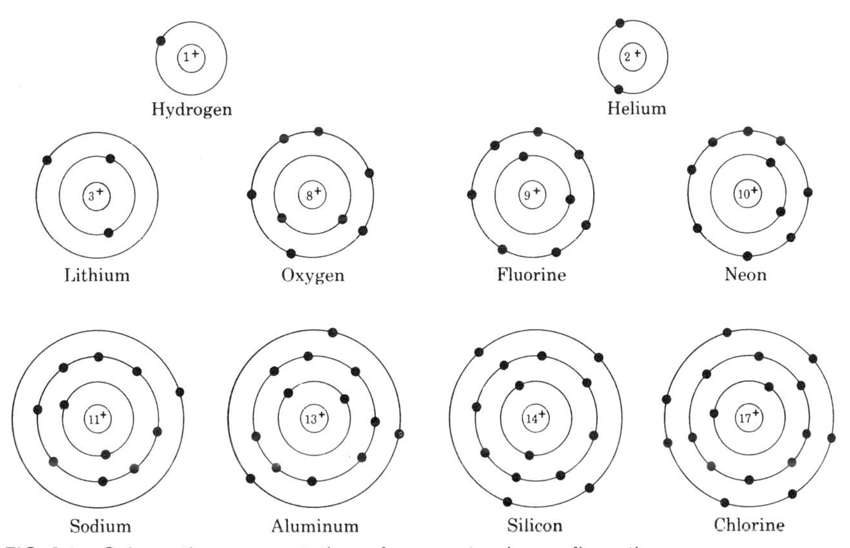

FIG. 3.1. Schematic representation of some atomic configurations.

Table 3.1
PERIODIC TABLE OF THE CHEMICAL ELEMENTS

Periods	Ia	IIa	IIIb	IVb	Vb	VIb	VIIb	VIIIb			Ib	IIb	IIIa	IVa	Va	VIa	VIIa	VIIIa
	METALS												NONMETALS					
			TRANSITION ELEMENTS															
1	1 H	2																2 He
2	3 Li	4 Be											5 B	6 C	7 N	8 O	9 F	10 Ne
3	11 Na	12 Mg											13 Al	14 Si	15 P	16 S	17 Cl	18 Ar
4	19 K	20 Ca	21 Sc	22 Ti	23 V	24 Cr	25 Mn	26 Fe	27 Co	28 Ni	29 Cu	30 Zn	31 Ga	32 Ge	33 As	34 Se	35 Br	36 Kr
5	37 Rb	38 Sr	39 Y	40 Zr	41 Nb	42 Mo	43 Tc	44 Ru	45 Rh	46 Pd	47 Ag	48 Cd	49 In	50 Sn	51 Sb	52 Te	53 I	54 Xe
6	55 Cs	56 Ba	57 *	72 Hf	73 Ta	74 W	75 Re	76 Os	77 Ir	78 Pt	79 Au	80 Hg	81 Tl	82 Pb	83 Bi	84 Po	85 At	86 Rn
7	87 Fr	88 R	89 **	104	105	106												

*LANTHANIDE SERIES	57 La	58 Ce	59 Pr	60 Nd	61 Pm	62 Sm	63 Eu	64 Gd	65 Tb	66 Dy	67 Ho	68 Er	69 Tm	70 Yb	71 Lu
**ACTINIDE SERIES	89 Ac	90 Th	91 Pa	92 U	93 Np	94 Pu	95 Am	96 Cm	97 Bk	98 Cf	99 Es	100 Fm	101 Md	102 No	103 Lw

Table 3.2

ATOMIC WEIGHTS, 1971

	Symbol	Atomic Weight
Aluminum	Al	26.9815
Antimony	Sb	121.75
Argon	A	39.948
Arsenic	As	74.9216
Barium	Ba	137.34
Beryllium	Be	9.0122
Bismuth	Bi	208.981
Boron	B	10.811
Bromine	Br	79.904
Cadmium	Cd	112.40
Calcium	Ca	40.08
Carbon	C	12.0111
Cerium	Ce	140.12
Cesium	Cs	132.905
Chlorine	Cl	35.453
Chromium	Cr	51.996
Cobalt	Co	58.9332
Copper	Cu	63.546
Dysprosium	Dy	162.50
Erbium	Er	167.26
Europium	Eu	151.96
Fluorine	F	18.9984
Gadolinium	Gd	157.25
Gallium	Ga	69.72
Germanium	Ge	72.59
Gold	Au	196.967
Hafnium	Hf	178.49
Helium	He	4.0026
Holmium	Ho	164.930
Hydrogen	H	1.0080
Indium	In	114.82
Iodine	I	126.9045
Iridium	Ir	192.2
Iron	Fe	55.847
Krypton	Kr	83.80
Lanthanum	La	138.91
Lead	Pb	207.19
Lithium	Li	6.941
Lutetium	Lu	174.97
Magnesium	Mg	24.305
Manganese	Mn	54.9380
Mercury	Hg	200.59
Molybdenum	Mo	95.94

	Symbol	Atomic Weight
Neodymium	Nd	144.24
Neon	Ne	20.179
Nickel	Ni	58.71
Niobium	Nb	92.906
Nitrogen	N	14.0067
Osmium	Os	190.2
Oxygen	O	15.9994
Palladium	Pd	106.4
Phosphorus	P	30.9738
Platinum	Pt	195.09
Potassium	K	39.102
Praseodymium	Pr	140.907
Protactinium	Pa	231.036
Radium	Ra	226.025
Radon	Rn	222.
Rhenium	Re	186.2
Rhodium	Rh	102.905
Rubidium	Rb	85.467
Ruthenium	Ru	101.07
Samarium	Sm	150.4
Scandium	Sc	44.956
Selenium	Se	78.96
Silicon	Si	28.086
Silver	Ag	107.868
Sodium	Na	22.9898
Strontium	Sr	87.62
Sulfur	S	32.064
Tantalum	Ta	180.947
Tellurium	Te	127.60
Terbium	Tb	158.9254
Thallium	Tl	204.37
Thorium	Th	232.038
Thulium	Tm	168.934
Tin	Sn	118.69
Titanium	Ti	47.90
Tungsten	W	183.85
Uranium	U	238.029
Vanadium	V	50.9414
Xenon	Xe	131.30
Ytterbium	Yb	173.04
Yttrium	Y	88.9059
Zinc	Zn	65.37
Zirconium	Zr	91.22

to 7, farthest from the nucleus. The maximum number of electrons that can occupy each of the shells is as follows: K, 2; L, 8; M, 18; and N, 32. The *O*, *P*, and *Q* shells are not completely filled.

With the exception of the *K* shell, each shell contains subshells designated as *s*, *p*, *d*, *f*. When the subshells are filled the number of electrons occupying them is as follows:

Shell:	*K*	*L*		*M*			*N*			
Subshell:	*s*	*s*	*p*	*s*	*p*	*d*	*s*	*p*	*d*	*f*
Electrons:	2	2	6	2	6	10	2	6	10	14

In the lighter elements between hydrogen and argon (atomic numbers 1–18) electrons are added in regular succession to the *K*, *L*, and *M* shells until all subshells including *M-p* are filled, yielding a total of 18 electrons. These 18 elements constitute the first three "short" periods of the periodic table.

Addition of one electron to the argon superstructure results in an atom of potassium and the addition of two electrons in an atom of calcium. However, these two electrons enter the *N-s* shell, not the *M-d*, which remains vacant. With the *N-s* shell now filled, additional electrons enter the *M-d*. A sequence of 10 elements intervenes before the *M-d* is filled and further addition to the *N* shell may continue. These 10 elements, scandium through zinc, are the transition metals and do not resemble the elements of the short periods. Starting with gallium, electrons again enter the *N* shell, forming a sequence from gallium through bromine which do resemble the elements of the three short periods. The fourth period, the first "long" period, ends with the noble gas krypton.

The fifth and sixth periods are somewhat analogous to the fourth with each containing transition elements, 39–48 in the fifth period and 57–80 in the sixth period. For a discussion of them as well as a complete treatment of the periodic table one is referred to a book on chemistry (see references at end of chapter). The above sketchy discussion is given here to take us to the fourth period with the introduction of the transition metals. They will be referred to later (see discussion of chromophores under *Color* in Chapter 6). For it is the transition metals and their electronic configurations that are a major factor in the color of gemstones.

The electrons in the outermost shell, called the *valence* electrons, largely determine the chemical properties of an element as well as its place in the periodic table. The number of valence electrons in groups Ia, IIa, IIIb, IVb, Vb, VIb, and VIIa is the same as the group number. In group VIII are the noble gases which, with the exception of helium with only one shell occupied by two electrons, have eight electrons in their outermost shells. This is the most stable configuration with the valence electron shells completely filled.

Ions and Chemical Bonding

There is a strong tendency for all atoms to achieve stable, completely filled outer shells by losing or gaining electrons. When this is done, the number of electrons no longer equals the number of protons and the atom has either a positive or negative charge and is called an *ion*. Positive ions, called *cations*, have lost electrons; negative ions, called *anions*, have gained electrons. The gain or loss of a single electron gives a unit charge. Thus if two electrons are gained, the charge is -2; if two electrons are lost, the charge is $+2$, etc. The charge of ions of the common elements found in gem minerals is given in Table 3.3.

Ionic Bond

Sodium in group I has a single valence electron that is easily lost, leaving the atom with a single positive charge. On the other hand, chlorine in group VII, with seven electrons in its valence shell, can easily attain the stable configuration by adding an electron, and thus assume a negative charge. A water solution of sodium chloride (common salt) contains positive ions of sodium (Na^+) and negative ions of chlorine (Cl^-). If the volume of the solution is sufficiently reduced by evaporation, the attractive forces of the opposing electrical charges cause the ions to join together as a crystal nucleus (Fig. 3.2*a*). On continued evaporation more and more ions attach themselves to the growing nucleus to form a

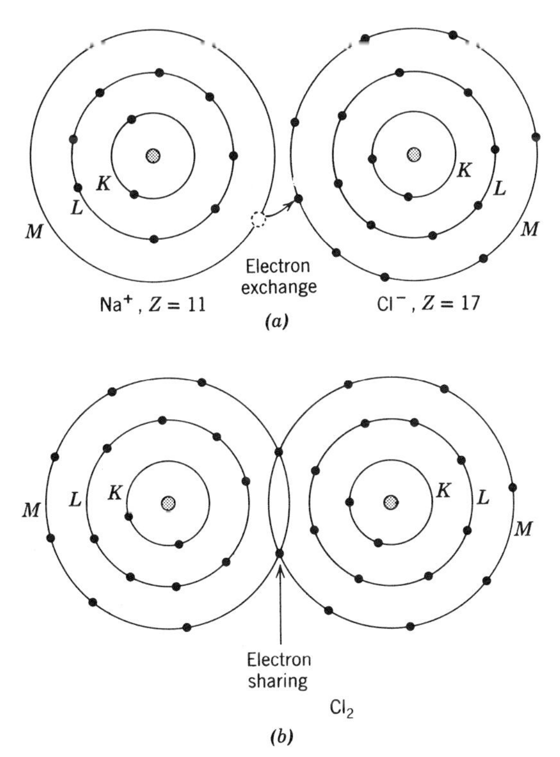

FIG. 3.2. (*a*) Schematic representation of electron exchange (ionic bond) between Na and Cl atoms to form Na^+ and Cl^-. (*b*) Electron sharing (covalent bond) between two Cl atoms to form a molecule of Cl_2. Z is the atomic number.

crystal that settles out of solution. This is the principal type of chemical bonding in minerals and is called the *ionic bond*.

Covalent Bond

Another way atoms can achieve the inert gas configuration is by sharing electrons. The outer regions interpenetrate so that some electrons do double duty in the joined atoms. This electron sharing, called the *covalent bond*, is the strongest of the chemical bonds. A simple example is illustrated by the element chlorine whose atoms, as we have seen, have seven electrons in the valence shell. Two chlorine atoms join to form a molecule of chlorine gas by one electron from each completing the outer shell of the other (Fig. 3.2*b*) and locking them firmly together. It is this same type of bonding that holds one carbon atom to another in diamond. Carbon in group IV of the periodic table has four valence electrons. In the diamond structure every carbon atom is joined to four others by sharing electrons, as shown in Figure 3.3.

Although unimportant in gem materials, there are two other bond types that must be mentioned to make the listing complete. The *van der Waals* bond, named after the Dutch physicist, is the weakest type of bonding. It is easily broken and, when present in minerals, defines planes of ready cleavage and gives a mineral a low hardness—too low for most gem materials. The soft mineral talc (steatite or soapstone) is an example. The *metallic bond* is the force to which metals owe their cohesiveness. Although it is present in the metals used in jewelry (gold, platinum, and silver), it plays no part as a bonding mechanism in gemstones.

Thus only the ionic and covalent bonds are important in gem materials, natural or synthetic. Very few minerals have only a single bond type and most are held together by a combination of the two. It is the weakest bond that determines the physical properties such as hardness and cleavage; but the stronger the bond, the harder the mineral.

Ionic Radii

In general we can consider minerals as ionic structures with essentially spherical ions packed together

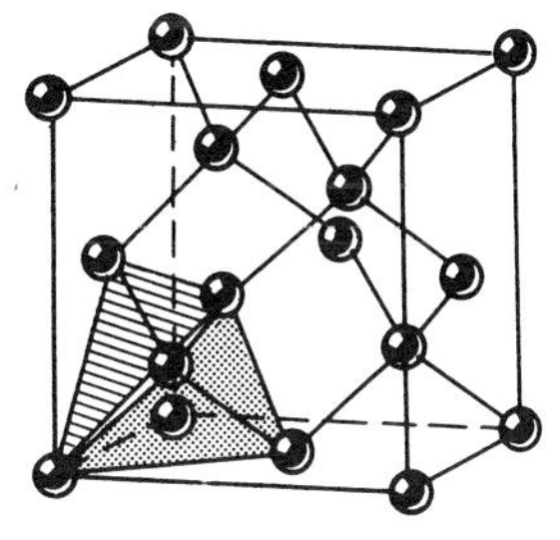

FIG. 3.3. Diamond structure showing carbon atoms joined to four other carbon atoms by a strong covalent bond. This arrangement is tetrahedral or 4 coordination. A single tetrahedron is outlined with a carbon atom at its center bonded to four others at its corners (see Fig. 3.6*c*).

Table 3.3
CHARGES AND RADII OF THE COMMON IONS

Atomic Number	Element	Symbol and Ionic Charge	Ionic Radius (Å)
4	Beryllium	Be^{2+}	0.35
5	Boron	B^{3+}	0.23
6	Carbon	C^{4+}	0.15
8	Oxygen	O^{2-}	1.40
9	Fluorine	F^{-}	1.36
11	Sodium	Na^{+}	0.97
12	Magnesium	Mg^{2+}	0.66
13	Aluminum	Al^{3+}	0.51
14	Silicon	Si^{4+}	0.40
16	Sulfur	S^{6+}	0.30
17	Chlorine	Cl^{-}	1.81
19	Potassium	K^{+}	1.33
20	Calcium	Ca^{2+}	0.99
22	Titanium	Ti^{4+}	0.68
24	Chromium	Cr^{3+}	0.63
25	Manganese	Mn^{2+}	0.80
		Mn^{3+}	0.66
26	Iron	Fe^{2+}	0.74
		Fe^{3+}	0.64
28	Nickel	Ni^{2+}	0.69
29	Copper	Cu^{+}	0.96
		Cu^{2+}	0.72
40	Zirconium	Zr^{4+}	0.79
50	Tin	Sn^{4+}	0.71

in an ordered three-dimensional array. The way atoms arrange themselves depends on both their charge and their relative size. Although it is impossible to measure directly the size of a single ion, the radii of most ions have been determined empirically.

When an anion and a cation approach each other under the influence of their opposite electrostatic charges, repulsive forces are set up. The distance at which the repulsive force, arising from the negatively charged electron clouds, balances the attractive force is the interionic distance between the pair of ions. Although not rigorously true, we can think of the ions as rigid spheres in contact and the interionic distance as the sum of their radii. Interionic distances can be determined by X-ray techniques; thus if we assume the radius of one ion, we can calculate the radius of the other. In this way ionic radii have been determined for most of the elements. In general anions are larger than cations and crystal structures can be thought of as closely packed anions with smaller cations arranged between them. In Table 3.3 are given the ionic radii of the elements that make up most of the gem minerals.

Coordination Principle and Radius Ratio

When cations and anions join in a crystal structure, each tends to coordinate, that is, to gather to itself, as many ions of opposite charge as size permits. The coordinated ions arrange themselves about the central coordinating ion in such a way that lines joining their centers outline a geometrical figure, a *coordination polyhedron*. The number of anions that can cluster about a central cation is the coordination number. Figure 3.4 is a model of the sodium chloride structure. In it each sodium ion is

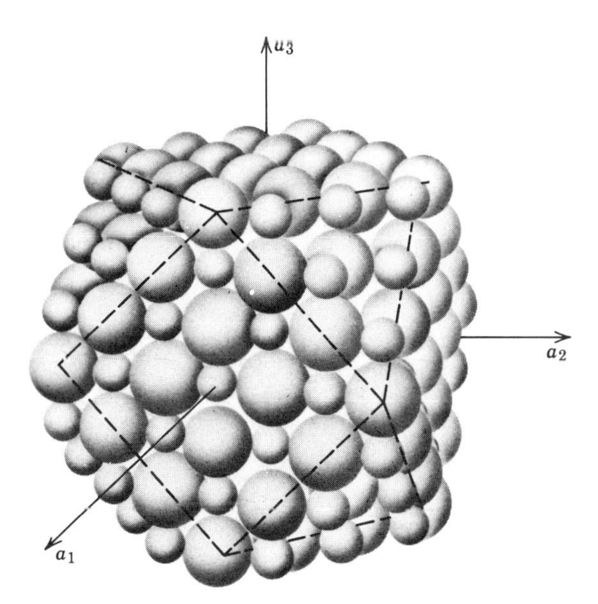

FIG. 3.4. Sodium chloride packing model. Sodium small, chlorine large. Dashed lines outline a cubo-octahedron. Note the cube "faces" (square) are underlain by sheets of equal numbers of Na^+ and Cl^- ions; the octahedral "faces" (triangular) are underlain by alternating sheets of Na^+ and Cl^- ions.

FIG. 3.5. Atomic packing model of gold. The model shows gold atoms in 12 coordination (cubic closest packing).

found to coordinate six of the larger chlorine ions and thus chlorine is in 6 coordination with respect to sodium (Fig. 3.6*b*).

The type of coordination polyhedron and hence the coordination number depend on the relative size of the coordinated ions. The relative size is expressed by a *radius ratio,* $R_c\!:\!R_a$, where R_c is the radius of the cation and R_a the radius of the anion in angstrom units. Take, for example, sodium chloride. From Table 3.3 we find the radius of chlorine (Cl^-) is 1.81 Å and the radius of sodium (Na^+) is 0.97 Å. Therefore, the radius ratio of sodium to chlorine is $R_{Na^+}\!:\!R_{Cl^-} = 0.97\!:\!1.81 = 0.54$.

There are two ways of packing identical spheres in three dimensions with a minimum of space between them, an array called *closest packing*. One way has an hexagonal arrangement, *hexagonal closest packing*; the other has a cubic arrangement, *cubic closest packing*. In either case each sphere is in contact with 12 closest neighbors, that is, in 12 coordination. Figure 3.5 is a photograph of a model of the gold structure with atoms

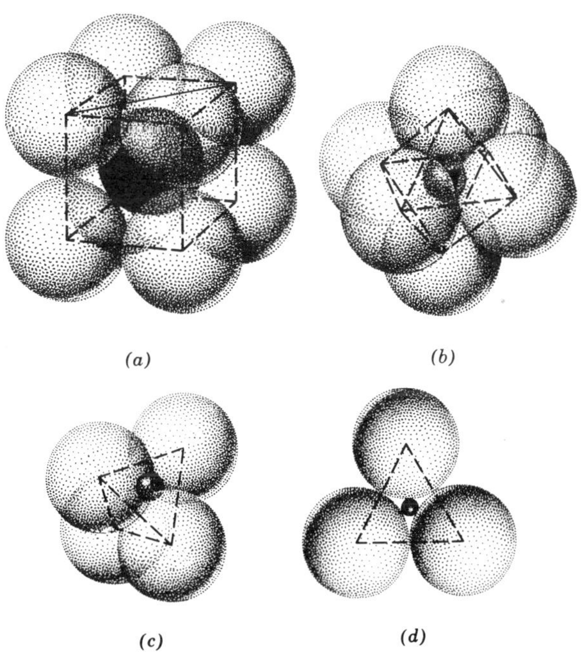

FIG. 3.6. Coordination polyhedra of anions about cation. (*a*) Cubic or 8 coordination with $R_c\!:\!R_a < 1 - 0.723$. (*b*) Octahedral or 6 coordination with $R_c\!:\!R_a = 0.732 - 0.414$. (*c*) Tetrahedral or 4 coordination with $R_c\!:\!R_a = 0.414 - 0.225$. (*d*) Triangular or 3 coordination with $R_c\!:\!R_a = 0.225 - 0.155$.

in cubic closest packing, a structure gold has in common with silver, platinum, and copper. Aside from the native metals, 12 coordination is rare in minerals.

If the radius ratio is slightly less than 1, the coordination polyhedron is a cube with the anions at its eight corners. This is 8, or cubic, coordination which is stable provided the radius ratio is 0.732 or greater. At this critical ratio, the anions are touching each other as well as the central cation. With a lesser radius ratio the cation is too small for 8 coordination, and 6 or octahedral coordination is stable. With decreasing radius ratios other coordinations become stable. Figure 3.6 illustrates 8, 6, 4, and 3 coordination and the range of radius ratios stable for each.

Although there are many other factors that influence the coordination pattern in a crystal, for most minerals the radius ratio principle is valid.

Illustrations of Crystal Structures

There are three types of models used to help one visualize crystal structures. The packing model, used to illustrate sodium chloride (Fig. 3.4), is most realistic, for spheres can be made proportional to the size of the ions represented. However, because one can not look inside the closely packed model, an open "ball and stick" model is frequently used, particularly for more complicated structures, to

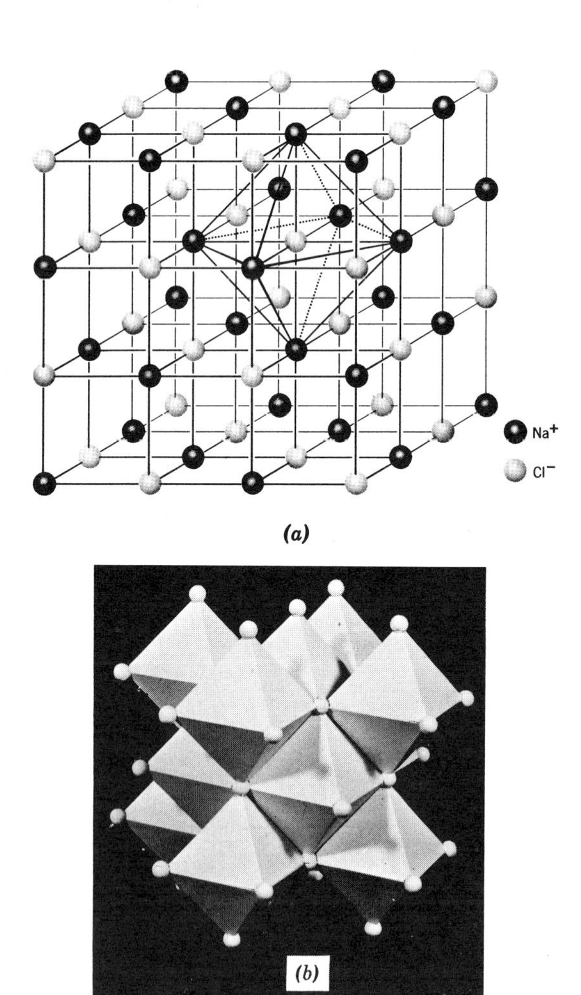

FIG. 3.7 Sodium chloride structure. (*a*) Open "ball and stick" model with octahedral polyhedron outlined. (*b*) Polyhedral model.

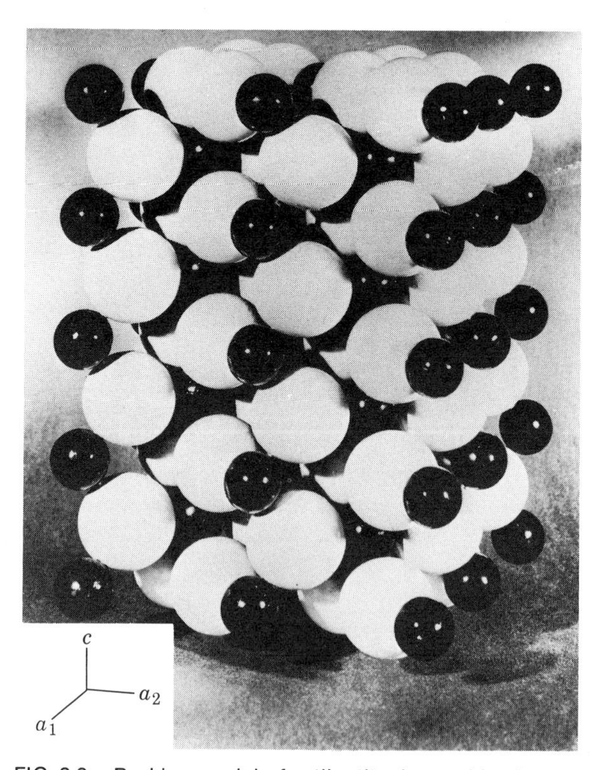

FIG. 3.8. Packing model of rutile, titanium oxide. Oxygen (white) is in 6 coordination about titanium (black) and titanium is in 3 coordination about oxygen. The ratio of titanium:oxygen = 3:6 or TiO_2.

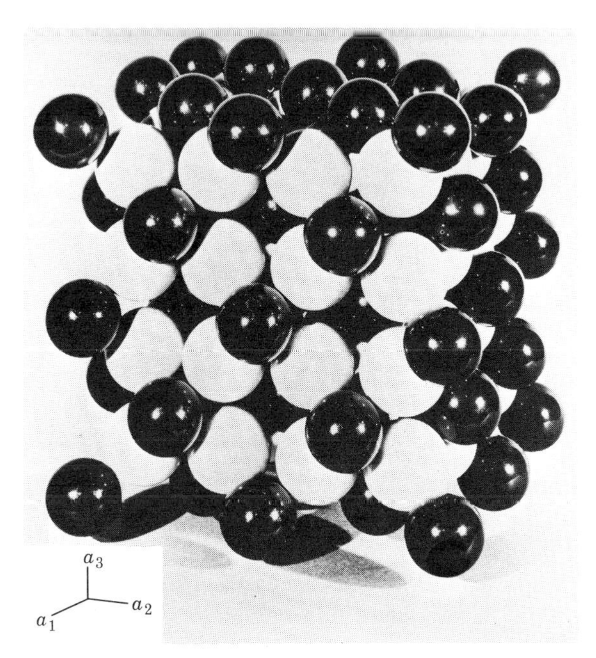

FIG. 3.9. Packing model of fluorite, calcium fluoride. Fluorine (white) is in 8 coordination about calcium (black). To maintain electrical neutrality only half the calcium sites are filled and calcium is in 4 coordination about fluorine. The formula is thus CaF_2.

show the relation of the component ions. Finally, the structure may be represented by coordination polyhedra, such as octahedra and tetrahedra, joined together at the points where they share ions without necessarily showing the individual coordinated ions. The polyhedral model is most useful in depicting the repeat units in complicated structures. The open and polyhedral models (Fig. 3.7) illustrate the same sodium chloride structure as the packing model (Fig. 3.4). Other examples of structure models are given in Figures 3.8–3.10.

Isostructure

There appears to be little similarity between sodium chloride (NaCl) and periclase, magnesium oxide (MgO), a synthetic gem material. Sodium chloride is water soluble with low density, hardness, and melting point: periclase is insoluble in water with high density, hardness, and melting point. Yet structure analysis shows a marked similarity. The magnesium ions in periclase are in 6 coordination with respect to oxygen and in sodium chloride each sodium ion coordinates six chlorine ions. The sodium

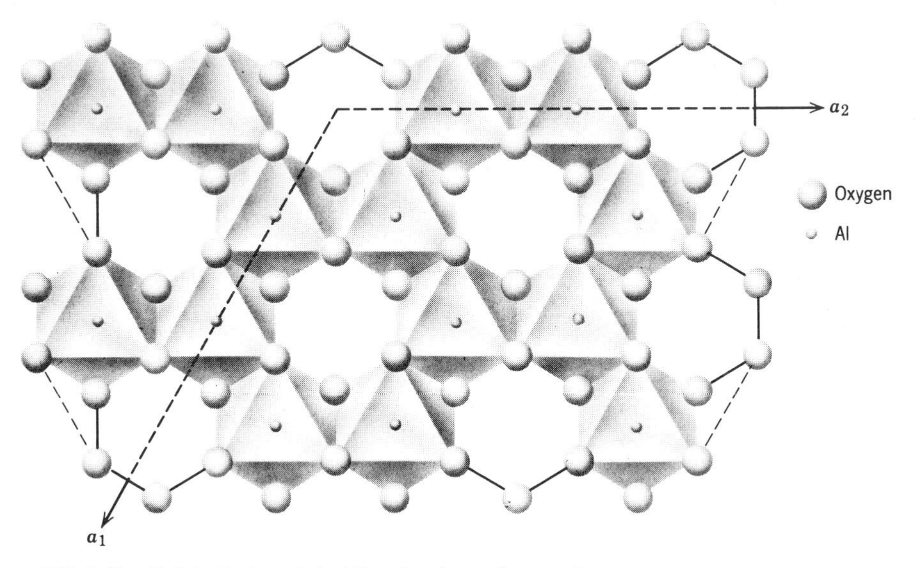

FIG. 3.10. Polyhedral model of the structure of corundum (ruby and sapphire) showing a basal sheet of octahedra with one octahedron vacant for every two octahedra with aluminum. The ratio of aluminum:oxygen = 2:3, giving the formula Al_2O_3. a_1 and a_2 are the directions of the crystal axes.

chloride models (Figs. 3.4 and 3.6*b*) could serve equally well for periclase. Minerals such as these with identical geometrical arrangements in their structures are said to be *isostructural*.

Minerals which contain the same anion often belong to "isostructural groups." For example, the garnet group and the spinel group of gems constitute two major isostructural groups, with large ranges in chemical composition within each group. The variation of the properties of group members results from different cations occupying a given structural site in the crystal.

Ionic Substitution and Solid Solution

Because in nature the solutions and melts from which minerals crystallize contain many elements, it is the rule rather than the exception for minerals to contain elements not required for their stable formation. Such additional elements are often present in minute amounts and may act as coloring agents, such as chromium or vanadium in emerald and iron in amethyst. These metals are not present as tiny bits of metallic chromium or iron but as ions (e.g., Fe^{2+}, Fe^{3+}, or Cr^{3+}) substituting for a major element in the mineral. For example, in sodium chloride small percentages of potassium (K^+) may randomly substitute for some of the sodium (Na^+) in the structure; similarly some fluorine (F^-) may replace some chlorine (Cl^-) anions. Figure 3.11 shows a packing model of the sodium chloride structure in which there is random ionic substitution within the structural framework.

FIG. 3.11. Random substitution of anions (large black) in the sodium chloride structure.

There are many mineral examples in which ionic substitution is extensive. Substitution of one ion for another takes place readily if the ions differ in radius by less than 15%, provided overall neutral charge of the mineral is maintained. For example, in the olivine group are the minerals forsterite, magnesium silicate (Mg_2SiO_4), and fayalite, iron silicate (Fe_2SiO_4). Iron (Fe^{2+}) and magnesium (Mg^{2+}) answer the requirements of similar ionic radius and have identical charges and thus can substitute for each other in all proportions. Such substitution is called *solid solution*. The properties of the intermediate members of the group change with composition. With no iron, forsterite is colorless but with increasing iron the mineral darkens, going from pale olive green to dark green to black in fayalite. The specific gravity also changes with a progressive increase with increase in iron. In peridot, gem olivine, about 10% of the magnesium in forsterite has been replaced by iron.

Exsolution

When minerals crystallize at high temperatures, their high internal thermal energy allows for less stringent space requirements on the structure than would be permissible at lower temperatures, and ionic substitution may be extensive. An example is found in the feldspars. Potassium feldspar, orthoclase, formed at a relatively high temperature can tolerate considerable amounts of sodium ions in place of potassium ions. When such an orthoclase cools, the poor fit of the sodium ions in the potassium position results in stresses forcing the sodium to migrate through the structure to form domains (small localized areas) of sodium feldspar, albite. This phenomenon is called *exsolution* and, in the example given, is responsible for the opalescent effect in some moonstone.

Chemical Formulas

The name of each of the elements listed in Table 3.2 is followed by its chemical symbol. We thus find Aluminum Al, Carbon C, Oxygen O, Silicon Si, etc. Of the gem minerals only diamond occurs as a native element and its formula is merely C for carbon. Other gem minerals are composed of two or more elements and their formulas indicate the proportions of the elements present. For example, quartz is SiO_2, indicating that there are two atoms of oxygen for each atom of silicon; in corundum, Al_2O_3, there are three atoms of oxygen for two atoms of aluminum. In Chapter 13 the description of each gem mineral includes the chemical formula. Thus a familiarity with chemical symbols gives one a ready means of telling what elements are present and the ratios of their atoms to each other.

References and Suggested Reading

Evans, R. C., 1966. *An Introduction to Crystal Chemistry*, 2nd ed. The University Press, Cambridge. 410 pp.

Dickerson, R. E. and I. Geis, 1976. *Chemistry, Matter, and the Universe*. W. A. Benjamin, Inc., Menlo Park, Calif. 669 pp.

Hurlbut, C. S. and C. Klein, 1977. *Manual of Mineralogy*, 19th ed. John Wiley & Sons, Inc., New York. Chapter 4.

4 CRYSTALLOGRAPHY

Introduction

Since the discovery that X-rays could be used to reveal the secrets of the interior of a crystal, X-ray study has been the principal means of crystallographic research. From it has come our knowledge of crystal structures, that is, the geometrical arrangements of atoms, their sizes, and the forces that hold them together.

Space Lattices

However, long before the use of X-rays, crystallographers were concerned with crystal structure. They correctly deduced from the external shapes and physical properties that crystals had an ordered interior and speculated on the arrangement of the building units. In 1849 the French physicist Auguste Bravais proved that there are only 14 ways of arranging identical points in space so that each one has the same number of neighbors at the same distances and directions. These arrangements, known as the 14 *space lattices* or sometimes as the *Bravais lattices,* are the frameworks on which crystals are built. The *unit cell* is the smallest parallelepiped of the lattice. These simplest lattice units, different for every mineral, vary in the lengths of the edges and, in many, in the angles between the edges. Some cells are called *primitive* with lattice points only at the corners; others, *multiple* cells, have additional points at face centers or along body diagonals (Fig. 4.1).

Point Groups and Space Groups

We now know that every crystalline substance is composed of atoms or ions joined together to form a geometrical arrangement—a motif—whose shape and size depends on the type and number of atoms involved. In real crystals it is these motifs that occupy the lattice points and are repeated in three dimensions, filling all space. It has been shown that there are only 32 ways of arranging atoms about a point to accomplish this. These are called the 32 *point groups* or the *crystal classes.* When the point groups are combined with the space lattices, there result 230 possible arrangements called the *space*

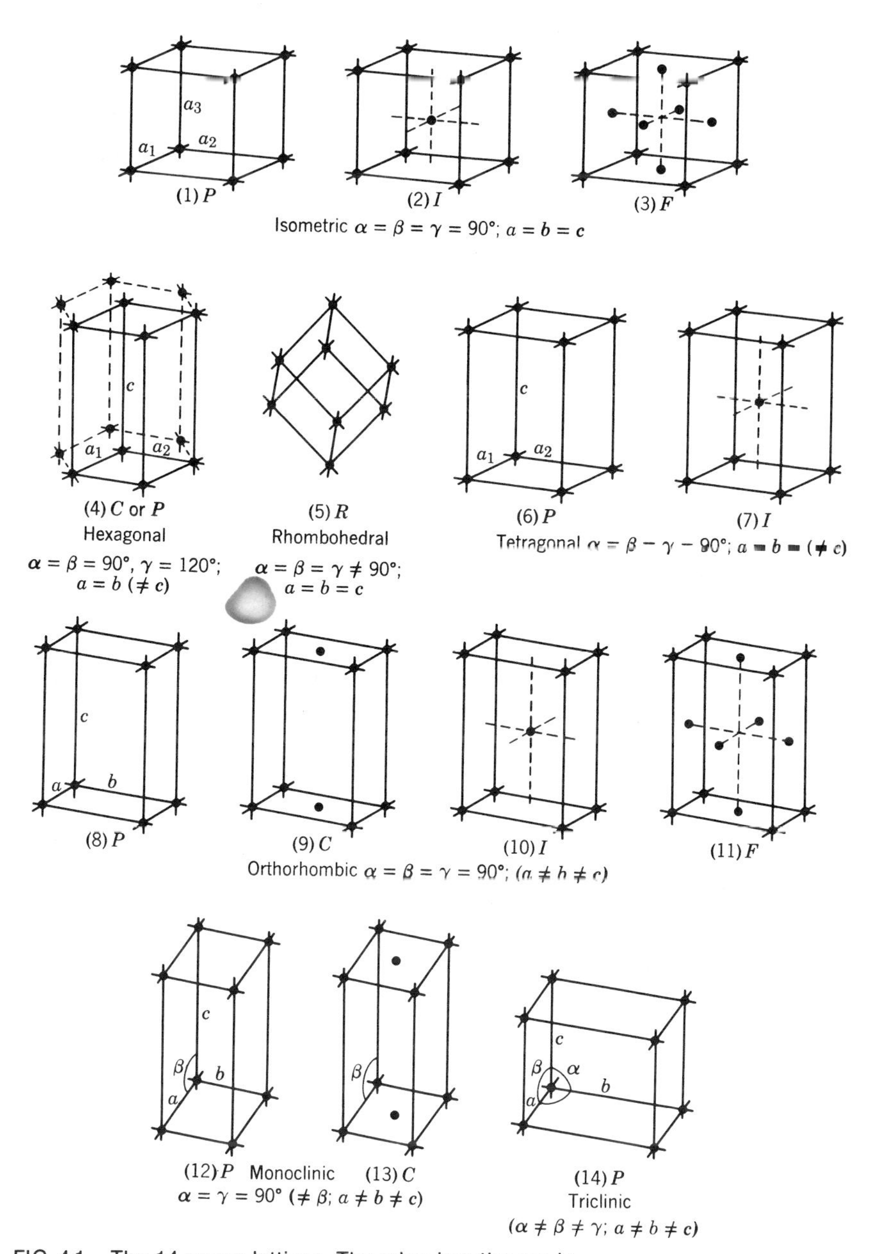

FIG. 4.1. The 14 space lattices. The edge lengths are indicated by *a*, *b*, and *c* and the angles between them by α, β, and γ. Each lattice has its own constraints regarding edge lengths and angles. In the above notation the nonquivalence of angles and edges that usually exist but are not mandatory are set off by parentheses. Primitive lattices are indicated by *P*. Multiple lattices are indicated by *I* (body centered), *F* (face centered), *C* (end centered). *R* is the rhombohedral lattice.

groups. To define a crystal structure both the space group and the arrangement of the atoms must be given.

Fortunately for normal work, the gemologist need not be concerned with space groups or even with point groups but only with the *crystal systems* into which the 32 point groups fall. The characterizing features of the crystal systems are discussed later in this chapter under *Morphological Crystallography*.

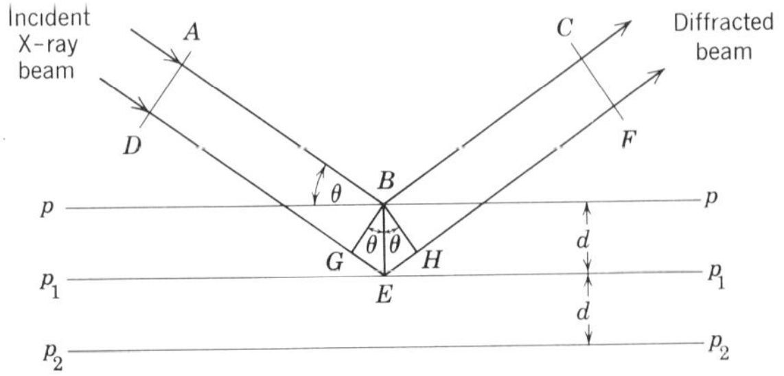

FIG. 4.2. Geometry of X-ray reflection.

X-Ray Crystallography

The Bragg Law

It was at the suggestion of Max von Laue of the University of Munich in 1912 that X-rays were first used in the study of crystals. In the initial experiments an X-ray beam was passed through a crystal plate and allowed to strike a photographic film. It was discovered that a regular array of dark spots appeared on the film resulting from X-rays diffracted from internal planes. Shortly thereafter, W. L. Bragg in England pointed out that, although X-rays are diffracted from atomic planes, they act as though they are reflected. The reflections are not continuous but take place only under certain conditions. The conditions are expressed by the equation $n\lambda = 2d \sin \theta$, known as the *Bragg law*. In the equation n is a whole number (1, 2, 3, ..., n), λ is the wavelength, d is the distance between successive parallel lattice planes, and θ is the angle at which the X-ray beam strikes a given atomic plane.

The X-ray beam penetrates the crystal so that the resulting reflection is not from a single plane but from many parallel planes. In order to be of sufficient intensity to be recorded, the reflections from all planes must be in phase and reinforce each other. This is the condition of the Bragg equation as illustrated in Figure 4.2. In the diagram the trace of parallel planes is represented by p_1, p_2, p_3, etc. with spacing d. The X-ray beam strikes p_1, then p_2, p_3, etc. In order that reflections from p_1 and p_2 be in phase, the distance represented by *GEH* must be equal to the wavelength or an integral number of wave lengths, $n\lambda$. This condition can be satisfied only at definite values of angle θ.

Crystal Structure and X-rays

There are many ways in which crystals are studied by means of X-rays but in those that yield the most information, a tiny crystal is rotated (the rotation may be a complete 360° or a limited number of degrees) about a known crystallographic direction in the path of an X-ray beam of known wavelength. As the crystal is slowly turned, various families of atomic planes, each with a characteristic d spacing, make the proper θ angles with the X-ray beam to satisfy the Bragg law. In some methods the diffracted beams are recorded on photographic film; in others both the position and intensity of the diffraction are recorded electronically. In either case the X-ray crystallographer can interpret the results in terms of crystal structure. He can determine the dimensions of the unit cell, that is, the length of the cell edges and the angles between them. With more exacting procedure the positions of the various atoms within the cell can be specified. Because of the complexity of the equipment, its operation, and interpretation of results, we must rely on the skilled specialist for information on crystal structures.

The Powder Method

A less complicated means of X-ray study of crystalline materials (one the gemologist might use) is known as the *powder method*. The amount of material required is small; scrapings from the girdle of a gemstone are sufficient. These powdered crystalline particles can be picked up and bonded together into a needlelike spindle. The spindle is mounted at the center of the powder camera, a cylindrical light-tight box, and a filmstrip is snugly fitted around the inside wall (Fig. 4.3*a*). When X-rays strike the powdered material composed of many randomly oriented grains, many particles are in position to satisfy the Bragg law. The diffractions that take place simultaneously from the crystal planes are recorded on the filmstrip as curved lines (Fig. 4.3*b*). Since the spacing and intensity of these lines are characteristic for every crystalline substance, the powder photograph has been called the "fingerprint" of a mineral. The powder method of X-ray analysis is thus a powerful means of mineral identification.

Morphological Crystallography

If we refer to the space lattices shown in Figure 4.1, we find that, although there are 14 lattices, there are only seven basic shapes. Each of these shapes can be thought of as a unit (unit cell) which repeated over and over in three dimensions builds up a crystal. Thus in the broadest sense all crystals fall into one of seven groups based on the shape of the unit. But two of these (hexagonal and rhombohedral)

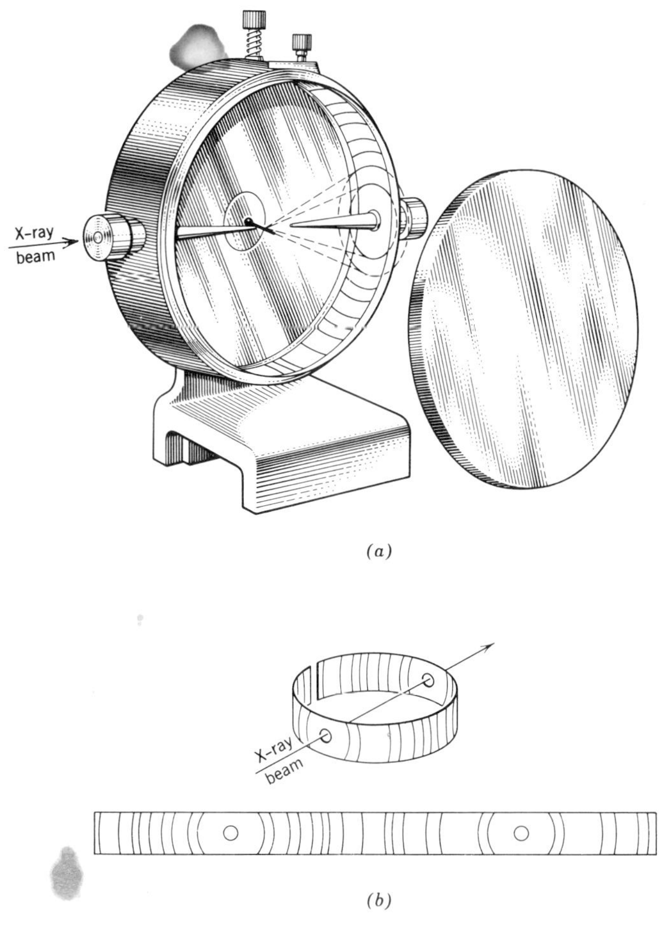

FIG. 4.3. (*a*) X-ray powder camera. (*b*) X-ray diffractions recorded on the cylindrical film appear as curved lines when the film is developed and laid flat.

have certain similar characteristics which permit them to be described as one. By combining these two, there result six groups known as the *crystal systems* into which the 14 lattices of Figure 4.1 are grouped. We discuss crystal morphology, that is, the outward shapes of crystals, in terms of the crystal systems.

Crystal Faces

The outward forms of crystals result partly from the shape of the building unit and partly from the way units are added to the growing crystal. It is easy to picture large crystals having the same shape as the tiny units of which they are built. For example, a large cube can be made by stacking together cubic building blocks in three dimensions, n units along each edge. The large cube will then have n^3 units. In Figure 4.4a $n = 5$ and $n^3 = 125$. However, in nature because of varying conditions of temperature, pressure, or nature and movement of the solution, the building units may not be added uniformly and a great variety of shapes results. Unequal stacking may result in a tabular crystal (Fig. 4.4*b*) or an elongated crystal (Fig. 4.4c). From the crystallographic point of view both of these are considered as cubes with six faces at right angles to each other. But when units are left off the corners (Fig. 4.4*d*) or off the edges (Fig. 4.4e) sloping faces develop. The resulting outward appearance is no longer cubic but the fundamental building unit is the same tiny cube.

As another example, let us consider the stacking of unit cells of peridot, the mineral olivine. The unit has much the shape of an ordinary brick (Fig. 4.5a) with the three edges at right angles, but all of different lengths. There can be many stacking arrangements including, as in the case of the cube, a large "brick" having the same relative dimensions as the unit. But consider only a "wall" of units (Fig.

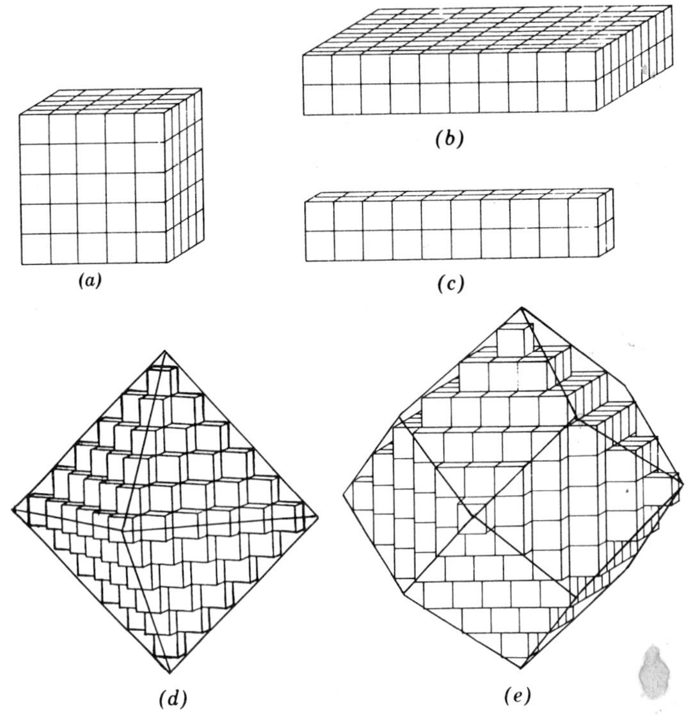

FIG. 4.4 Different external shapes from stacking identical cubic unit cells. (*a*) Geometrical cube. (*b*) and (*c*) Malformed cubes. (*d*) Octahedron. (*e*) Dodecahedron.

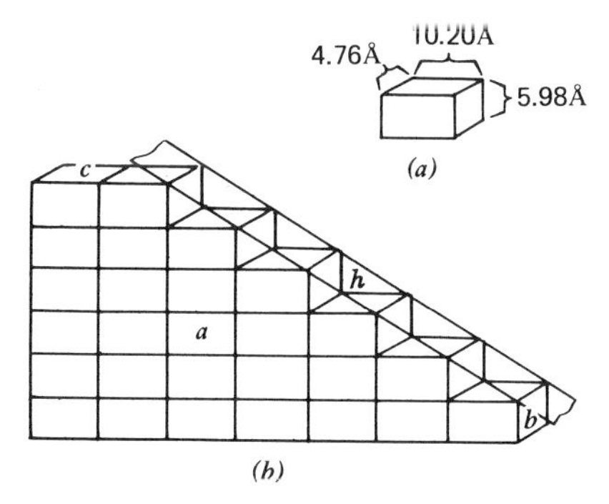

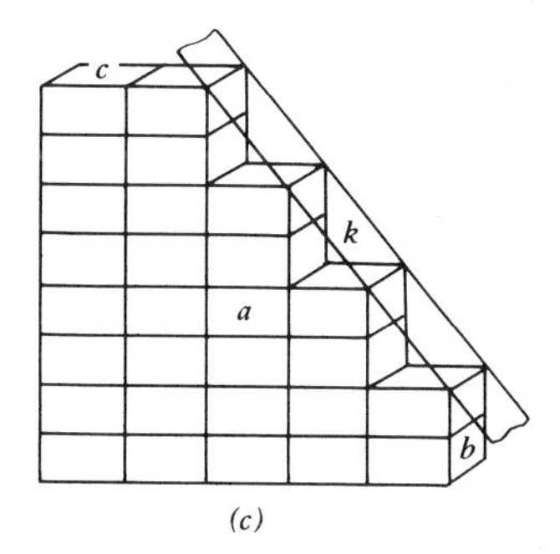

FIG. 4.5. (a) Unit cell of olivine. (b) Stacking of unit cells to produce face *h*. (c) Different stacking produces face *k*.

4.5*b*) with the bottom row the longest and the length of each higher row one unit less than the row below. The sloping "crystal face" *h* is thus formed. A different stacking of the same units is shown in Figure 4.5*c*. Here with the bottom pair of rows the same length and one unit longer than the next higher pair of rows, etc., "face" *k* is developed at a steeper angle than "face" *h* (Fig. 4.5*b*). In crystals the building units are so small that instead of seeing steplike surface, we see only smooth planes.

Because the units are identical with known dimensions, the angles that "faces" *h* and *k* make with c can be calculated: $c \wedge h = 149°37'$ and $c \wedge k = 129°25'$. The crystal drawing of olivine (Fig. 4.6) shows *h* and *k* with other faces that have developed in the same manner as those described.*

Since the unit cell dimensions of olivine of a given composition are constant, the angles between like faces are also constant no matter where the crystal is found or how it is formed. This observation of the constancy of interfacial angles was first made on another mineral, quartz, in 1669 by Nicolas Steno. Today his name is given to the generalization known as Steno's law, which states that *the angles between equivalent faces of crystals of the same substance are constant.* For this reason the measurement of interfacial angles is a valuable means of mineral identification. A mineral may be found in crystals differing greatly in size and shape, but the angles between corresponding faces are always the same.

The smooth plane surfaces of a well-formed crystal resemble in many respects the facets of a cut gemstone. Although both have a geometrical regularity, there is a fundamental difference between them. The symmetrical disposition of facets on a cut gem results from the design of the lapidary, whereas the crystal faces, as an outward expression of internal order, reflect the symmetry of the crystal structure. Symmetry is an important concept in crystallography for by its recognition a crystal can be assigned to one of the six crystal systems.

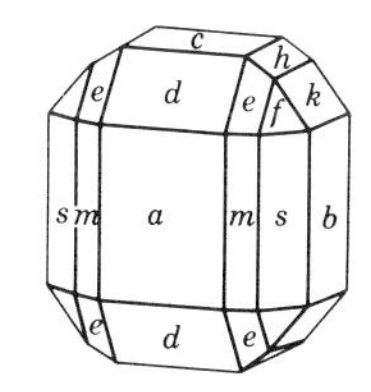

FIG. 4.6. Olivine crystal.

* In Figure 4.6 and in most of the other crystal drawings appearing later, the crystal faces are lettered to make reference to them easy. Conventionally the front face is *a*, the side face *b*, and the top face *c*. Most of the other letters have been arbitrarily chosen by the person describing the crystal.

Crystal Symmetry

Close examination of well-formed crystals of different minerals reveals that the quality of the faces differs; even on the same crystal some may be brilliant, others dull and pitted. Furthermore, it will be observed that the faces and angles between them

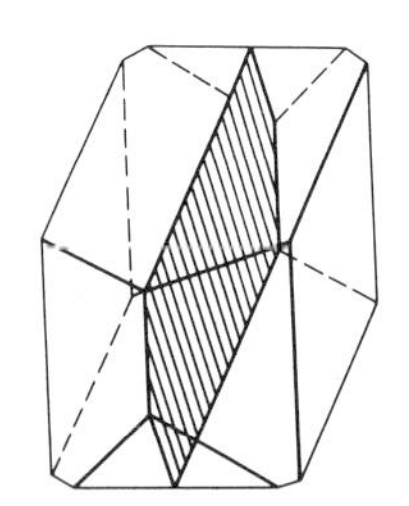

FIG. 4.7. Symmetry plane.

are not haphazard but have a regularity of arrangement. Note this regular distribution on the olivine crystal (Fig. 4.6). On one crystal all faces may be similar, whereas on another only two faces are similar and these may be located on opposite sides of the crystal. In short, it will be found that like faces have a symmetrical arrangement, the same for all crystals of a given mineral. Externally crystals exhibit three types of symmetry, known as the *symmetry elements:* symmetry across a plane, symmetry about a line (axis), and symmetry about a point.

Symmetry Plane

An imaginary plane passed through a crystal that divides the crystal into two halves so that each half is the mirror image of the other is a symmetry plane. It is sometimes called a *mirror plane,* for if a crystal could be split in two and one-half held against a mirror, the reflected image appears to restore the complete crystal (Fig. 4.7). Some crystals have as many as nine planes of symmetry but others have none.

Symmetry Axis

A symmetry axis is an imaginary line through the crystal about which it can be rotated and repeat itself two or more times during a complete rotation. Symmetry axes may be 1-, 2-, 3-, 4-, or 6-fold. Figure 4.8 illustrates both 4-fold and 2-fold axes. If rotated about CC', the crystal assumes the identical appearance four times during a 360° rotation; however, if rotated on AA^3 or A^2A^4, it repeats itself only twice during a complete rotation. A 1-fold axis is included in the list for completeness, but we do not consider it, for rotation on a 1-fold axis merely returns the crystal to its initial position.

Symmetry Center

A crystal is said to have a center of symmetry if an imaginary line can be passed from any point on its surface through its center to an identical point on the opposite side. Symmetry centers may be found in crystals having both axes and planes of symmetry as in Figures 4.7 and 4.8. Figure 4.9 has a center as its only symmetry element. Similar and parallel faces on opposite sides of the crystal indicate a symmetry center.

Crystal Classes

Crystals are grouped by their symmetry, that is, by the combination of planes, axes, and center, into crystal classes. Surprisingly the number of such groups is not large for the presence of one symmetry element places restrictions on the others. There are

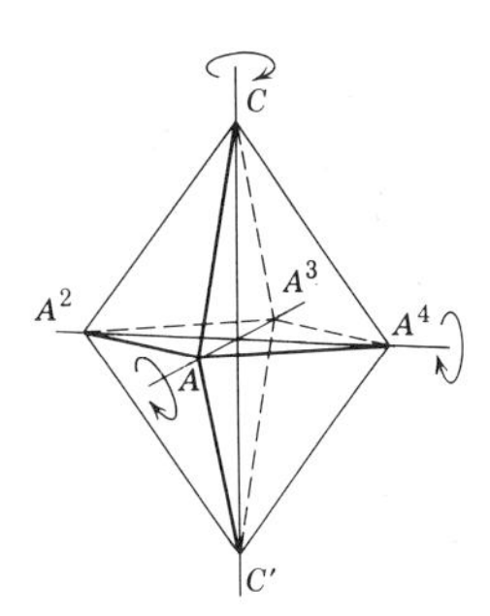

FIG. 4.8. Symmetry axes. AA^3 and A^2A^4 are 2-fold axes. CC' is a 4-fold axis.

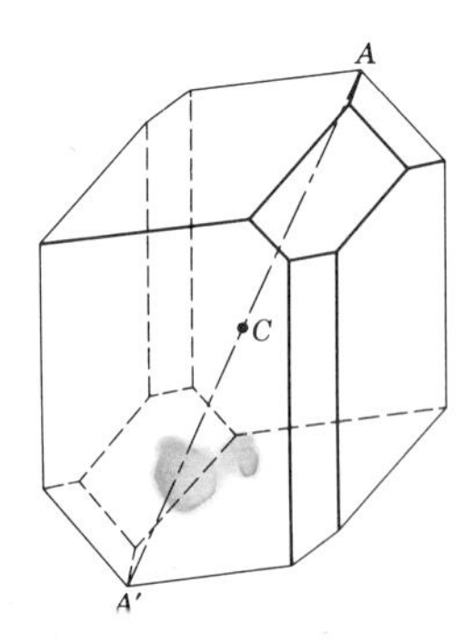

FIG. 4.9. Symmetry center.

only 32 crystal classes, the same 32 symmetrical arrangements in which we saw atoms could be arranged about a point (see *Introduction* to this chapter). Thus the crystal classes and point groups have like symmetry and are designated by the same names or symbols. All the crystal classes whose building units (unit cells) have the same basic shape have certain symmetry aspects in common and are thus grouped into one of the six crystal systems. Only one class in each crystal system is considered, the one having the highest symmetry as represented by the shape of the unit cell.

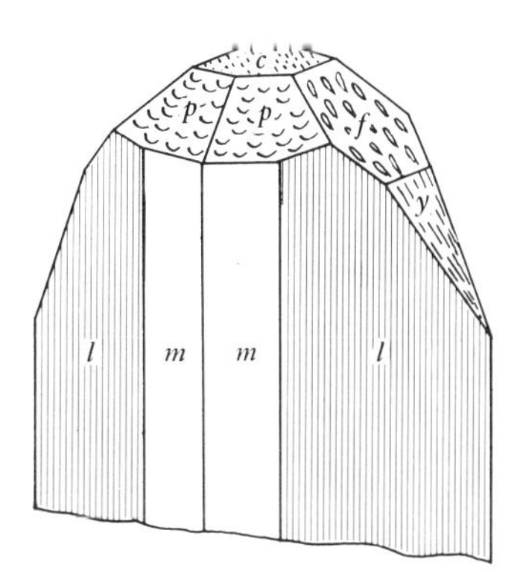

FIG. 4.11. Topaz crystal showing different appearance of different forms.

Crystal Axes

Crystals are conventionally described in relation to lines passing through the crystal that serve as axes of reference. These imaginary lines, the *crystallographic axes,* are ideally parallel to the edges of the unit cell and proportional to their lengths. However, in any given crystal when the unit cell dimensions and orientation are unknown, the axes are usually assumed to be parallel to the intersection of major faces. The symmetry may also be of help, for crystallographic axes are commonly either symmetry axes or perpendicular to symmetry planes.

All crystals with the exception of those belonging to the hexagonal system (discussed below under crystal systems) are referred to three crystallographic axes designated *a, b,* and *c*. In the general case (triclinic system) all axes are of different lengths and at oblique angles to each other. But for simplicity in describing their orientation consider those illustrated in Figure 4.10. Here three axes of different lengths are mutually perpendicular and when in the conventional orientation are as follows: the *a* axis is horizonal in a front–back position, the *b* axis is horizontal in a right–left position, and the *c* axis is vertical. The ends of each axis are designated plus or minus; the front end of *a,* the right-hand end of *b,* and the upper end of *c* are positive; the opposite ends are negative.

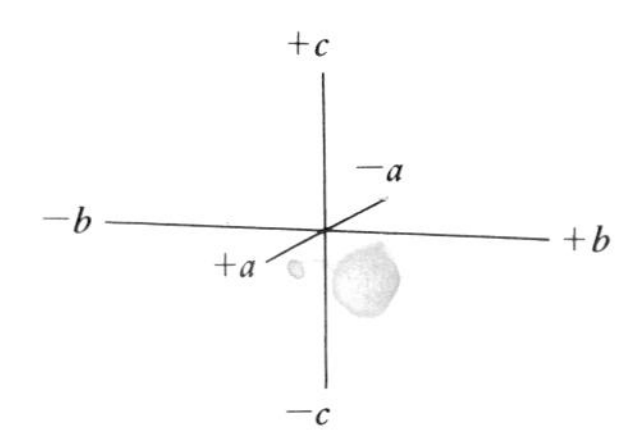

FIG. 4.10. Orthorhombic crystal axes.

Crystal Forms

Form in its most familiar meaning indicates the general outward appearance. In crystallography the term *habit* is used to express the external shape and *form* is used in a special and restricted sense. A crystal form comprises a group of crystal faces all of which have the same relation to the symmetry elements and to the crystallographic axes. Since the faces of a form are underlain by the same arrangement of atoms, they all have the same appearance. On a malformed crystal the faces of a form are of different shape and size but the similarity is evidenced by luster, etchings, or striations. When several forms are present on a single crystal, the faces of each form have a like appearance, as in the drawing of topaz (Fig. 4.11).

Crystal Systems

A rigorous description of a mineral requires that its crystal class be given, but for our study, a gem mineral is sufficiently characterized by indicating its crystal system. Consequently in the following discussion when we speak of symmetry and forms

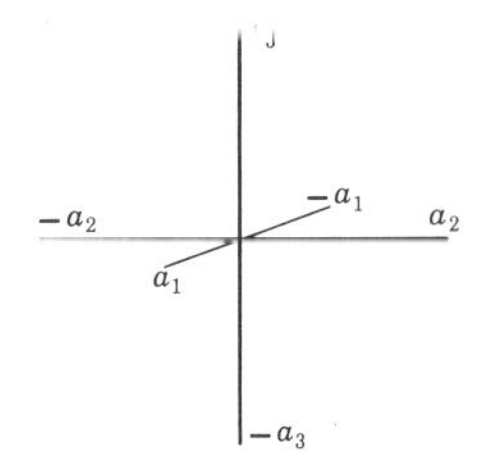

FIG. 4.12. Isometric crystal axes.

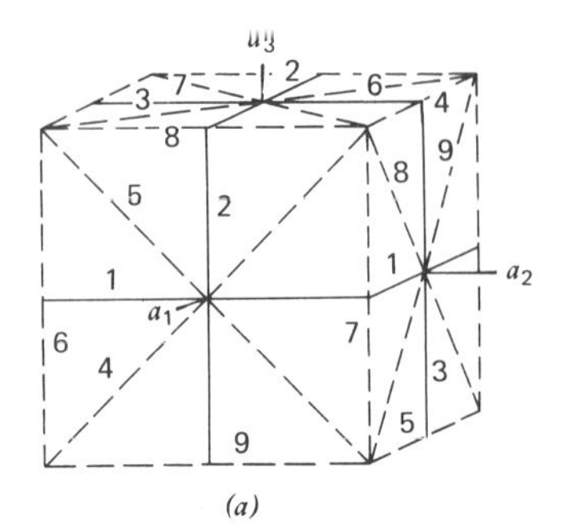

(a)

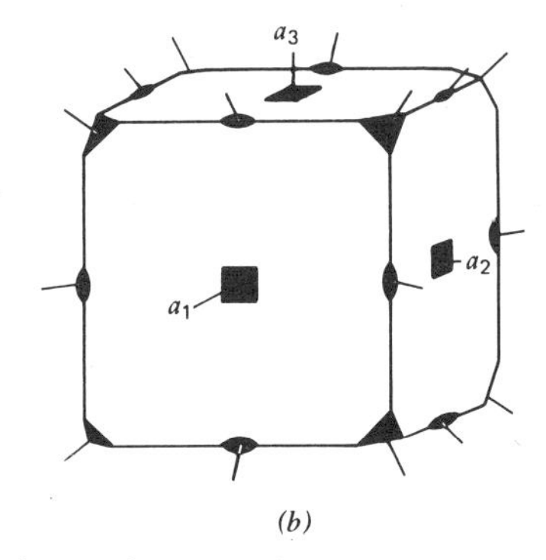

(b)

FIG. 4.13. Isometric crystal symmetry. (*a*) Symmetry planes. The traces of symmetry planes are shown on cube faces and edges. Solid lines (1–3) are axial planes with the crystal axes at their intersections. Broken lines (4–9) are diagonal planes, two of which are perpendicular to each cube face. (*b*) Symmetry axes. ■ = 4-fold axis, ▲ = 3-fold axis, ⬬ = 2-fold axis.

of a crystal system, we refer specifically to the symmetry and forms of the highest symmetry class in each system. Less than 15% of the gem minerals crystallize in lower symmetry classes.

Isometric System

Crystallographic Axes and Symmetry.

The unit cell of the isometric system is a cube. There are thus three equal crystallographic axes at right angles to each other (Fig. 4.12). Since they are interchangeable, they are all designated by the same letter, a; a_1 front–back, a_2 right–left, a_3 vertical. The positive and negative ends of the axes are indicated in Figure 4.12.

The isometric system has the highest crystal symmetry and the following symmetry elements: nine planes, three 4-fold axes (the crystallographic axes), four 3-fold axes, six 2-fold axes, and a center (Fig. 4.13).

Forms

All the forms of the isometric system are *closed forms;* that is, every form by itself encloses space and can appear alone on a crystal. This is in contrast to some forms in the other crystal systems that are *open forms* and must be combined with faces of other forms to enclose space. The description and illustration of each form are those of the perfect geometrical solid, but it must be remembered that in real crystals the faces are frequently unequally developed.

Cube. The cube (Fig. 4.14) has six square faces at right angles to each other. Because of the six faces it is sometimes called *hexahedron*. The cube is the commonest form shown by fluorite.

Octahedron. The octahedron is composed of eight equilateral triangular faces (Fig. 4.15). This is the form most frequently shown by diamond and spinel.

Dodecahedron. The dodecahedron has 12 diamond-shaped faces (Fig. 4.16). Because each face is a rhombus, it is sometimes called the rhombic dodecahedron to distinguish it from another 12-

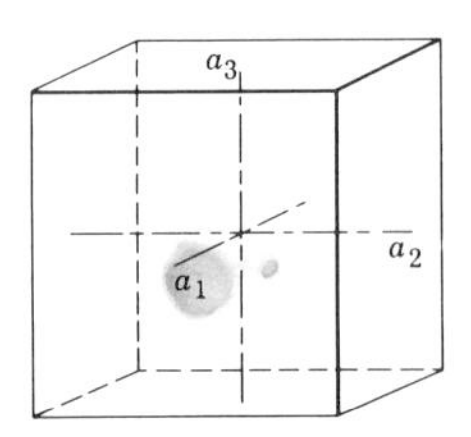

FIG. 4.14. Cube.

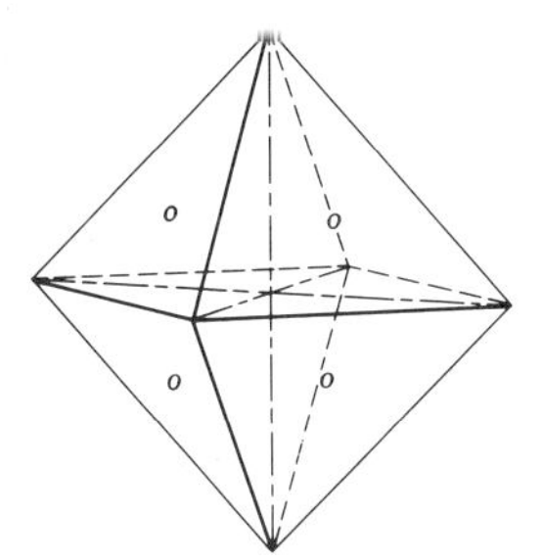

FIG. 4.15. Octahedron.

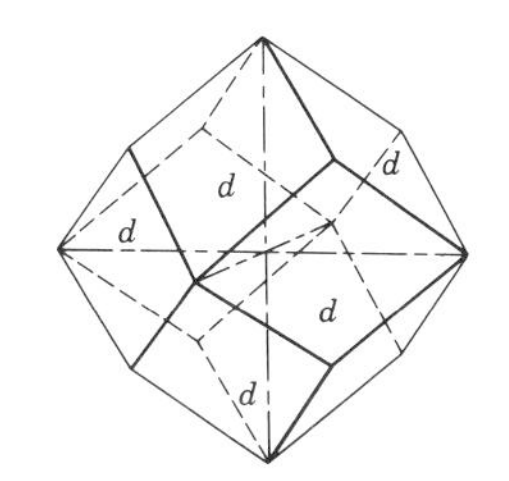

FIG. 4.16. Dodecahedron.

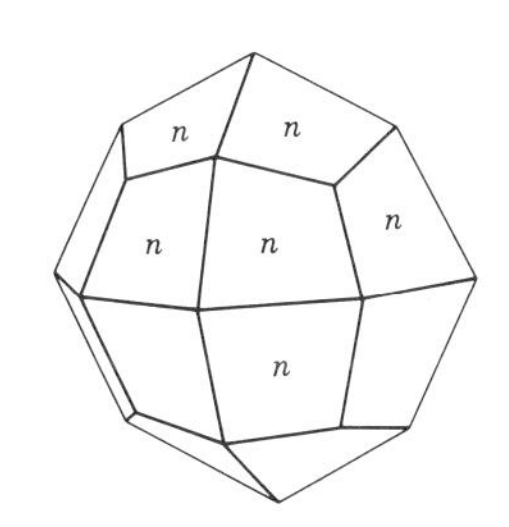

FIG. 4.17. Trapezohedron.

sided form of lower symmetry. The dodecahedron is the commonest form of garnet.

Trapezohedron. This form (Fig. 4.17) is composed of 24 faces and receives its name because each face is trapezium-shaped. The trapezohedron is frequently present on garnet crystals.

Each of the isometric forms mentioned may occur by itself or in combination with other forms as shown in Figure 4.18. It is interesting to note that in spite of their greatly differing shapes, the symmetry of all forms is the same.

Gem minerals crystallizing in the isometric system include diamond, spinel, garnet, and fluorite.

Hexagonal System

The hexagonal system is divided into the *hexagonal division* and the *rhombohedral division.* Crystals in the two divisions can be distinguished by their symmetry: hexagonal crystals have one 6-fold symmetry axis; rhombohedral crystals have one 3-fold axis.

Crystallographic Axes

The crystals of both divisions of the hexagonal system are referred to four crystallographic axes rather than to three as in the other crystal systems. Three of these axes, designated a (a_1, a_2, and a_3) are of equal length and lie in a horizontal plane at 120° to each other. The fourth axis, c, is vertical and either shorter or longer than the a axes. The conventional orientation of the axes is given in Figure 4.19 with plus and minus ends indicated.

Hexagonal Division, Symmetry

A 6-fold symmetry axis, coincident with the c crystallographic axis, is the most significant symmetry

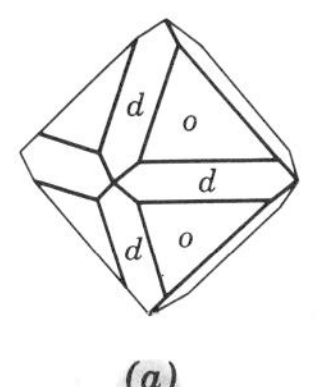

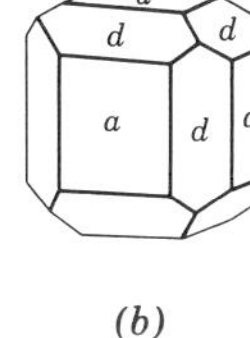

(c)

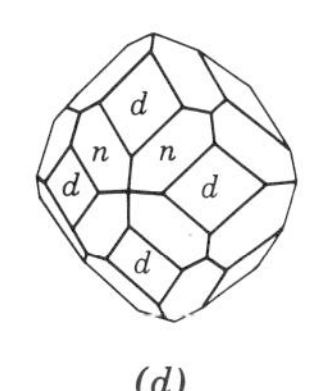

FIG. 4.18. Combinations of isometric forms. (*a*) Dodecahedron, (*d*) and octahedron, (*o*). (*b*) Cube, (*a*) and dodecahedron, (*d*). (*c*) Octahedron, (*o*), dodecahedron, (*d*), and cube, (*c*). (*d*) Trapezohedron, (*n*) and dodecahedron, (*d*).

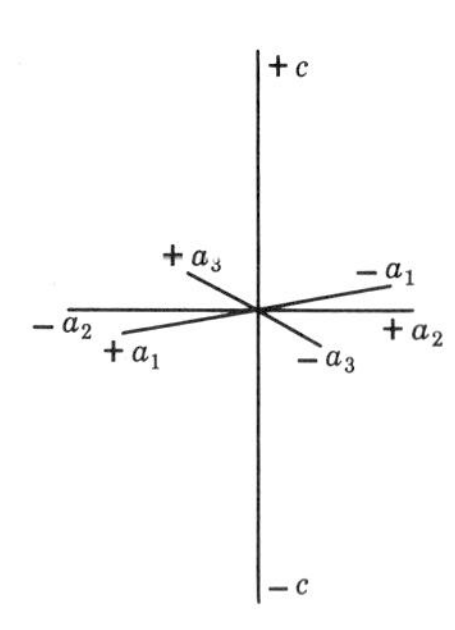

FIG. 4.19. Hexagonal crystal axes.

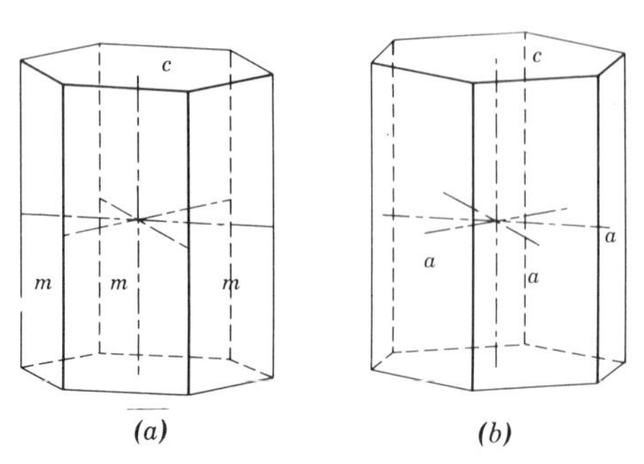

FIG. 4.21. Hexagonal prisms. (*a*) First-order. (*b*) Second-order.

element. In addition there are six 2-fold axes, three of which coincide with the *a* axes; the other three lie midway between them in a horizontal plane. There are seven symmetry planes; six are vertical and each includes a 2-fold axis. The seventh symmetry plane is horizontal (Fig. 4.20). A symmetry center is also present.

Hexagonal Division, Forms

Prisms. A prism is defined as a form composed of three or more faces whose intersection edges are mutually parallel. The *hexagonal prism* consists of six vertical faces. It is called first-order if each face is parallel to an *a* crystallographic axis (faces *m* Fig. 4.21*a*) and second-order if the faces are perpendicular to the *a* axes (faces *a* Fig. 4.21*b*). The dihexagonal prism has 12 faces parallel to the *c* axis none of which is parallel or perpendicular to an *a* axis (faces *n* Fig. 4.22). All these prisms may be present on the same crystal, but since they are open forms they can exist only in combination with other forms.

Dipyramids. A pyramid is defined as a form composed of three or more faces that intersect at a point. A dipyramid can be considered to be two identical pyramids base to base. There are two hexagonal dipyramids, first- and second-order, each composed of 12 faces, six above and six below (Figs. 4.23*a* and *b*). The dihexagonal dipyramid has 24 faces, 12 above and 12 below (Fig. 4.24). The faces of all the dipyramids intersect the *c* axis, but the intersections with the *a* axes correspond to those in prisms.

Pinacoid. A pinacoid is a form composed of two parallel faces. In the hexagonal system there is a *basal pinacoid* made up of two horizontal faces. This is an open form and is shown, faces *c*, in combination with prisms in Figures 4.21 and 4.22.

Rhombohedral Division, Symmetry

There are one 3-fold axis coinciding with the *c* crystallographic axis, three 2-fold axes coincident with the horizontal *a* axes, three symmetry planes

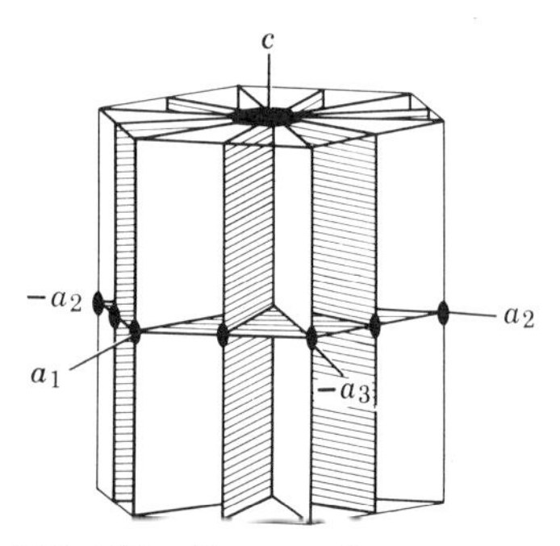

FIG. 4.20. Hexagonal symmetry.

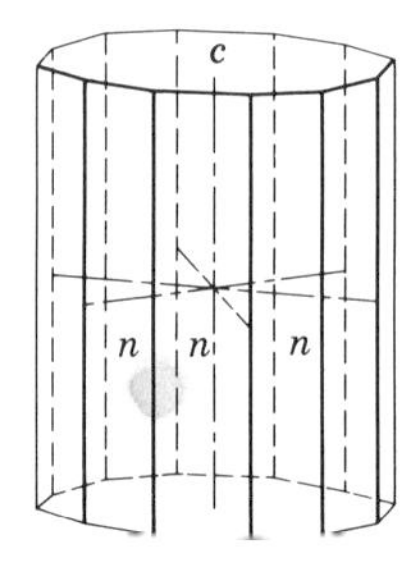

FIG. 4.22. Dihexagonal prism.

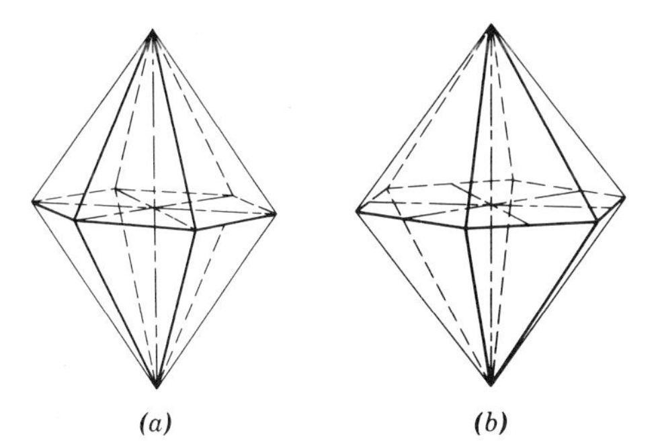

FIG. 4.23. Hexagonal dipyramids. (*a*) First-order. (*b*) Second-order.

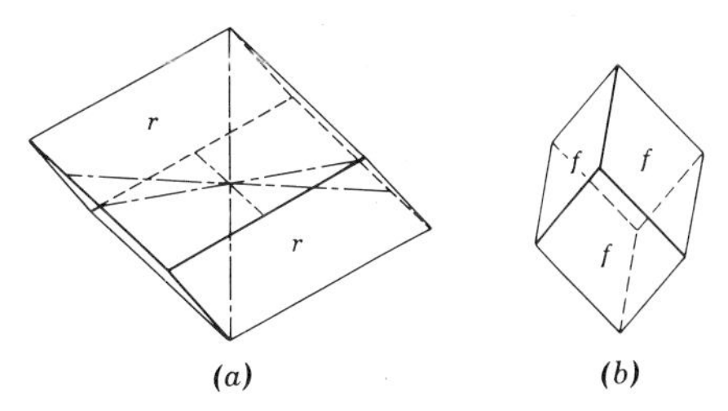

FIG. 4.26. Rhombohedrons.

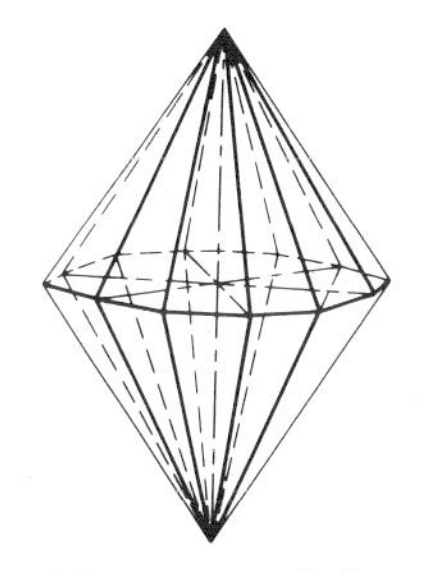

FIG. 4.24. Dihexagonal dipyramid.

perpendicular to the 2-fold axes, and a symmetry center (Fig. 4.25).

Rhombohedral Division, Forms

Rhombohedron. The rhombohedron is a six-sided form, each face of which is a rhombus (diamond-shaped). The rhombohedron resembles a cube that has been compressed (Fig. 4.26*a*) or elongated along one of its 3-fold axes (Fig. 4.26*b*).

Scalenohedron. The scalenohedron (Fig. 4.27) is a 12-sided form somewhat resembling a dipyramid. However, each face is a scalene triangle, and the six upper faces and six lower faces meet forming a zigzag edge.

All the forms of the hexagonal division can be

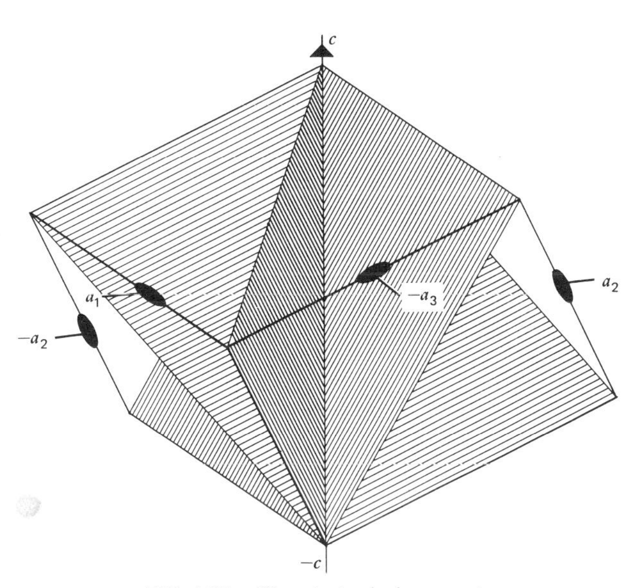

FIG. 4.25. Rhombohedral symmetry.

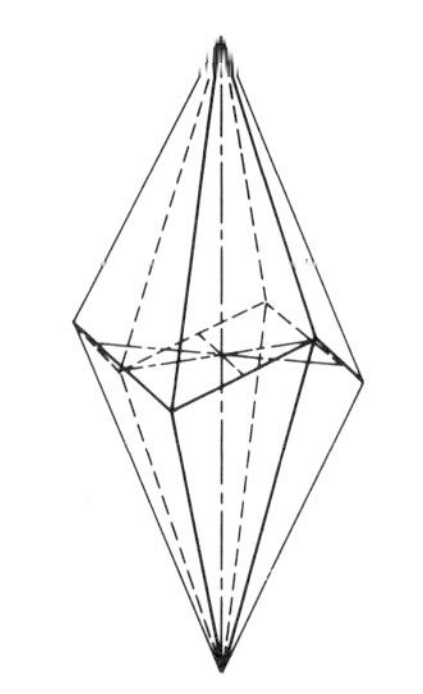

FIG. 4.27. Scalenohedron.

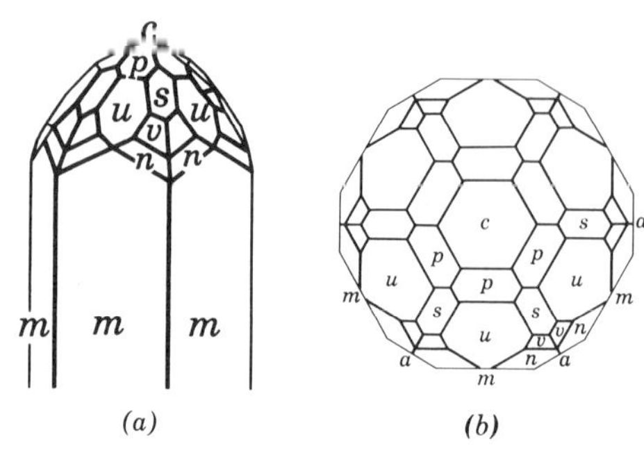

FIG. 4.28. Beryl. (*a*) Multifaced crystal. (*b*) Top view showing the forms of (*a*) and in addition the second-order prism, *a*.

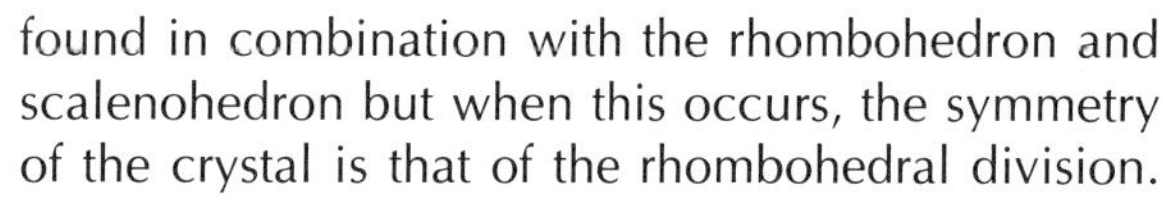

found in combination with the rhombohedron and scalenohedron but when this occurs, the symmetry of the crystal is that of the rhombohedral division.

Outstanding examples of hexagonal gem minerals are beryl (Fig. 4.28) (hexagonal division) and corundum (Fig. 4.29) (rhombohedral division).

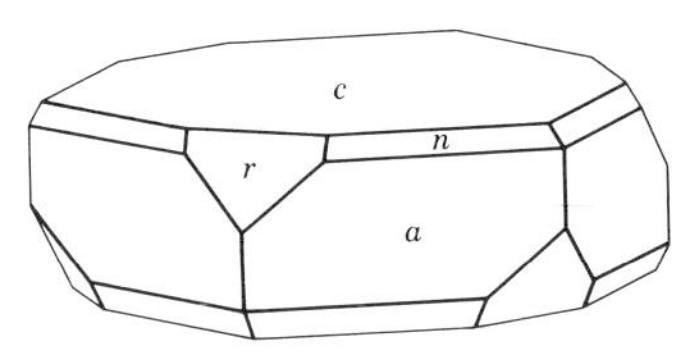

FIG. 4.29. Corundum crystal (typical of ruby).

Tetragonal System

Crystallographic Axes and Symmetry

Tetragonal crystals are referred to three crystallographic axes at right angles to each other. The horizontal axes, designated as a_1 and a_2, are equal and interchangeable, but the vertical axis, *c*, is of different length (Fig. 4.30). The symmetry is as follows: one 4-fold axis, coincident with the *c* axis; four 2-fold axes two of which coincide with the *a* axes, the other two at 45° to them in the horizontal plane; five planes, of which four are vertical and include a 2-fold axis, and the fifth is horizontal; and a symmetry center (Fig. 4.31).

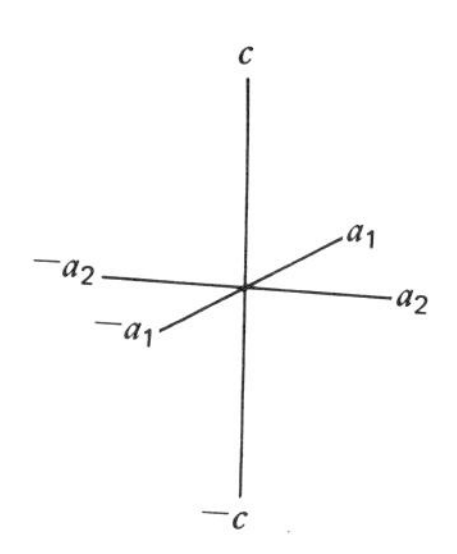

FIG. 4.30. Tetragonal crystal axes.

Forms

Prisms. There are two geometrically identical tetragonal prisms each with four vertical faces at right angles to each other. The first-order prism faces intersect both horizontal *a* axes; the second-order prism faces are perpendicular to one *a* axis and are parallel to the other (Figs. 4.32*a* and *b*). The ditetragonal prism is a form with eight vertical faces all of which intersect both *a* axes (Fig. 4.32 *c*). Since all

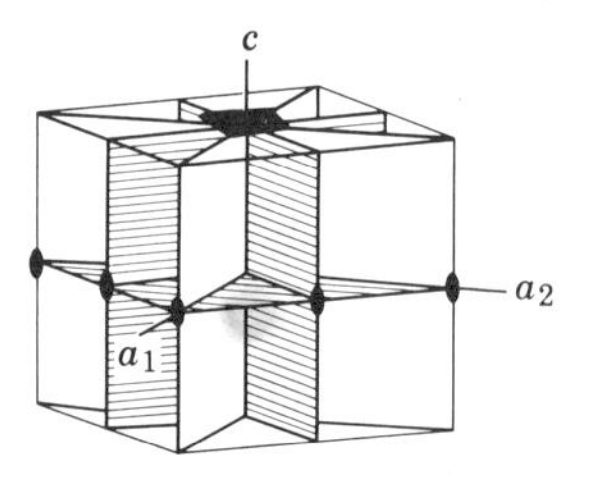

FIG. 4.31. Tetragonal symmetry.

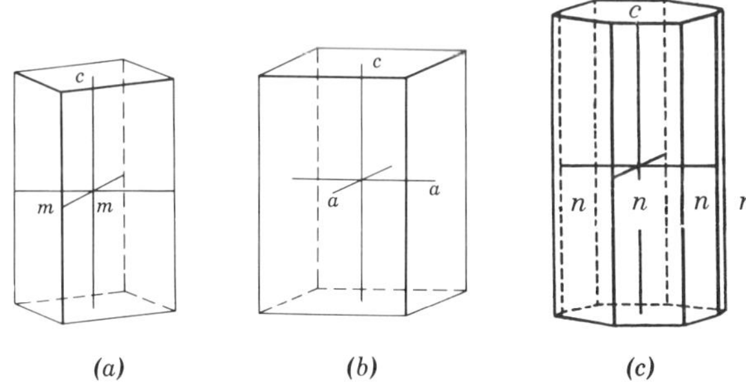

FIG. 4.32. Tetragonal prisms. (*a*) First-order. (*b*) Second-order. (*c*) Ditetragonal prism.

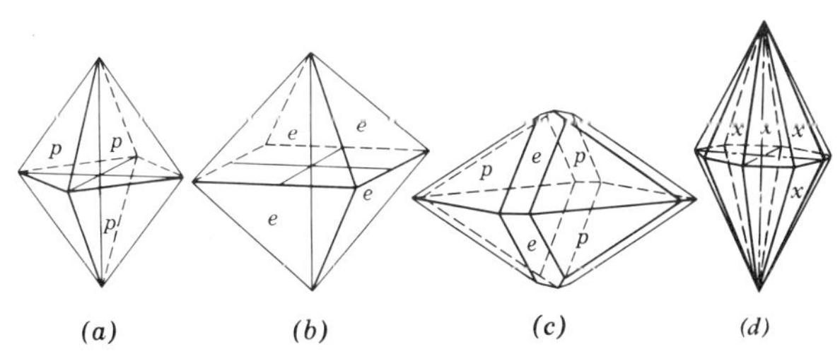

FIG. 4.33. Tetragonal dipyramids. (*a*) First-order. (*b*) Second-order. (*c*) Combination of first- and second-order. (*d*) Ditetragonal dipyramid.

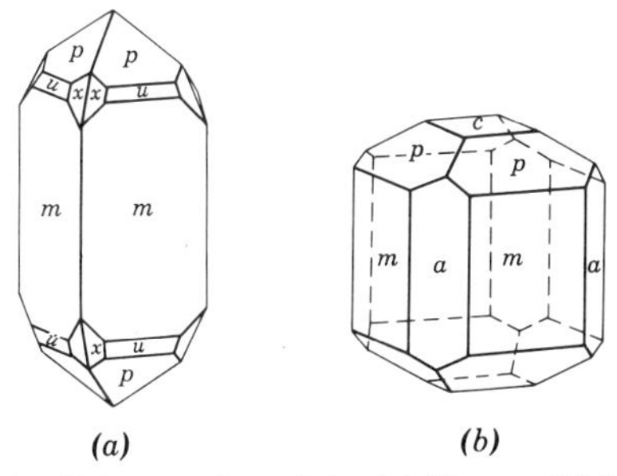

FIG. 4.34. Tetragonal crystals. (*a*) Zircon. (*b*) Idocrase.

the prisms are open forms, they occur on crystals only in combination with other forms.

Dipyramid. There are two eight-faced tetragonal dipyramids, first- and second-order, each with four faces above and four below (Figs. 4.33*a* and *b*). The ditetragonal dipyramid has 16 faces, eight above and eight below (Fig. 4.33*d*). The faces of all dipyramids intersect the *c* axis, but the intersections on the *a* axes are the same as the corresponding prisms.

Pinacoid. As in the hexagonal crystal system there is a basal pinacoid composed of two horizontal faces. It is shown (faces *c*) in combination with prisms in Figures 4.32*a* and *b*.

The gem minerals zircon (Fig. 4.34*a*) and idocrase (Fig. 4.34*b*) are found frequently in well-formed tetragonal crystals.

Orthorhombic System

Crystallographic Axes and Symmetry

The forms of orthorhombic crystals are referred to three unequal crystallographic axes at right angles to each other. The conventional orientation of these axes, designated *a, b, c,* is shown in Figure 4.35. There is no fixed rule as to which axis is the longest or the shortest. The symmetry is as follows: three 2-fold axes each coinciding with a crystallographic axis, three planes each perpendicular to a 2-fold axis, and a symmetry center (Fig. 4.36).

Forms

Prisms. Orthorhombic prisms have four faces which are parallel to one axis and intersect the other

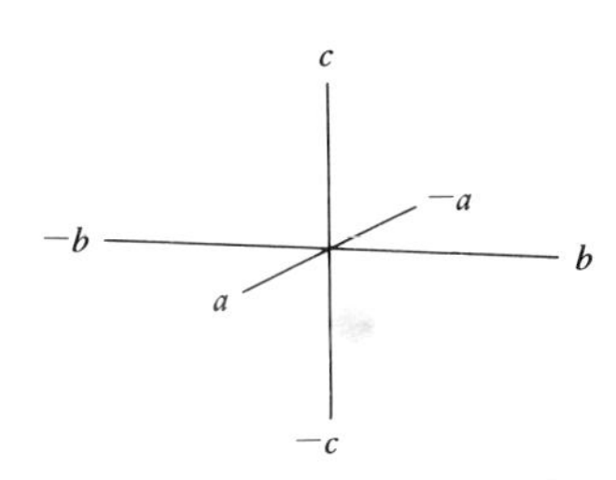

FIG. 4.35. Orthorhombic crystal axes.

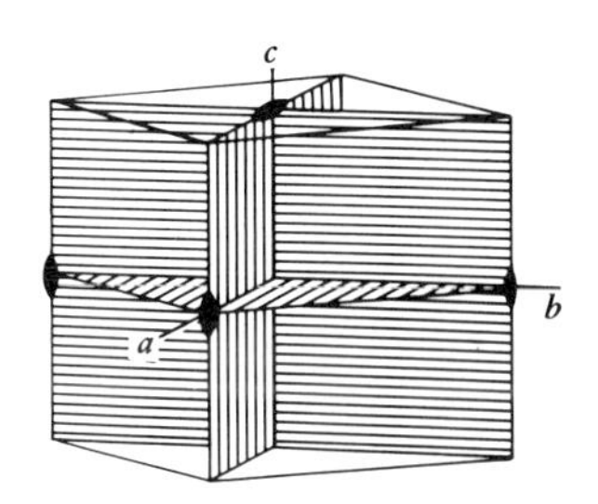

FIG. 4.36. Orthorhombic symmetry.

two. There are thus three types (Fig. 4.37) as follows:

First-order: parallel to *a* (first axis); intersects *b* and *c*.
Second-order: parallel to *b* (second axis); intersects *a* and *c*.
Third-order: parallel to *c* (third axis); intersects *a* and *b*.

The first- and second-order prisms are sometimes called horizontal prisms and the third-order a vertical prism. Since all prisms are parallel to one axis and intersect the other two, one prism can be transformed to another by a different choice of axes.

Dipyramids. The orthorhombic dipyramid has eight triangular faces, each of which intersects all the crystallographic axes (Fig. 4.38).

Pinacoids. There are three pinacoids whose faces are perpendicular to one axis and parallel to the other two. They are shown in combination with the three prisms (Fig. 4.37) and are as follows:

Front pinacoid. Intersects *a;* parallels *b* and *c*.
Side pinacoid. Intersects *b;* parallels *a* and *c*.
Basal pinacoid. Intersects *c;* parallels *a* and *b*.

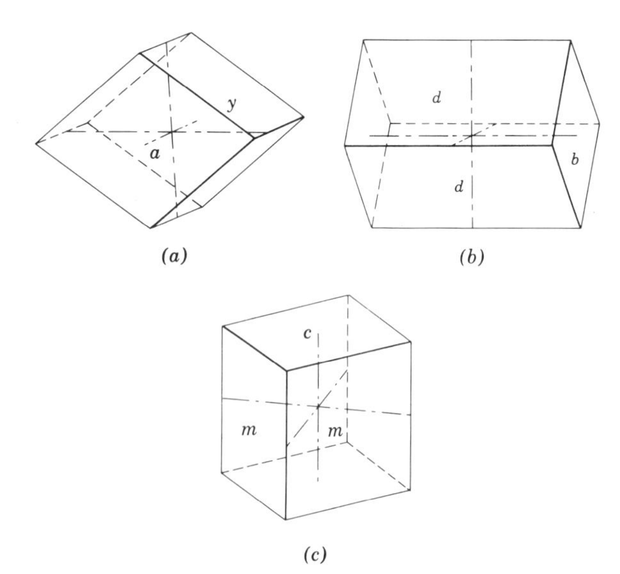

FIG. 4.37. Orthorhombic prisms and pinacoids. (*a*) First-order prism, *y*, front pinacoid, *a*. (*b*) Second-order prism, *d*, side pinacoid, *b*. (*c*) Third-order prism, *m*, basal pinacoid, *c*.

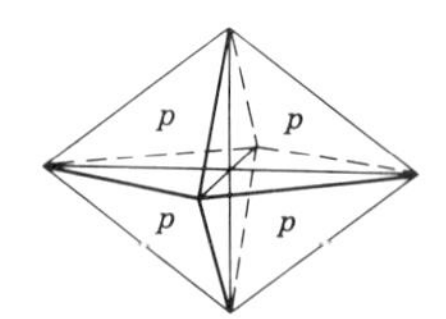

FIG. 4.38. Orthorhombic dipyramid.

As is the case with the prisms, the pinacoids can be interchanged depending on the choice of axes.

Orthorhombic representatives of gem minerals are topaz and staurolite (Figs. 4.39*a* and *b*) and peridot (see Fig. 4.6).

Monoclinic System

Crystallographic Axes and Symmetry

Monoclinic crystals are referred to three unequal axes, *a, b,* and *c*. The *c* axis is vertical, *b* is right–left at right angles to *c,* and *a* is also at right angles to *b* but is inclined downward toward the front. The angle between +*a* and +*c* is known as β (Fig. 4.40). The sloping *a* axis reduces the symmetry to one 2-fold axis coincident with *b,* one plane (the *a–c* plane) perpendicular to the 2-fold axis, and a symmetry center (Fig. 4.41).

Forms

Only two types of forms are present in the monoclinic system, prisms and pinacoids.

Prisms. In the monoclinic system prisms are four-faced forms. The first- and third-order prisms are similar to those of the orthorhombic system: first-order parallels *a,* intersects *b* and *c;* third-order

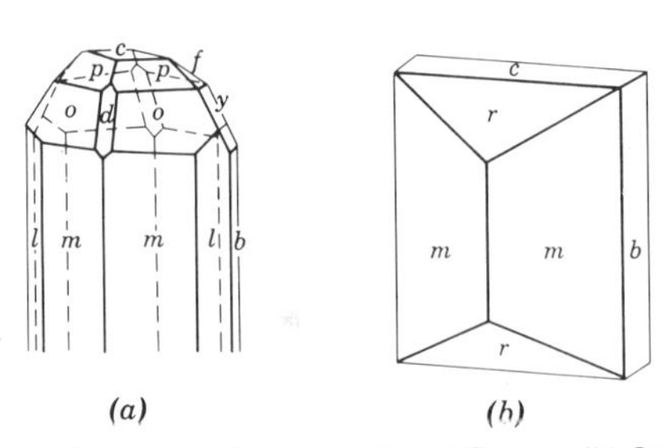

FIG. 4.39. Orthorhombic crystals. (*a*) Topaz. (*b*) Staurolite.

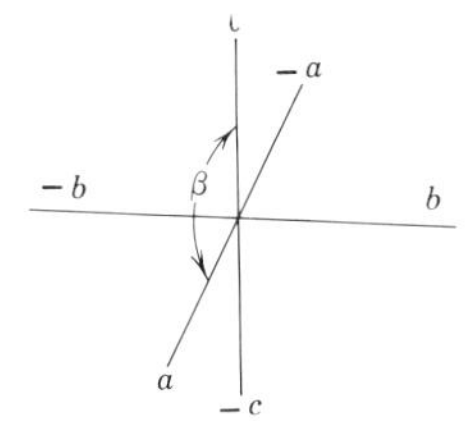

FIG. 4.40. Monoclinic crystal axes.

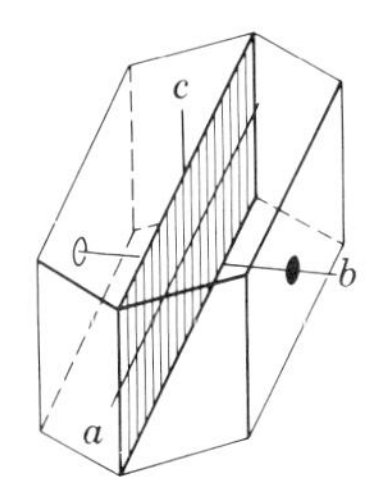

FIG. 4.41. Monoclinic symmetry.

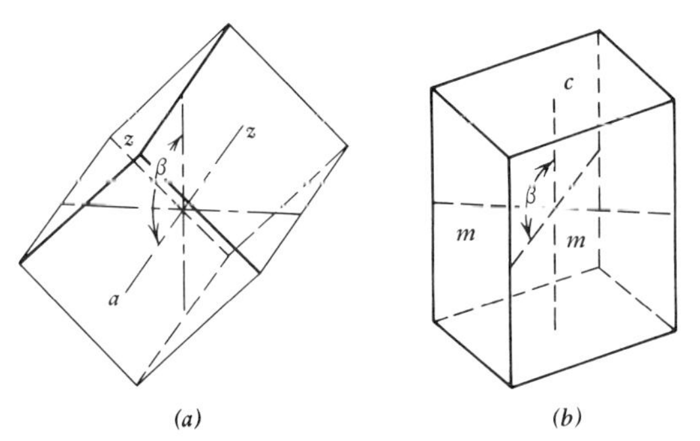

FIG. 4.42. Monoclinic prisms. (*a*) First-order, *z*, with front pinacoid, *a*. (*b*) Third-order, *m*, with basal pinacoid, *c*.

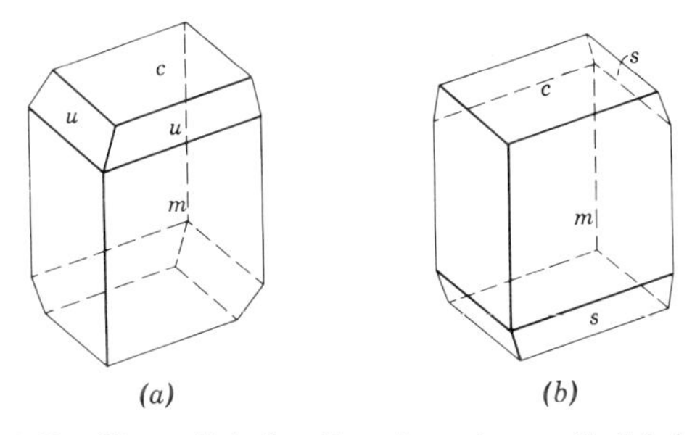

FIG. 4.43. Monoclinic fourth-order prisms with third-order prisms, *m*, and basal pinacoids, *c*. (*a*) Positive, *u*. (*b*) Negative, *s*.

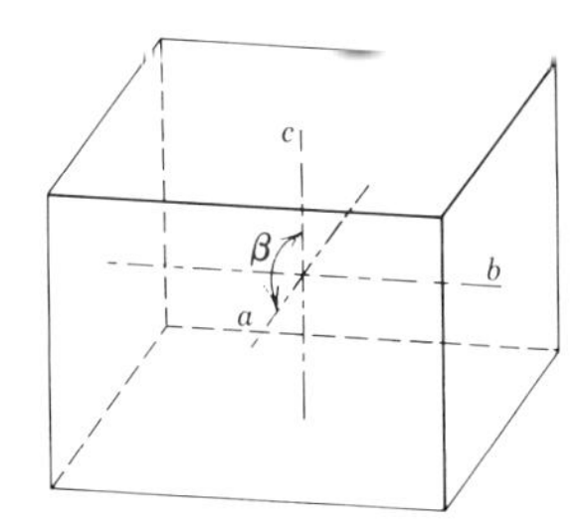

FIG. 4.44. Monoclinic pinacoids: *a* front, *b* side, *c* basal.

parallels *c*, intersects *a* and *b* (Fig. 4.42). Because the opposite ends of the *a* axis are not interchangeable, there is no second-order prism. For this same reason there can not be an eight-faced pyramid but there are two four-faced prisms instead. The faces *u* (Fig. 4.43*a*) at the top front of the crystal belong to the positive fourth-order prism, whereas faces *s* (Fig. 4.43*b*) at the top back of the crystal are of the negative fourth-order prism.

Pinacoids. As in the orthorhombic system there exist the three pinacoids (Fig. 4.44):

Front pinacoid. Intersects *a*; parallels *b* and *c*.
Side pinacoid. Intersects *b*; parallels *a* and *c*.
Basal pinacoid. Intersects *c*; parallels *a* and *b*.

Because of the sloping *a* axis, the four-faced second-order prism of the orthorhombic system becomes two pinacoids: positive second-order pinacoid parallel to *b* (second axis), intersecting +*a* and +*c*; negative second-order pinacoid parallel to *b*, intersecting −*a* and +*c* (Fig. 4.45).

Many gem minerals crystallize in the monoclinic system including nephrite, epidote (Fig. 4.46*b*), orthoclase (Fig. 4.46*a*), sphene, and diopside.

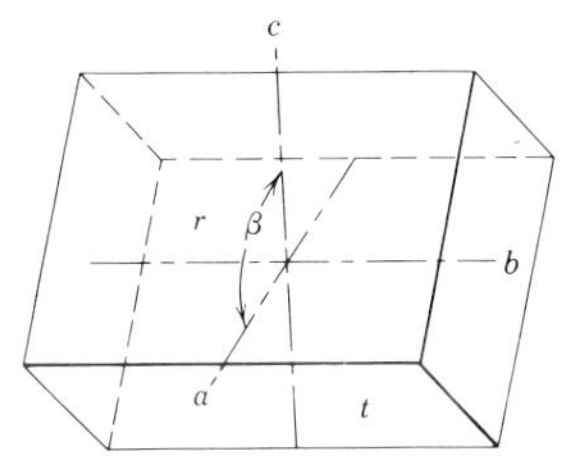

FIG. 4.45. Second-order pinacoids and side pinacoid. Faces *r* positive, faces *t* negative.

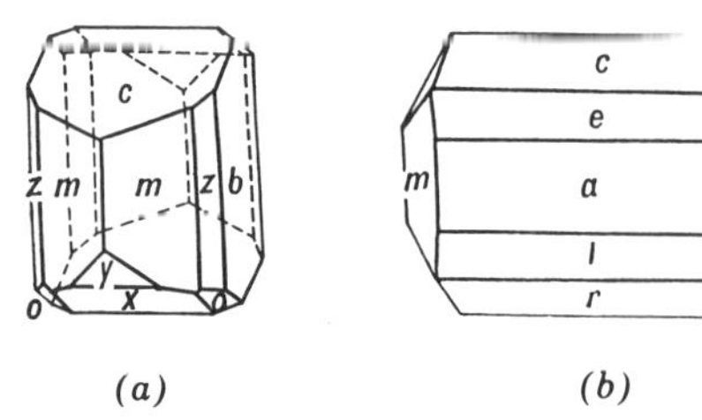

FIG. 4.46. Monoclinic crystals. (a) Orthoclase. (b) Epidote.

Triclinic System

Crystallographic Axes and Symmetry

In the triclinic system the crystallographic axes are all of unequal length and at oblique angles to each other. Symmetry planes and axes are absent and a center is the only symmetry element.

Forms

The pinacoid is the only form. Thus faces are arranged in pairs on opposite sides of the crystal as shown in the drawings of axinite and albite (Fig. 4.47). Kyanite, rhodonite, and microcline are also triclinic gem minerals.

Twin Crystals

In our discussion of morphological crystallography only well-formed single crystals were considered. In nature such ideal geometrical forms are rare and it is far more common to find crystals in aggregates, sometimes in parallel growths, or more frequently in completely random orientation to each other. Occasionally one finds two or more crystals of the same mineral that have grown together in such a way that certain lattice directions are parallel whereas other lattice directions are in reverse position (Fig. 4.48). These are known as *twins* or *twin crystals*.

The members of a twin are related to each other in two ways: (1) as though one individual were derived from the other by reflection over a plane (Fig. 4.49), the *twin plane;* or (2) as though one part had been rotated 180° with respect to the other about a crystal direction, common to both, the *twin axis.* The twin plane is usually parallel to a common crystal face but never parallel to a symmetry plane. A twin axis is commonly a symmetry axis but never one of even-fold symmetry (2-, 4-, 6-fold). Rotation on such an axis would bring all corresponding lattice planes into parallel position. In addition to defining a twin by twin axis or twin plane, called the *twin law,* it is further designated as either a contact twin or a penetration twin. In the former the individuals are in contact, usually on the twin plane (Fig. 4.50*a*), whereas in a penetration twin the two individuals interpenetrate one another (Fig. 4.50*b*).

When three or more individuals are twinned according to the same law, the aggregate is called

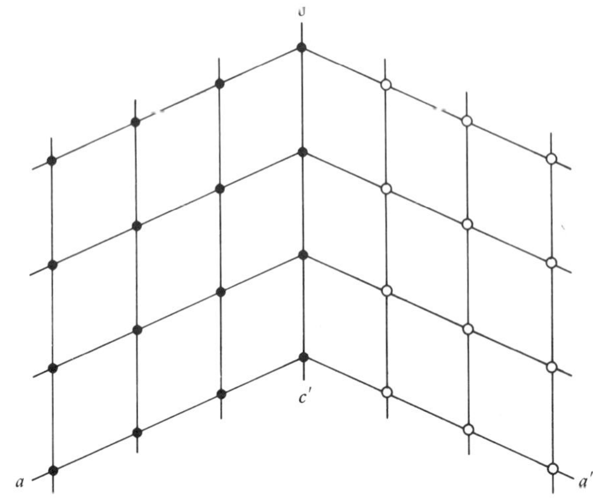

FIG. 4.48. Cross section through a monoclinic lattice showing a twin relation along *cc*′. On this line lattice points are common to the two individuals. The twin relation can be considered as 180° rotation about *cc*′ or as a mirror reflection along *cc*′.

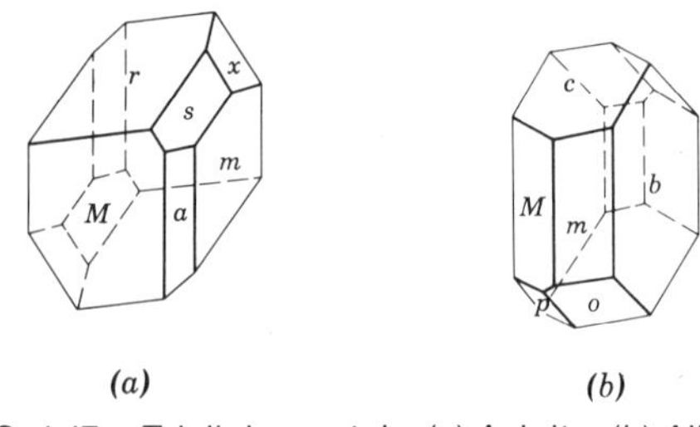

FIG. 4.47. Triclinic crystals. (*a*) Axinite. (*b*) Albite.

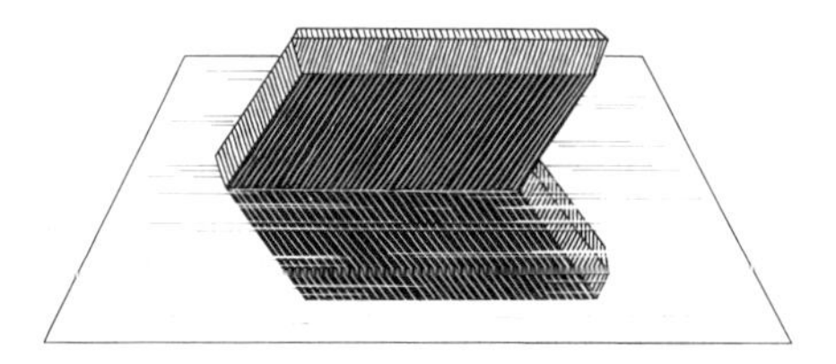

FIG. 4.49. Twinning by reflection.

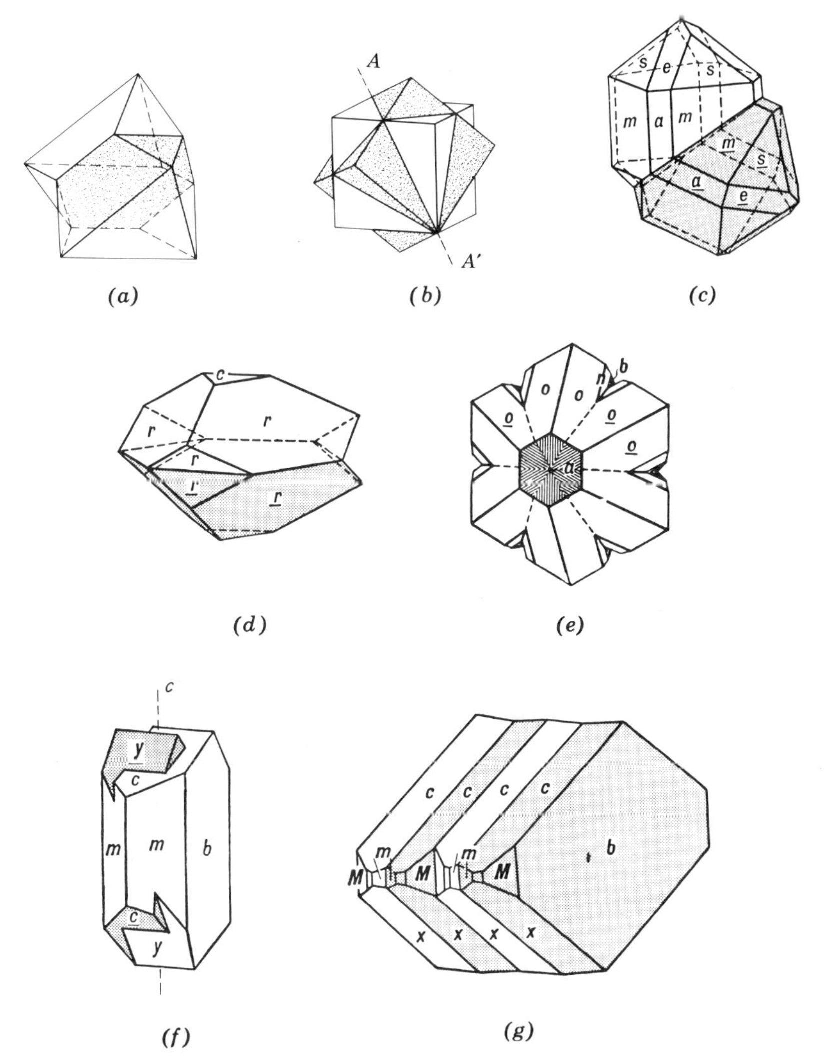

FIG. 4.50. Examples of twinned crystals in the different crystal systems. (*a*) Isometric: diamond, contact twin (called spinel twin). (*b*) Isometric: fluorite, penetration twin, *AA'* twin axis. (*c*) Tetragonal: cassiterite, contact twin. (*d*) Hexagonal: calcite, contact twin. (*e*) Orthorhombic: chrysoberyl, cyclic twin. (*f*) Monoclinic: orthoclase (Carlsbad) penetration twin. (*g*) Triclinic: albite, polysynthetic contact twins.

a *repeated* or *multiple twin*. If the twin planes are all parallel, the resulting group is called *polysynthetic* (Fig. 4.50g). Such closely spaced twin planes give rise to parallel striations as are characteristic of the plagioclase feldspars. It the twin planes are not parallel, but parallel to other faces of the same crystal form, a *cyclic* twin results (Fig. 4.50e). Chrysoberyl is frequently found in cyclic twins.

Twinning is found in crystals of all the crystal systems, but is most common in those of low symmetry. Figure 4.50 illustrates twins in each of the crystal systems.

References and Suggested Reading

Hurlbut, C. S. and C. Klein, 1977. *Manual of Mineralogy*, 19th ed. John Wiley & Sons, Inc., New York. Chapters 2 and 3.

5 PHYSICAL PROPERTIES OF GEM MATERIALS

The elements that make up gemstones and the way their atoms are packed together in a crystal structure impart to the gem various physical properties that characterize it. It is on the determination of these properties that gem identification largely rests, for only rarely must the gemologist resort to chemical or X-ray tests. The various ways in which gems react to light, although physical properties, are of such importance in gem study and identification that they are considered separately in Chapter 6.

Cleavage, Parting, and Fracture

Cleavage is the propensity of many crystalline substances to break parallel to certain atomic planes yielding smooth surfaces. Its presence indicates a weaker bonding across the planes or a greater spacing between them than elsewhere in the structure. Because cleavage is the splitting of a crystal between atomic planes, any parallel plane through the crystal is a potential cleavage. Moreover, it is always parallel to a possible crystal face, usually a face of a common form.

Cleavage is described by its quality, ease of production, and crystallographic plane. Quality is expressed as perfect, good, imperfect, poor, etc.; and plane is indicated by naming the crystal form to which it is parallel. Thus cleavage in diamond is perfect octahedral, in calcite perfect rhombohedral, in diopside imperfect prismatic, and in beryl poor pinacoidal. Cleavage is always consistent with the symmetry. Therefore, in octahedral cleavage there are four equivalent planes, in rhombohedral cleavage three, in prismatic cleavage (orthorhombic and monoclinic) two, but in pinacoidal cleavage only one (Fig. 5.1).

In rough material cleavage is easily observed, but in a cut stone there may be little or no evidence of it. The gem should be examined with a lens or microscope for smooth interior reflecting surfaces indicating incipient cleavage cracks. Also, if a stone is mounted, it should be examined carefully near the points of contact with the prongs; the pressure of mounting may have produced tiny cleavages. Since cleavage is a constant property, its detection is an excellent diagnostic criterion. For example, quartz and topaz resemble one another but close examination frequently reveals the single pinacoidal cleavage of topaz, whereas quartz has no cleavage.

Care should be exercised in wearing a gem possessing a good cleavage. For, although it may appear to have all the attributes of a durable stone, being dropped or being dealt an inadvertent sharp blow may cause it to cleave.

Parting is the breaking of a crystal along planes of structural weakness usually resulting from polysynthetic twinning. Parting is sometimes called *false cleavage* for it resembles cleavage in that both are parallel to atomic planes. However, parting takes place only along discrete twin planes with an appreciable distance between them, whereas the distance between potential cleavage planes is their interatomic spacing. For example, corundum has no cleavage but is commonly polysynthetically twinned on both the rhombohedron (Fig. 5.2) and the pinacoid and may thus have four planes of parting. Between these parting planes, the mineral breaks with a conchoidal fracture.

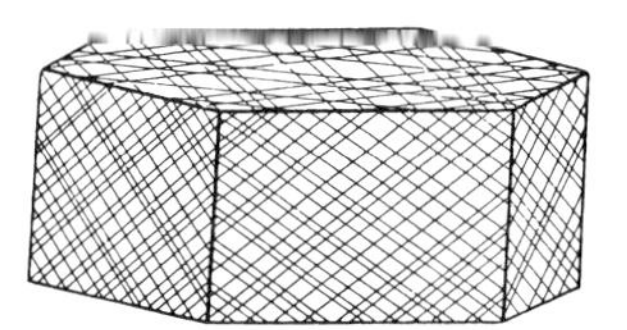

FIG. 5.2. Rhombohedral parting, corundum (ruby).

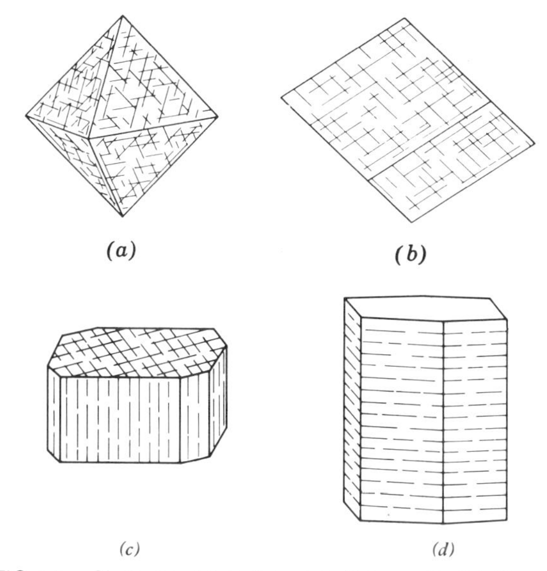

FIG. 5.1. Cleavage. (*a*) Octahedral (diamond). (*b*) Rhombohedral (calcite). (*c*) Prismatic (diopside). (*d*) Pinacoidal (beryl).

Fracture is the way a substance breaks other than along cleavage or parting planes. The most distinctive type of fracture is *conchoidal* (shell-like), that is, a break yielding smooth concave surfaces. This is the familiar type of fracture in glass and thus is characteristic of natural glass, obsidian (Fig. 5.3). Quartz, chalcedony, opal, and olivine all break with a conchoidal fracture. Other types of fracture are described by common adjectives such as even, uneven, rough, and splintery.

Hardness

The resistance a gem mineral offers to scratching or abrasion is its hardness which, in the description of individual gems, is designated by **H**. The resistance

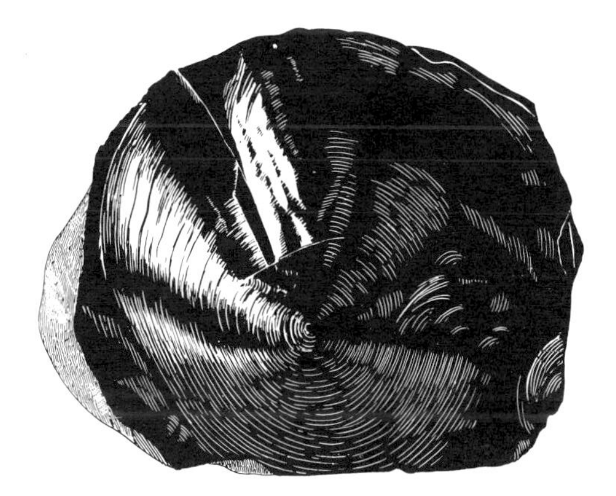

FIG. 5.3. Conchoidal fracture, obsidian.

depends on the strength of the bonding forces holding the atoms together. To put it simply, the stronger the bonding, the harder the stone. Hardness is thus an important property, for durability, one of the principal attributes of a fine gem, largely depends on it.

Several quantitative schemes for measuring hardness have been developed, but the *scratch hardness*, long used by mineralogists, serves well for the gemologist. In this method one determines whether the sharp edge of a mineral of known hardness will scratch the smooth surface of an unknown. In 1824, the Austrian mineralogist, F. Mohs, selected 10 relatively common minerals and arranged them in order of increasing hardness 1 to 10, given as follows:

Mohs Scale of Hardness

1. Talc	6. Orthoclase
2. Gypsum	7. Quartz
3. Calcite	8. Topaz
4. Fluorite	9. Corundum
5. Apatite	10. Diamond

Each mineral can scratch those with a lower number on the scale but can not scratch those with a higher number. For example, diamond readily scratches corundum, corundum scratches topaz, and topaz scratches quartz. Two minerals of the same hardness can scratch each other.

In addition to the minerals of the hardness scale, it is frequently convenient to use the following as test materials: the hardness of a copper coin is about 3, the steel of a pocketknife a little over 5, window glass $5\frac{1}{2}$, and the steel of a file 7. With practice one can estimate the hardness of a stone less than 7 by the relative ease or difficulty with which a scratch is made by the sharp point of a broken file. For example, glass (**H** $5\frac{1}{2}$) is scratched with comparative ease, whereas considerable pressure is required to produce a faint scratch on idocrase (**H** $6\frac{1}{2}$).

The hardness of cut stones can be best determined by using a set of *hardness points* (Fig. 5.4). These are metal pencils set with sharp-pointed fragments of minerals of known hardness. The set usually consists of seven points with hardnesses of 10, 9, $8\frac{1}{2}$, 8, $7\frac{1}{2}$, 7, and 6.

Using rough material a scratch can be attempted on any smooth surface but to avoid injury to a faceted gem, the test must be used with extreme caution. In fact, identification should be made by other means if possible, and a hardness test used only as a last resort. When it is necessary to check hardness on a transparent stone, the scratch should be attempted on a back facet near the girdle or on the girdle itself and made as short as possible. Observing the process under a low-power microscope or with a hand lens is desirable for it enables one to see a scratch only a small fraction of a millimeter in length. After a scratch has been made, the stone should be wiped clean and again examined under magnification to determine whether a scratch has indeed been made. A small streak of powder from a test mineral softer than the stone may resemble a true scratch.

If the gem is not in a setting, one can and should avoid scratching it to determine its hardness. This can be done by using the gem girdle as the "scratcher" and working up from low to higher hardness until no scratch is obtained on the scale mineral. Such a procedure requires relatively large pieces of minerals in the hardness scale, much larger than those in hardness points.

The bonding in a crystal may vary from one set of atomic planes to another; thus hardness, which depends on bond strength, may also vary with crystallographic direction. As a result, a crystal may show varying degrees of hardness depending on the direction in which it is scratched. However, in most gems the difference is too slight to be detected with the scratch test. Kyanite is an outstanding exception; in one direction **H** = 5, and at right angles **H** = 7. Variation in the hardness of diamond has long been recognized by diamond cutters. Planes parallel to the cube face can be polished with relative ease, but it is difficult to cut and polish a facet exactly parallel to an octahedron face.

Although the Mohs scale is useful in characterizing the hardness of gems, it is not linear; that is, the intervals between adjacent pairs of minerals are unequal. The hardness of the minerals of the scale measured by more quantitative techniques

FIG. 5.4. Set of hardness points manufactured by Rayner. Courtesy of Gemological Institute of America.

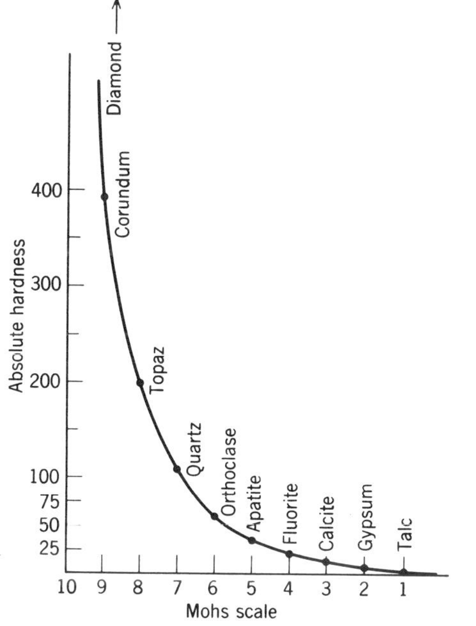

FIG. 5.5. Comparison of the Mohs hardness scale with absolute measurements of hardness.

than a scratch test leads to an absolute hardness scale shown in Figure 5.5. Equal positions of the minerals in the Mohs scale are preserved, but corundum can be seen to be four times as hard as quartz, and diamond, if the diagram would permit, over four times as hard as corundum.

Tenacity

The cohesiveness of a mineral, that is, the resistance it offers to crushing or breaking, is its tenacity. Tenacity of a gemstone is usually described as *brittle* or *tough* and is not necessarily related to hardness. Most diamonds are brittle and, because of the excellent cleavage, shatter when struck a sharp blow. However, the black variety *carbonado*, made up of many tiny intergrown crystals, is tough as well as hard. Jade (**H** 6–7) is a gem material that offers great resistance to breaking or crushing for, like carbonado, it is composed of intergrowths of many small

Table 5.1
SPECIFIC GRAVITY CHANGE WITH CHANGE IN CATION

Mineral	Composition	Atomic Weight of Cation	Specific Gravity
Calcite	$CaCO_3$	Ca 40	2.7
Rhodochrosite	$MnCO_3$	Mn 55	3.5
Smithsonite	$ZnCO_3$	Zn 65	4.4

crystals and a cleavage in one crystal is usually not continuous with cleavages in adjacent particles.

Specific Gravity

Specific gravity (**S.G.**) or relative density* is a number expressing the ratio between the weight of a substance to the weight of an equal volume of water at 4°C. Thus if a ruby weighs 1 gram (5 carats) and an equal volume of water weighs 0.250 grams (1.25 carats), the specific gravity is 4. The specific gravity of a gemstone not only is easily determined but is one of the most important characterizing properties. Furthermore, unlike hardness, it is a nondestructive test.

In crystalline materials specific gravity depends chiefly on two factors, the kinds of atoms present and the manner in which they are packed and bonded together. In the table of atomic weights (Table 3.2) the relative weights of the atoms are shown to range from 1 for hydrogen to 238 for uranium. In general we find that minerals composed of the heavier elements have higher specific gravities than those composed of lighter elements, if the crystal structure is the same. This is well illustrated by the carbonate minerals (Table 5.1) in which the only difference is in the cation.

In considering hardness it was pointed out that the stronger the bonding of the constituent atoms, the harder the mineral. In a similar manner bonding influences specific gravity; the more closely packed and more strongly bonded the atoms, the higher the specific gravity. In comparing polymorphic minerals (minerals composed of the same atoms in the same proportions) it is generally true that the mineral with the higher hardness also has the greater specific gravity. Anatase and rutile, polymorphs of TiO_2, offer an example: anatase—**H** $5\frac{1}{2}$, **S.G.** 3.88; rutile—**H** $6\frac{1}{2}$, **S.G.** 4.20. The most dramatic illustration is given by the polymorphic forms of carbon, diamond and graphite. Diamond with closely packed and strongly bonded atoms has **H** 10, **S.G.** 3.52; graphite with widely spaced layers of identical atoms held together by a weak bond has **H** $1\frac{1}{2}$, **S.G.** 2.23.

* Density and specific gravity are sometimes used interchangeably. When used properly, *density* requires the citation of units, such as grams per cubic centimeter.

Determination of Specific Gravity

Hydrostatic Weighing

We have seen that specific gravity is determined by dividing the weight of a gemstone by the weight of an equal volume of water. Although it is easy to weigh the stone, it is more difficult to determine the weight of an equal volume of water. To do this one makes use of the principle that a substance immersed in water is buoyed up and weighs less than in air; the loss of weight is equal to the weight of water displaced. Thus to determine specific gravity, the gemstone is first weighed in air and then in water. The difference in the two readings is the weight of the equal volume of water and specific gravity can be expressed as follows:

$$\frac{\text{weight in air}}{\text{loss of weight in water}} = \textbf{S.G.}$$

For example, a topaz-appearing gem weighs 9.75 carats* (1.950 grams) in air and 6.07 carats (1.214

* One carat equals 0.200 grams; or there are 5 carats/gram.

grams) when immersed in water. The specific gravity is calculated:

$$\frac{9.75}{9.75 - 6.07} = \frac{9.75}{3.68} = 2.65 = \textbf{S.G.}$$

The stone is therefore not topaz, but probably citrine quartz.

Chemical Balance

A chemical or diamond balance (Fig. 5.6) used to accurately weigh a gemstone can also be used for the determination of specific gravity by placing a bridge over one of the pans as shown in Figure 5.7. A bridge can be purchased but it can also be made by cutting out the long sides of a cardboard box of appropriate size. A beaker filled about three-fourths full of water is placed on the bridge. At one end of a fine wire (stainless steel is best but copper or brass will do) a spiral coil is made to hold the stone when immersed in water; the other end is bent to form a hook to suspend the wire from the balance arm. The hook must be at the proper distance so that the spiral basket when completely immersed does not touch the bottom or side of the beaker. The balance must then be brought to zero reading by counterbalancing the weight of the wire with another wire, or any appropriate weights, in the other pan. The stone is then placed on the pan beneath the bridge and weighed in air. It is then transferred to the spiral basket and weighed in water.

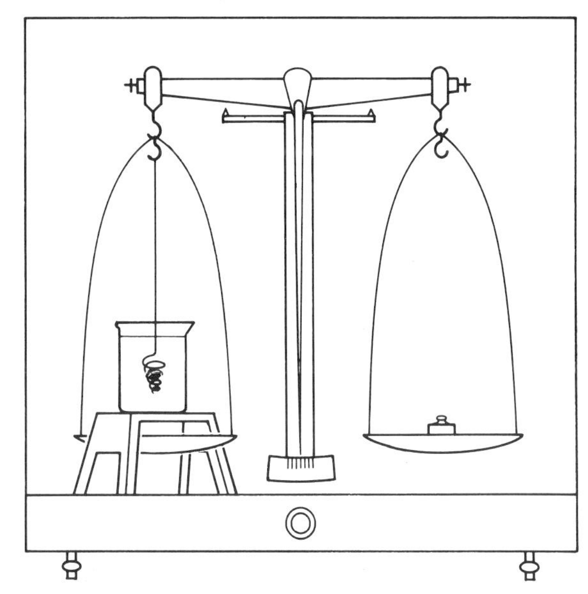

FIG. 5.7. Diagram of chemical balance arranged for hydrostatic weighing.

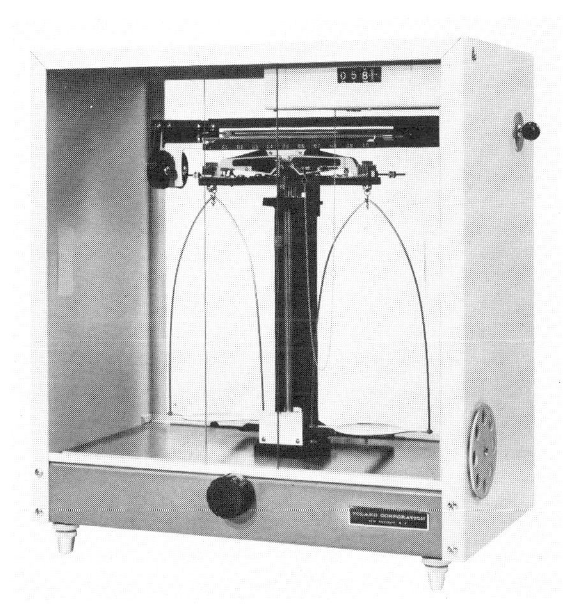

FIG. 5.6. Chemical balance. Courtesy of Voland Corporation, New York.

If ordinary tap water is used, the air which it contains will form bubbles on the wire and stone resulting in erroneous measurements. This can be avoided by using distilled water or boiled tap water.

A single pan balance (Fig. 5.8) with digital reading of weight (either in carats or grams) today has replaced in many laboratories the older chemical type balance. Weighing with it is fast, easy, and accurate (to 0.001 carat, Fig. 5.9) and it can be used for specific gravity determination with a bridge spanning the single pan to support the beaker of water. However, in using it an initial reading is made of the wire with the spiral basket immersed in water. This reading must then be subtracted from the readings made when the stone is weighed both in air and water.

The surface tension of water has a damping effect on the free movement of the balance. If a thin wire is used, the effect is negligible in weighing a large stone but may introduce an appreciable error

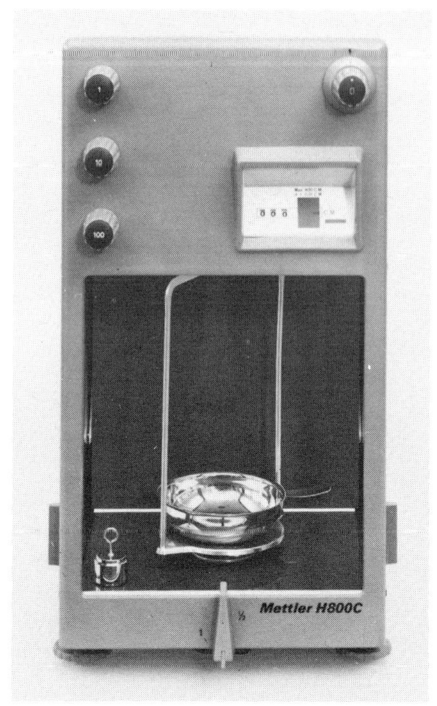

FIG. 5.8. A Mettler direct reading balance. Courtesy of Mettler Instrument Corporation, Princeton, New Jersey.

in specific gravity determination of a small stone. More accurate results can be obtained by adding a drop of liquid detergent to the water to lower the surface tension.

For even more precise measurements, particularly with small stones, it is desirable to use a liquid with a low surface tension. Toluene is such a liquid and is easily obtained. Although a volatile liquid which evaporates if left unstoppered, it maintains a constant density for a given temperature.

Water has a density of 1 only at 4°C (37.2°F); and for very precise work, a correction should be made if measurements are made at higher temperatures. However, at room temperature, 22°C (71.6°F), water has a density of 0.998, so close to 1 that no correction is necessary. However, the density of toluene, which at 22°C is 0.865, changes 0.001 with a 1°C change in temperature (see Fig. 5.10). For example, at 17°C it is 0.870, at 30°C 0.857. Thus for accurate work using toluene, temperature should be considered. The weighing procedure is the same with toluene (or any other liquid) as with water but the result must be multiplied by

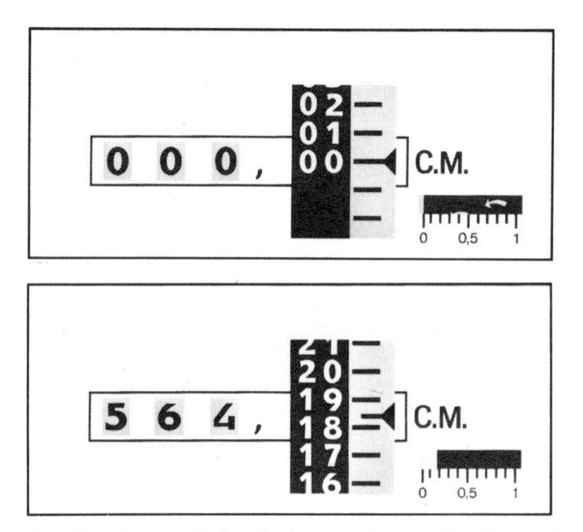

FIG. 5.9. Readout of the balance shown in Fig. 5.8. (*a*) Zero position. (*b*) Result, 564.184 carats. Courtesy of Mettler Instrument Corporation, Princeton, New Jersey.

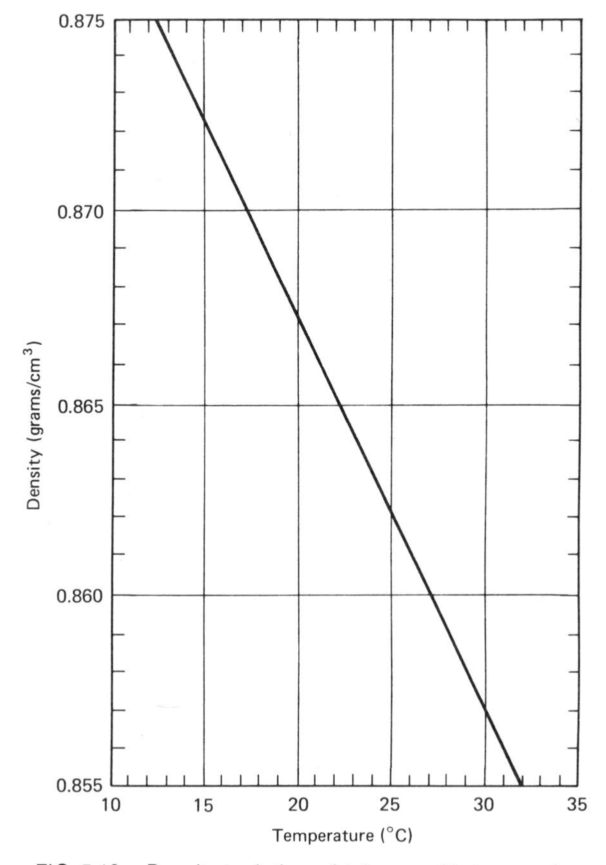

FIG. 5.10. Density variation of toluene with temperature.

the density of the liquid used. For example, the specific gravity of a stone weighing 2.1 carats in air and 1.6 carats in toluene at 27°C (82.4°F) is determined as follows:

$$\frac{2.1}{2.1 - 1.6} = \frac{2.1}{0.5} = 4.2 \times 0.86 = 3.612 = \textbf{S.G.}$$

Beam Balance

For those not having a chemical or diamond balance, specific gravity can be determined by an inexpensive and easily constructed *beam balance* (Fig. 5.11). The beam, of either wood or metal, is supported on a knife-edge, *a*, (such as a razor blade) which permits it to swing freely. The long arm, *b*, is divided in a decimal scale with zero at the fulcrum. The short arm carries two pans so arranged that one is in air and the other, suspended by a single thin wire, is in water. When the pans are empty, the beam is balanced by placing a piece of lead, *c*, on the short arm. A number of counterbalances of different weights are needed to hook over the long arm. They may be made of coiled wire or a wire hook can be used to carry metal washers of different weight, but their actual weight need not be known.

After the beam is adjusted to the "zero" position, a gemstone is placed in the upper pan and a counterbalance selected that will balance the beam when placed near the end of the longer arm. The position W_a, which is proportional to the weight of the stone in air, is read on the decimal scale and recorded. The stone is then transferred to the lower pan and the *same* counterbalance is moved closer to the fulcrum until the beam is again balanced. The second position W_w, read on the scale, is proportional to the weight in water. Since specific gravity is merely a ratio, it can be determined as $W_a/W_a - W_w$ = **S.G.** Although not as sensitive as a chemical balance, a well-constructed beam balance will give the specific gravity of a stone, 5 carats or more, correct to the second decimal place.

Heavy Liquids

Several liquids with relatively high densities can be used to determine specific gravity. The liquids most easily obtainable are bromoform (**S.G.** 2.89) and methylene iodide (**S.G.** 3.33). Both are miscible with toluene (**S.G.** 0.865) and acetone (**S.G.** 0.792) and by dilution with them any specific gravity below 3.33 may be obtained. *Caution*: the two heavy liquids must not be mixed with each other, for the combination will turn black.

If a gemstone floats when introduced into a heavy liquid, its specific gravity is obviously less than that of the liquid. The liquid can then be diluted with acetone or toluene until the stone neither rises nor sinks. The specific gravities of liquid and stone are then the same and that of the liquid can be determined by means of a Westphal balance or a set of glass density standards. The Westphal balance (Fig. 5.12) has a graduated arm, *a*, at the end of which is a sinker, *b*. When the sinker is immersed in water, an adjustable counterpoise, *c*, is moved to bring the beam to a horizontal position. The sinker floats when placed in a heavy liquid but sinks as the beam is brought to a horizontal position by placing calibrated riders, *r* 1 and *r* 0.1, on the grad-

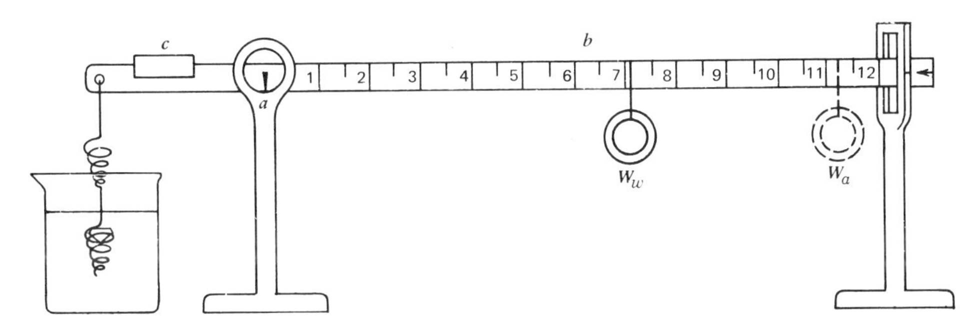

FIG. 5.11. Beam balance.

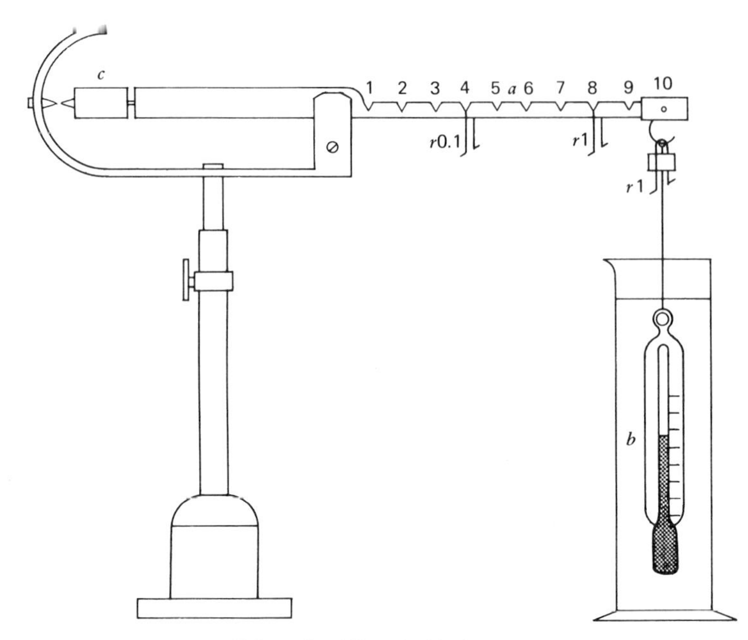

FIG. 5.12. Westphal balance.

uated arm. If the volume of the sinker is exactly 1 ml, and immersed in a liquid of **S.G.** 2.0, a unit rider (*r* 1) placed at the end of the beam will balance it; but if balanced with the unit rider at 8 on the beam, the **S.G.** is 1.8. In Figure 5.12, unit riders are on the beam at 10 and 8, and rider *r* 0.1 ($\frac{1}{10}$ the weight of *r* 1) is at 4. The specific gravity is 2.84 determined thus: 1 (weight of sinker) + 1 (*r* 1 at 10) + 0.8 (*r* 1 at 8) + 0.04 (*r* 0.1 at 4).

Although most gemologists do not have a Westphal balance, they do use heavy liquids for quickly obtaining approximate specific gravities. The pure liquids and others diluted to definite specific gravities are kept in tightly capped bottles. If a gem sinks when dropped into one of the liquids, it has a greater specific gravity than that of the liquid; but if it floats, its specific gravity is less than that of the liquid. Thus by making successive tests one can bracket the specific gravity of a stone. To avoid contamination, the stone (and tweezers) should be well cleaned before transferring from one liquid to another.

Convenient liquid densities are 2.67 (bromoform + acetone) in which quartz just floats and beryl sinks slowly; 2.89 (bromoform) in which quartz floats and topaz sinks rapidly; 3.10 (methylene iodide + acetone) in which tourmaline just floats, spodumene and fluorite sink slowly; and 3.33 (methylene iodide) in which nephrite floats and jadeite is suspended or sinks slowly.

Clerici Solution (**S.G.** 4.15 maximum) is another heavy liquid used by some gemologists. It is a water solution of thallium formate and thallium malonate and can thus be diluted with distilled water to any specific gravity lower than 4.15. Also if too much water is added, the density can be increased by evaporating the water at a low temperature. It has, however, the disadvantages of being not only expensive but also poisonous. Therefore it must be used with care. If it is available, dilutions to specific gravities of 4.00 and 3.50 are suggested for bracketing the specific gravities of gemstones.

Electrical and Magnetic Properties

Electroconductivity

The conduction of electricity in crystals is related to the type of bonding. Minerals with a pure metallic bond such as the native metals gold, silver, and copper are excellent electrical conductors. Other minerals in which the bonding mechanism is partially metallic are semiconductors; the less the metallic bond, the poorer the conduction. Because gem minerals lack metallic bonding, they are with few exceptions nonconductors of electricity.

Piezoelectricity

In Chapter 4, the crystal classes of highest symmetry were described, two for the hexagonal system and one for each of the other crystal systems. All of these classes have a symmetry center. However, in each crystal system there are classes (or a class) that lack a center of symmetry and have polar axes. That is, they have crystal axes with different properties at opposite ends. Pressure exerted at the ends of a polar axis of a nonconducting crystal causes electrons to flow to one end producing a negative charge, while a positive charge is induced at the opposite end. This is *piezoelectricity* (pressure electricity). Because they are strongly piezoelectric, two important gem minerals, quartz and tourmaline, have interesting uses.

Plates cut from quartz have been used since 1921 as oscillators to control radio frequencies. Subjected to the alternating current of a radio circuit, the quartz is caused to vibrate at a frequency dependent on the plate thickness and type of cut. The frequency of radio transmission or reception is determined by the frequency of the quartz oscillator. Tiny quartz plates used in quartz watches serve the same function. That is, they vibrate at a constant predetermined frequency to control the radio frequency of an electronic circuit which provides the time display.

Tourmaline is hexagonal with c a polar axis. Plates cut at right angles to this axis generate an electric current when pressure is exerted on their broad surfaces. The current generated is proportional to the area of the plate and the pressure. Pressure gauges made from tourmaline were developed in 1945 to record the blast pressure from the first atomic bomb and since then have been used with each atomic explosion.

Pyroelectricity

The property of pyroelectricity (heat electricity) is observed, as is piezoelectricity, only in crystals with polar axes. It is evidenced by the development of positive and negative electric charges at opposite ends of a polar axis when a temperature change takes place in the crystal. The property has no practical application but, because of it, tourmaline gemstones in the jeweler's window may attract dust particles when heated by the sun.

Frictional Electricity

The electrostatic charge induced in a substance by rubbing is frictional electricity, a phenomenon shown by most minerals. The charge is evidenced in a substance by its ability to attract light objects such as bits of paper. Whether the charge is positive or negative depends on the substance and the material with which it is rubbed. Frictional electricity was early (600 BC) observed in amber and much later the effect was called "electrification" from *electron*, the Greek name for amber. The development of a frictional electric charge is not diagnostic for amber, for it can be produced in materials made to simulate amber.

Magnetism

A few minerals are attracted to a small magnet and are said to be *magnetic*; lodestone, a variety of magnetite, is itself a natural magnet. If attraction to a small magnet is used as the test, most minerals are nonmagnetic. However, many iron-containing minerals are attracted by powerful electromagnets and in this way are separated from those that are iron-free. The magnetic test is unimportant in gemology for it is the rare gemstone that is even feebly attracted to a strong magnet.

References and Suggested Reading

Anderson, B. W., 1971. *Gem Testing*, 8th ed. Van Nostrand Reinhold Co., New York. Chapter 5.

Hurlbut, C. S. and C. Klein, 1977. *Manual of Mineralogy*, 19th ed. John Wiley & Sons, Inc., New York. Chapter 5.

Liddicoat, R. T., 1975. *Handbook of Gem Identification*, 10th ed. Gemological Institute of America, Santa Monica, Calif. Chapter 4.

Webster, R., 1975. *Gems*, 3rd ed. Butterworth and Co., Ltd., London. Chapters 27 and 28.

6 GEMS AND LIGHT

Of the many properties of gems, those that depend on light are of greatest importance, for they are responsible for all the attributes that make gems things of beauty. They include not only those properties we see when light is reflected from a gemstone surface but also those less easily observed that depend on the effect crystal structure has on transmitted light. The latter are referred to as *optical properties* and their determination is paramount in gem identification.

PART I. GENERAL

Luster

The luster of a mineral is the quality of shining in reflected light. In general, minerals are divided into two groups on the basis of luster. Those that reflect light as a metal have *metallic luster*; whereas those of the second group, which includes most of the remaining minerals, have a *nonmetallic luster*. There is no sharp division between them and the few minerals with an intermediate luster are said to be *submetallic*. Metallic minerals have few representatives among gems, but hematite and pyrite are examples. Our concern, then, is mostly with nonmetallic gem minerals, the luster of which is described by the following terms:

Vitreous. The luster of glass. This luster is the most common among gem minerals. Quartz and tourmaline are examples.

Adamantine. The brilliant luster like that of a diamond. It is found only in gems with high refractive index such as synthetic rutile and to a lesser degree in zircon and demantoid.

Resinous. The luster of resin. Sphalerite is an example.

Silky. A silklike luster caused by reflection of light from parallel fibers as in the satin spar variety of gypsum.

Pearly. An iridescent pearl-like luster shown by mother-of-pearl and by some minerals on surfaces parallel to a good cleavage.

Light Transmission

Most gemstones are light-colored and usually transmit some light, if not through the thick part of the stone at least through thin edges. The amount of transmitted light depends on the amount reflected from the surface or absorbed or scattered in passing through the stone. The following terms are used to express the degree to which light is transmitted:

Transparent. If the outline of an object seen through the stone is perfectly distinct.

Semitransparent or *subtransparent.* If an object is seen but with indistinct outline.

Translucent. If light is transmitted but objects cannot be distinguished.

Semitranslucent or *subtranslucent.* If light is transmitted only on thin edges.

If no light is transmitted, even on thin edges, the stone is *opaque*. Most minerals with metallic luster are opaque.

Color

Of the several properties that contribute to the beauty of a gem, color is a major factor and in some gems, beauty rests in color alone. Of all the properties possessed by gem minerals, color, for some, is the most constant; yet in others it is the most fickle and changeable and perhaps the least understood.

Nature of Light

To account for all light phenomena two theories are necessary: the *quantum* or *corpuscular theory* and the *wave theory*. The quantum theory regards light as discontinuous bundles of energy called *quanta* or *photons*. The wave theory considers light as electromagnetic waves that transmit energy by oscillating or vibrating at right angles to the direction of propagation. The wave motion is similar to that seen in a pond in which a floating leaf merely rises and falls with the passing wave, but the wave front moves forward. The distance between successive wave crests (or troughs) is the wavelength, λ (lambda), and the displacement either up or down from the position of rest is the amplitude (Fig. 6.1). The optical properties treated later in this chapter are best explained by the wave theory but the concept of the photon is used in the explanation of color.

Visible light is but a small part of the electromagnetic spectrum (Fig. 6.2) with wavelengths a billion times shorter than some radio waves and a billion times longer than some cosmic rays. Color is the response of the eye to this limited portion of the whole spectrum, from about 7500 to 3500 Å. However, some individuals are responsive to longer and shorter wavelengths than others. Thus for most people we can consider the longest visible wave-

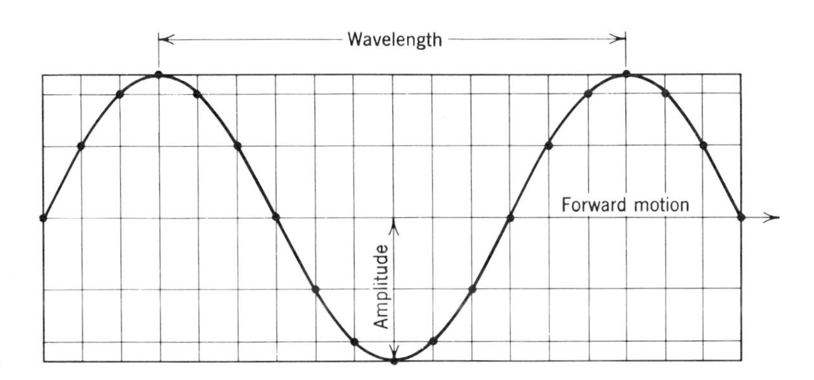

FIG. 6.1. Simple wave motion.

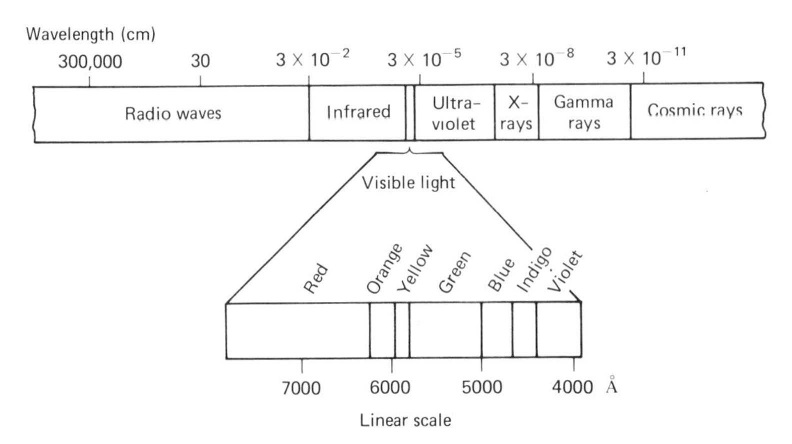

FIG. 6.2. The electromagnetic spectrum.

lengths as 7000 Å (red) and the shortest as 4000 Å (violet). As the wavelengths decrease from 7000 Å, the color varies continuously through red, orange, yellow, green, and blue to violet. White light is composed of all visible wavelengths.

When white light strikes the surface of a gemstone, part of it is reflected and part is refracted into the stone. If none of the reflected or refracted light is absorbed, the stone is colorless. If certain wavelengths are absorbed, color results from a combination of the wavelengths that reach the eye. For example, a stone in which blue has been absorbed is red, the complementary color. The eye is not equally sensitive to all colors and is most sensitive to green. Thus two minerals may produce the same color sensation by absorption of different wavelengths. Dioptase and emerald are both deep green, but dioptase absorbs most of the red wavelengths that are transmitted by emerald.

Chromophores

A major factor in absorption, and hence in color, lies in the electron configuration of the ions present in the crystal structure. In the discussion of atoms (Chapter 3) we learned that the electrons are in shells or orbitals at different energy levels, the outer shells at higher energy levels than the inner shells. If the radiant energy (photons) of visible light that shines on a crystal is sufficiently great (the shorter the wavelength, the greater its energy) it may elevate an electron to a higher, partially filled energy level and be absorbed in the process. Electronic transitions take place in the transition metal elements (see discussion in Chapter 3 under *Atoms, Ions, and the Periodic Table*) in which the *d* orbitals are only partially filled. These elements—titanium, vanadium, chromium, manganese, iron, cobalt, nickel, and copper—(usually present in only trace amounts) are thus the most important coloring agents in minerals and gems. Ions of such elements that produce color are called *chromophores.*

The absorption of light by an ion is also influenced by its nearest neighboring ions. The electron orbits of the chromophore are modified by the number and geometry of these neighbors. Further, in some crystals the transition metal ion occupies several different coordination sites each with its own geometry, and thus the effect on the chromophore is different in each site.

The oxidation state of a transition element is another factor that influences absorption. For example, the presence of ferrous iron (Fe^{2+}) characteristically produces a green color as in peridot, whereas ferric iron (Fe^{3+}) in a similar structural site in chrysoberyl results in a yellow color.

Idiochromatic Minerals

In some minerals, color is a fundamental property related to one of the major constituent elements and is thus constant. These are called *idiochromatic* and the characteristic color is an important means of identification. For example, the manganese minerals rhodochrosite and rhodonite are always red or pink; the copper minerals malachite and dioptase are always green, and the copper mineral azurite is always blue. Although the three copper minerals all contain Cu^{2+}, the azurite structure affects the absorption of the copper ions differently from the others, producing the blue color. Cuprite, copper oxide, contains Cu^{+} ions and is always red.

Most metallic minerals have a constant color and thus pyrite can be recognized by its brass-yellow metallic surface.

Allochromatic Minerals

The major elements composing most gem minerals produce no characteristic color and, when pure, the minerals are colorless. Yet many of them have color varieties that may be even more common than the colorless mineral. For example, colorless nephrite jade is less common than green varieties. As increasing amounts of iron enter the nephrite crystal structure, taking the place of magnesium, the mineral becomes darker and darker green and finally black. Minerals such as this that show a variation in color are called *allochromatic*. But the substitution of one major element for another, as in nephrite, is only one of the causes of color in allochromatic minerals.

The strongly light-absorbing ions of the transition elements, in small amounts as impurities substituting for a major element, may give a deep color to the mineral. Thus corundum, which is colorless when pure, is rendered red (ruby) by small amounts of chromium and blue (sapphire) by the presence of titanium and iron. Because of the influence of the crystal structure on the absorption effects of the chromophores, the same element (chromium) responsible for red in corundum gives rise in beryl to a deep green (emerald). And purple quartz (amethyst) contains trace amounts of iron, the same impurity ion that (in part) colors corundum blue.

Color Centers

The color of some minerals having no chromophores is attributed to *color centers* (F centers) which are lattice imperfections. The imperfections may be due to ions of one element in excess or deficiency of the amount required by the chemical formula, substitutional impurity ions, or mechanical deformation of the lattice. A color center is an anion vacancy that can trap a single electron. Certain wavelengths of light can excite the electron to a higher energy state and be absorbed in so doing.

Color Change

Color of a mineral may be changed by heat treatment or by exposure to X-rays or other high-energy radiation. There are many examples in which the color of gems is enhanced by such treatment. By proper heat treatment, some yellowish topaz turns pink, and brownish-red zircon becomes a lovely blue. Colorless diamonds when subjected to a certain radiation are colored green but a blue color results when exposed to another type of radiation. With heat treatment, the green diamond will turn yellow (see diamond under *Important Gemstones* in Chapter 13).

Color Caused by Inclusions

Minerals may be colored by the presence of mechanically incorporated inclusions of other minerals. This type of coloring is particularly common in quartz and is responsible for the large number of color varieties of chalcedony as well as the color in some coarsely crystallized quartz (q.v. under *Important Gemstones* in Chapter 13). Hematite is the most common pigmenting impurity and imparts its red color to many minerals including some feldspar and calcite as well as quartz.

Variation in Color

Color may not be constant even within a single crystal, as is well illustrated by tourmaline (Figs. 6.3*a* and *b*). An elongated crystal may be red at one end, green at the other end, and nearly colorless in between. In other tourmaline crystals there is a zonal arrangement of colors in concentric shells. This is best observed when a crystal is sliced at right angles to the length; it may show a green exterior

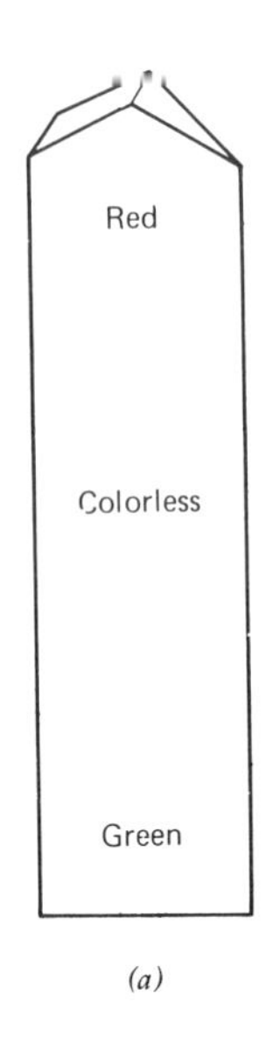

(a)

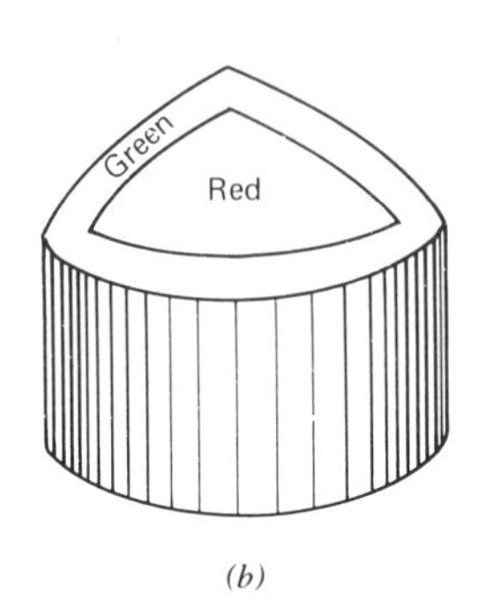

(b)

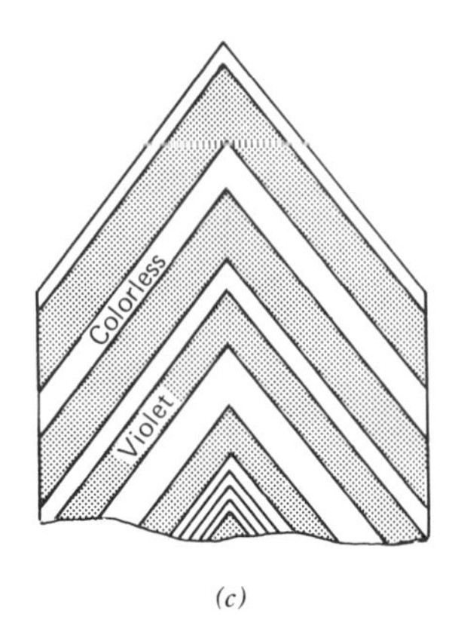

(c)

FIG. 6.3. Color zoning. (*a*) Tourmaline, longitudinal section. (*b*) "Watermelon" tourmaline, cross section. (*c*) Amethyst, longitudinal section.

enveloping a red interior. In some gems, although only one color is present, it is irregularly distributed. For example, in most amethyst the pigmented portions alternate with colorless bands parallel to rhombohedron faces of the crystal (Fig. 6.3c). This color arrangement is so characteristic that a purple stone evenly colored can be suspected of not being amethyst. The different colors of the zones or bands are related to slight changes in chemical composition.

Streak

The color of a mineral powder is its *streak*, so called because a streak of powder is produced by rubbing a specimen on a plate of unglazed porcelain (streak plate). In most nonmetallic minerals the streak is colorless and thus is of little importance in gem identification. In metallic minerals the streak is frequently colored and is characteristic, as, for example, the red streak of hematite produced by a black gem.

Color Filters

We have seen that the color of gemstones is due to the absorption of some of the wavelengths of the white light in which they are viewed. The resulting color is produced by the combination of the unabsorbed wavelengths (colors) that reach the eye. The colors of two gems may be very similar but the mixture of wavelengths giving rise to the color in one stone may be quite different from the mixture in the other. *Color filters* offer a quick and easy means of revealing this basic difference and thus aid in gem identification.

The most common filter is the "Chelsea" or "emerald filter" used particularly to distinguish emerald from other green stones and imitations. The filter absorbs most of the visible light but transmits the long red wavelengths and a band in the yellow-green portion of the spectrum. Since emerald transmits red light and absorbs some of the yellow-green, it appears red when illuminated by a bright incandescent lamp and viewed through the filter held close to the eye. Most other green stones and glass imitations absorb red light and appear green through the filter.

Unfortunately this test for emerald is not unequivocal, for a few other green stones, as well as synthetic emeralds, also appear red, whereas the color of emeralds from some localities appears unchanged. These exceptions, which include demantoid and green zircon, will be noted in the descriptions of specific gems.

Opalescence

The light entering some translucent to transparent gemstones is scattered internally by tiny particles or structural irregularities. The scattered light gives the stone a pearly or milky sheen, known as opalescence. It is best seen in cabochon cut moonstone and opal. The opalescence of moonstone is produced by thin lamellae formed as a product of exsolution (q.v. in Chapter 3). For the cause of the effect in opal, see the next section.

Play of Colors

The interference of light reflected from the surface or from the interior of a mineral may produce different colors as the angle of incident light changes. Surface iridescence, similar to that produced by soap bubbles, results from the interference of light as it is reflected from thin surface films. Because gemstones have been cut and polished, they rarely show this phenomenon and their play of color is caused by light diffracted internally from closely spaced fractures, twin or exsolution lamellae, or cleavage planes. It is this internal iridescence that produces the color effects in opal and in several gem varieties of feldspar.

Gem opal, although amorphous, is not a homogeneous material as is glass. It is composed of tiny uniform spheres (1500–3000 Å in diameter) of amorphous silica arranged in a closely packed, orderly, three-dimensional array with the voids between them occupied by air and water (Fig. 6.4). When white light passes through the essentially colorless opal and strikes planes of voids between spheres, certain wavelengths are diffracted and flash out of the stone as nearly pure spectral colors. This phenomenon is analogous to the diffraction of X-rays by crystals. In X-ray diffraction the interplaner spacings are of the same order of magnitude as the

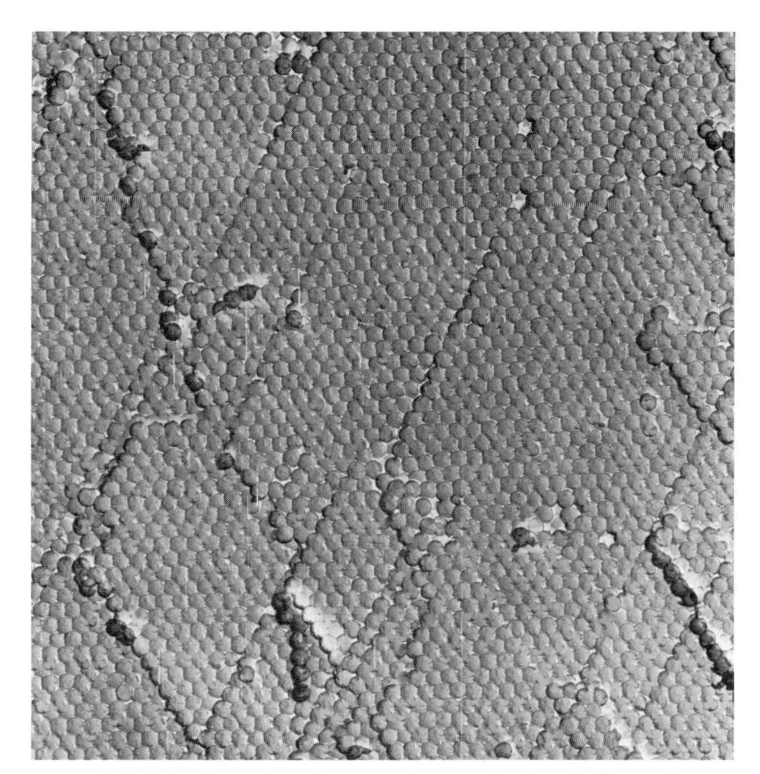

FIG. 6.4. Scanning electron micrograph of a chalky opal (8,500×) showing hexagonal packing of silica spheres. Because of the weak bonding between spheres, they are intact; in precious opal many of the spheres are broken. Courtesy of J. V. Sanders, University of Melbourne, Australia.

wavelengths of X-rays. The spacings in opal, determined by the sphere diameters, are far greater but so are the wavelengths of visible light. As the angle of incident light changes, different wavelengths satisfy the Bragg law (see under *X-Ray Crystallography* in Chapter 4), to be diffracted as flashes of brilliant colors. In common opal the domains composed of spheres of uniform size and packing are so small they merely scatter the light to give an opalescent effect.

A play of color, also called "schiller" or "labradorescence," is shown in several varieties of feldspar. Most familiar are labradorite with a color change from yellow to blue or green and peristerite with a delicate blue iridescence appearing on a gray or brownish body color as the angle of incident light changes. Evidence indicates that these feldspars, initially homogeneous, have by exsolution formed lamellar intergrowths of two feldspars with slightly different compositions (Fig. 6.5). If the thickness and periodicity of the lamellae are within certain limits, white light reflected from them will produce an *iridescence* due to the interference of some wavelengths. The phenomenon is similar to the color flashes in opal but, unlike opal, the periodicity of the intergrowths is rarely sufficiently regular to produce monochromatic colors. The resulting interference colors thus contain a range of wavelengths. Because the lamellae causing the reflection are all nearly parallel to a single crystal plane, a polished surface must be oriented with respect to this plane for the best color effects.

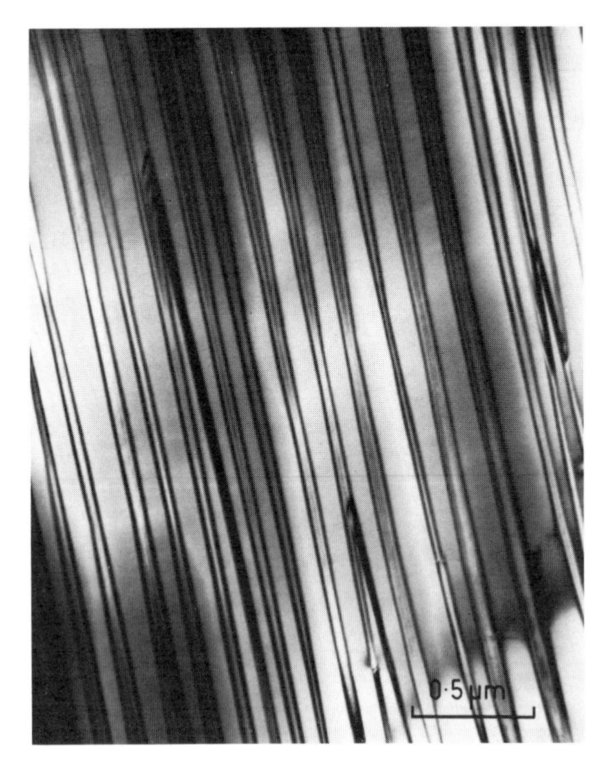

FIG. 6.5. Transmission electron microscope photograph of peristerite from Bancroft, Ontario. The alternate set of coarser lamellae is essentially pure albite; the finer set is oligoclase (about 25% anorthite). 1 μm = 1000 Å. Courtesy of Alex C. McLaren, Monash University, Clayton, Victoria, Australia.

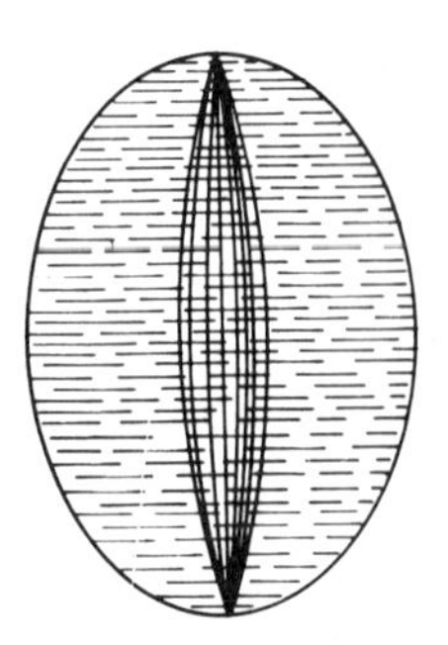

FIG. 6.6. Chatoyancy. Orientation of inclusions is at right angles to light beam crossing the stone.

Chatoyancy

Some minerals have a silky sheen that results from closely packed parallel fibers or parallel needlelike inclusions or cavities. When such a mineral is cut with a smooth rounded surface (a cabochon), its surface displays a band of light at right angles to the length of the fibers, inclusions, or cavities (Fig. 6.6). The effect is best seen in sunlight or under a single narrow light source. As the stone is turned, the narrow beam moves from side to side as does a cat's eye and the name *chatoyancy* is appropriately given to the property. The phenomenon is excellently shown in the "satin spar" variety of gypsum and were it not such a soft mineral, it would make a lovely gem. Many minerals sometimes have chatoyancy but most commonly seen in jewelry is the golden-brown "tiger's eye." This is quartz that has replaced crocidolite asbestos but preserved the fibrous nature of the asbestos. The most valuable chatoyant gem is the *cat's eye* variety of chrysoberyl. This is a greenish-yellow mineral that, as a beam of light moves across its polished surface, truly resembles a cat's eye.

Asterism

We have seen that chatoyancy results from inclusions oriented parallel to a single crystallographic direction. If there are three identical crystallographic directions, as in hexagonal crystals, needle-like inclusions may be present parallel to each of the like directions. A cabochon stone cut from such a crystal shows a beam of light on its surface perpendicular to each parallel set of inclusions (Fig. 6.7). This triple chatoyancy is called *asterism* and is best known in star rubies and star sapphires (see Plate II). In these gems the microscopic rutile needles that cause the effect have been constrained to grow throughout the crystal structure at 120° to each other. Since the length of all the observed rutile needles is perpendicular to the c crystallographic axis, to obtain a centered star, care must be taken in cutting to have the base of the cabochon also perpendicular to c.

FIG. 6.8. Rose quartz sphere (6 cm in diameter) showing asterism.

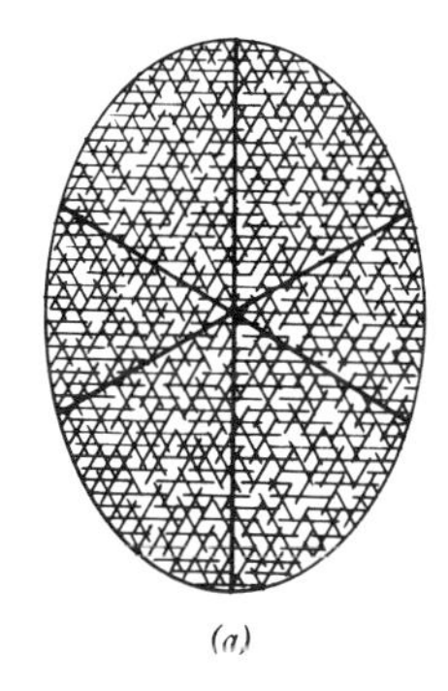

(a)

FIG. 6.7. Asterism. (*a*) Beams of light from three sets of inclusions at 120° to each other result in a six-pointed star. (*b*) A star produced by phlogopite mica in transmitted light.

Rose quartz frequently shows asterism which, as in ruby and sapphire, probably results from tiny rutile needles oriented at 120° by the hexagonal structure (Fig. 6.8). Almandine garnet may also display asterism; a six-pointed star is seen if the base of the cabochon is perpendicular to a 3-fold symmetry axis; other orientations may result in four-pointed stars. A nearly black variety of diopside gives a sharp four-pointed star made up of two bands that intersect at an angle slightly less than 90°.

Some phlogopite (brown mica) shows a remarkable asterism in both reflected and transmitted light. The six-pointed star (sometimes 12-pointed) results from light scattered by rutile needles oriented by the pseudohexagonal structure (Fig. 6.7*b*).

Luminescence

Substances that glow in the dark when crushed, heated (but not to incandescence), or exposed to short-wavelength radiation are *luminescent*. Of the three ways of producing luminescence only the last is of practical interest to the gemologist. A gemstone that emits visible light during exposure to ultraviolet light (also X-rays or cathode rays) is *fluorescent*. If luminescence continues after removal of the exciting rays, the stone is *phosphorescent*.

Luminescence commonly results when the energy of the shortwave radiation is absorbed by impurity ions (activators) and released as visible light. In some minerals structural defects may act as activators. Because different ions vary in their ability to absorb a given wavelength, two sources of ultraviolet light (u.v.) are used. Some stones fluoresce only in shortwave u.v. (2500 Å), others only in longwave u.v. (3500 Å), whereas still others fluoresce in either wavelength.

The cause of fluorescence is similar to the cause of color, and ions of the transition metals, particularly iron and manganese, are effective activators. Electrons in the impurity ions, excited by the invisible short radiation, are raised to higher energy levels. When they fall back to their initial (ground) state, they emit visible light. Because some of the exciting energy is lost as heat, the radiation emitted by the electrons in returning to the ground state is less energetic and thus of longer (visible) wavelength. In phosphorescent minerals there is a time lag between the excitation of electrons to a higher energy level and their return to the ground state.

The small amount of impurity ions on which fluorescence depends may have little or no effect on the color or general appearance of a gemstone. Thus one stone may show a vivid color under ultraviolet light, whereas another of the same mineral shows none. For example, only some fluorite, the mineral from which the property receives its name, fluoresces; and many diamonds, but not all, fluoresce blue in longwave u.v. Few minerals have a constant, and therefore diagnostic fluorescence. Of these that are occasionally cut as gems, scheelite fluoresces a pale blue, benitoite a bright blue, and willemite a yellow–green in shortwave u.v.

Under shortwave u.v. natural ruby commonly gives a red fluorescence but synthetic ruby usually yields a much more vivid color and appears to glow like a red-hot coal. Under both short- and longwave u.v. natural emerald may become a pale orange-red, but it more commonly fails to show any fluorescent color, whereas synthetic emerald frequently fluoresces a dull red. Although the difference in fluorescence of natural and synthetic emerald is a valuable distinguishing test, it is not 100% reliable. Therefore fluorescence should be used, as should most other tests, as an indicator of the stone's identity and not as a definitive test.

PART II: OPTICAL PROPERTIES

We have thus far considered the light-dependent properties of gems that can be observed merely by looking at them. The so-called *optical properties*, although less easily determined, are more informative and are of the greatest importance in gem identification. Their determination requires the use of special instruments and accessories which are described in the following discussion.

Reflection and Refraction

When a ray of white light passes from a rare medium (air) to a denser medium (gemstone), part of it is reflected at the surface but part of it enters the stone (Fig. 6.9). For the reflected ray the angle of incidence, i, equals the angle of reflection, r', both measured from a normal to the surface. The portion

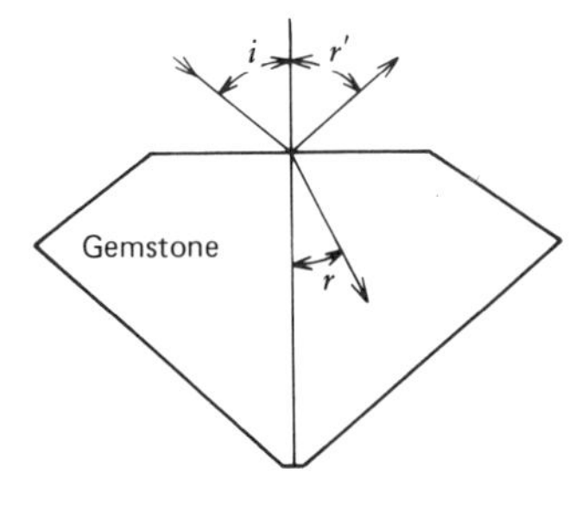

FIG. 6.9. Reflected and refracted light. For reflected light $r' = i$. For refracted light $r < i$.

of the light that enters the stone is bent or *refracted* from the path of the incident ray toward the surface normal. This *angle of refraction, r,* as measured from the normal, is always smaller than the angle of incidence, *i* (Fig. 6.9). The amount of refraction depends on the difference in density of the two media, in our example air and gemstone, and on the obliquity of the incident ray. The greater the difference in density and the greater the obliquity, the greater the refraction.

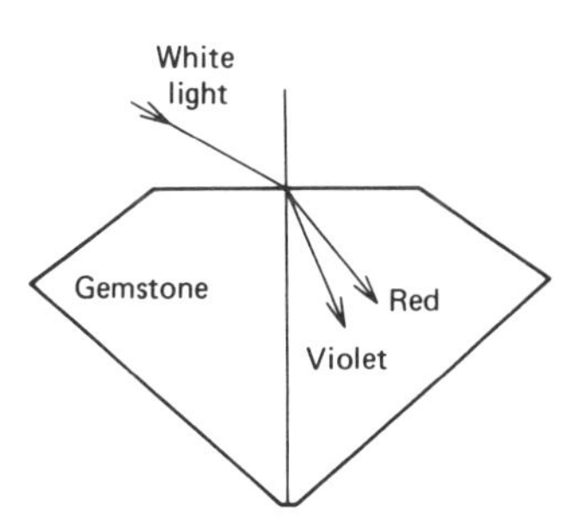

FIG. 6.11. Different refraction for different wavelengths of light.

Index of Refraction

Probably the most important law in optics is *Snell's law* that relates the angle of incidence *(i)* to the angle of refraction *(r)*. It states that *for the same two media the ratio of sin i:sin r is a constant.* The constant is called the *refractive index* (***R.I.***) and is usually expressed as ***R.I.*** = sin *i*/sin *r*. For example, consider light rays, *a* and *b*, striking the surface of a diamond (Fig. 6.10). For *a*, $i = 60°$ and $r = 21°$; for *b*, $i = 30°$ and $r = 11.92°$.

a

$$\frac{\sin 60° = 0.8660}{\sin 21° = 0.3583} = 2.42 = \textbf{\textit{R.I.}}$$

b

$$\frac{\sin 30° = 0.5000}{\sin 11.92° = 0.2066} = 2.42 = \textbf{\textit{R.I.}}$$

Light travels with a maximum velocity of about 300,000 km/sec in vacuum. Its velocity decreases when moving through any other medium an amount proportional to the optical density of the medium. It is thus possible to express refractive index as the ratio between the velocity of light in vacuum (*V*) to its velocity in the denser medium (*v*), that is, V/v = ***R.I.*** As a standard for comparison, the velocity of light in vacuum is taken as unity and other light velocities expressed in terms of $V = 1$. Since for air $v = 0.9997$ (almost as great as in vacuum), it also may be considered unity. We can thus state that when measured in air, ***R.I.*** = $1/v$, or the reciprocal of the velocity.

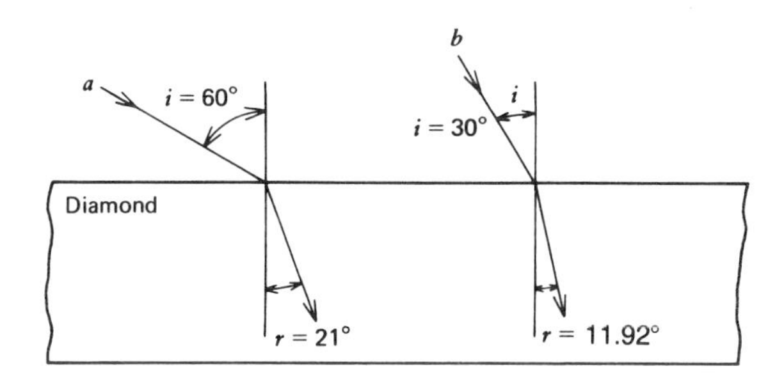

FIG. 6.10. Light rays incident to a diamond surface at different angles are refracted differently but in each case the ratio of sin *i*:sin *r* = 2.42.

Dispersion

When light passes from air into a denser medium, its velocity varies with the wavelength. Red light with the longest wavelength has the greatest velocity and is refracted least. Violet light with the shortest wavelength and least velocity is refracted most (Fig. 6.11). Thus the refractive index of a gemstone will be less for red than for violet light. This separation of the spectral colors, known as *dispersion*, is a phenomenon commonly observed when sunlight strikes a glass prism. But of more interest to us, it is dispersion that gives a diamond its fire causing it to emit red, yellow, and blue flashes of color as the stone is turned.

Dispersion in some gems is high, in others low, and thus this is another property by which gems are characterized. For a precise value of dispersion the wavelengths used to obtain the indices of refraction should be specified. However, usually (and for the values given here) dispersion is the difference in the refractive index measured in red light (λ = 6867 Å) and in blue-violet light (λ = 4308 Å).* Thus the dispersion of a diamond is 0.044 and of fluorite 0.0007.

Because gems have different refractive indices for each wavelength, ***R.I.*** data are usually given as measured in *monochromatic light*, that is, light of a single wavelength. Sodium light with λ = 5890 Å is most commonly used and is readily available in small portable lamps. However, the average values obtained with white light are close to those obtained with sodium light.

For most gemstones the index of refraction is the most important of the several properties that are used in their identification. It not only can be measured accurately but, unlike specific gravity, also can be determined with the stone in a setting.

Determination of Refractive Index

Critical Angle and Total Reflection

In the discussion of refraction, we saw that light in passing from a rare to a denser medium is refracted toward the normal (Fig. 6.9). If, however, the conditions are reversed and light passes from the denser to the rarer medium, it is refracted away from the normal. The angle of incidence (*I*) is thus less than the angle of refraction (*R*).† Consider Figure 6.12 in which light rays moving through glass (***R.I.*** 1.52) strike the glass–air interface at *O*. The greater the obliquity of the incident ray, the greater the angle of refraction. When the angle of refraction is 90° (ray *C*), the refracted ray moves parallel to the interface. This is known as the *critical angle* (**C.A.**). Rays with greater angles of incidence, such as *D* and *E*, do not pass into the air but are totally reflected back into the glass. For glass (***R.I.*** 1.52) this angle is 41°28′, but for diamond (***R.I.*** 2.42) the angle is 24°24′. The measurement of the critical angle thus offers a quick, easy, and accurate means of determining index of refraction.

* Precise measurements of dispersion are given between these wavelengths which are respectively the B and G Fraunhofer lines of the solar spectrum. See Table 8.1.

† *I* is here used for angle of incidence and *R* for angle of refraction when light moves from the denser to the rarer medium.

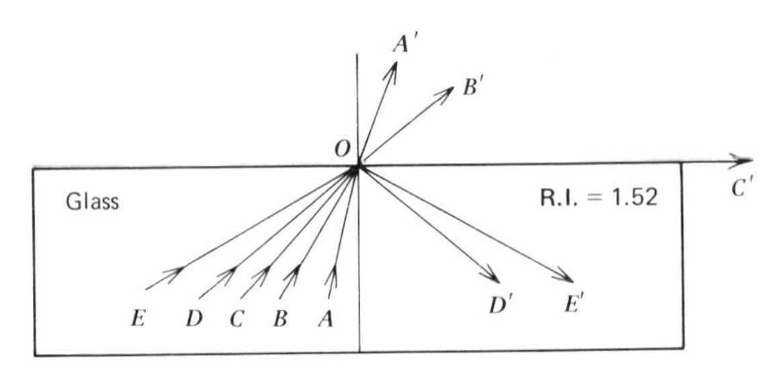

FIG. 6.12. The critical angle and total reflection.

Refractometer

The instrument by which refractive index is determined using the critical angle is the *refractometer*. The several different makes available to the gemologist operate on the same underlying principles (Figs. 6.13 and 6.14). Central to them all is a hemisphere of glass of high refractive index that is known. In some instruments hemicylinders are employed rather than hemispheres. In using the instru-

FIG. 6.13. Rayner refractometer with which the refractive index is read on the external dial. Courtesy of the Gemological Institute of America.

FIG. 6.14. Refractometer of Gemological Institute of America. Courtesy of GIA.

ment a flat facet of the gemstone is placed on the flat polished surface of the hemisphere but separated from it by a thin film of liquid. The liquid is to ensure optical contact by excluding the air and must have a refractive index lower than that of the hemisphere but higher than that of the stone. Light passing through a ground glass enters one quadrant of the hemisphere to strike the gemstone at varying angles of incidence. When the angles are less than the critical angle, the light is partly reflected and partly refracted through the stone and into the air. But when the angles of the light rays are greater than the critical angle, they are totally reflected back into the other quadrant and can be observed by an appropriately positioned telescope. The boundary between the brighter and darker portions of the field is at the critical angle (Fig. 6.15). Although light is refracted on entering the liquid film, it is refracted an equal amount on leaving and the effect can be disregarded. If the critical angle (**C.A.**) and the refractive index of the hemisphere (*N*) are known, the ***R.I.*** of the stone can be calculated as ***R.I.*** = sin ***C.A.*** × *N*.

In early refractometers, the critical angle was measured on a graduated circle and the refractive index of the stone calculated. Modern jeweler's refractometers are so constructed that the rays, totally reflected through the hemisphere, fall on a scale calibrated in index of refraction. The image of the scale is then reflected by a mirror or prism to a focusing eyepiece. If the refractive index of the stone is within the range of the refractometer, part of the scale is brightly illuminated and part is relatively dark. The position of the boundary, called a shadow edge, separating light from darker portions is determined by the refractive index, which can be read directly from the scale (Fig. 6.16).

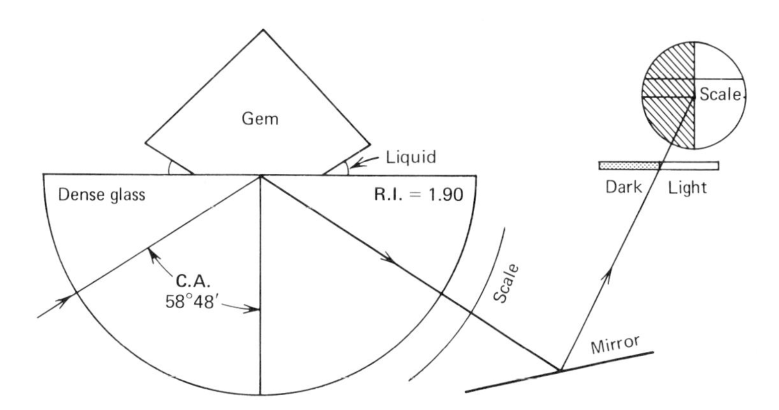

FIG. 6.15. Simplified diagram of light rays striking a gemstone at the critical angle and passing through the refractometer. **C.A.** = 58°48′. **R.I.** gem = sin 58°48′ (0.855) × 1.90 = 1.625. See Fig. 6.16.

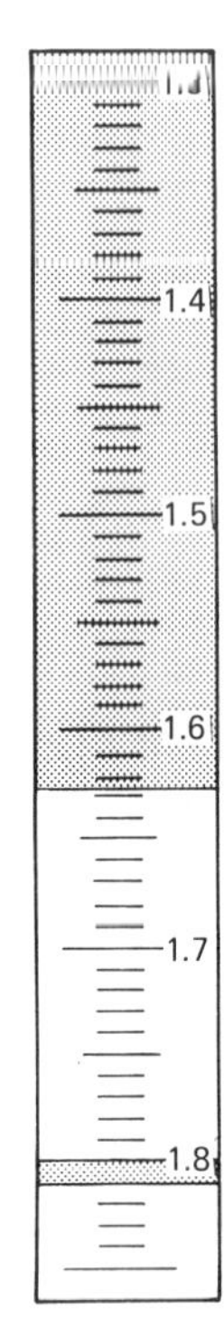

FIG. 6.16. Scale as seen through eyepiece of refractometer with shadow edge at 1.625. The faint edge at 1.81 indicates the refractive index of the liquid.

An improved refractometer has been developed by Rayner in which refractive index readings are made on an external dial (Fig. 6.13). Shadow edges are observed as in other instruments but on a plain field. On rotation of the dial, an opaque screen moves up and down on the right side of the illuminated area and its edge can be accurately aligned with the shadow edge. The refractive index is then read directly from the calibrated dial.

The refractive indices of the glass hemisphere and of the contact liquid are limiting factors in the use of the refractometer. The glass with a refractive index of about 1.90 sets an absolute upper limit to the ***R.I.*** determinations. Since the refractive index of the liquid must be greater than that of the stone, it usually determines the upper limit. Although liquids with ***R.I.*** = 2.0 or greater are available, those with ***R.I.*** > 1.81 are corrosive and may etch the glass hemisphere. The liquid most easily available is methelyene iodide saturated with sulfur, with ***R.I.*** = 1.78. If about 10% tetraiodoethylene is added to the methelyne iodide–sulfur mixture, the ***R.I.*** is increased to 1.81. It is this latter liquid that is usually supplied with a refractometer.

Using such a liquid, and if spinel were the test stone (***R.I.*** 1.72), one would see a dark–light boundary at a scale reading of 1.72 but would also see another at 1.81, the index of refraction of the liquid. If in making the test only the reading of 1.81 is observed, it indicates that the refractive index of the stone is higher than that of the liquid.

Lead incorporated in glass to make it highly refractive reduces its hardness, making it susceptible to damage. To avoid scratching the polished surface of the relatively soft glass, the gemstone must be placed on and removed from the hemisphere with care. It should be lifted, not slid, off and the liquid cleaned from the glass with lens paper. When not in use, the glass should be covered to protect it from settling dust; and, if the liquid has been allowed to dry, it should be cleaned with xylol or toluene.

Spinel (***H*** 8), used as the hemisphere in some refractometers, has several advantages over the lead-glass hemisphere. It is much less easily damaged, has low dispersion, and the readings can be made more accurately because of a more open scale. However, it does not permit determinations of refractive index greater than 1.72.

Because of dispersion, gems have a different refractive index for each wavelength of light. Thus when a white light source is used for the refractometer, the boundary between the light and dark portions of the scale is not sharp but is a band showing the spectral colors. For example, the red edge of the band for spinel might give a ***R.I.*** of 1.718 and at the violet end 1.738, a dispersion of 0.02. Readings can be made using white light with sufficient accuracy for most identifications, but for precise measurements a monochromatic source, such as a sodium light, is necessary. Some gems, as we see later, do not have a single refractive index but yield two shadow edges on the scale.

Spot Method

It has been stressed that for normal ***R.I.*** determination with the refractometer a plane facet of the gem must be placed in contact with the polished hemisphere or hemicylinder. However, in 1948* Lester B. Benson of the Gemological Institute of America described a method by which approximate ***R.I.*** readings could be obtained with a refractometer on cabochon stones and on faceted stones too small to give a distinct shadow edge. This is called the *spot method* and in Britain the *distant vision method*. Refractive index determinations by this method can be carried out most easily using a refractometer from which the eyepiece can be easily removed, such as the GIA Duplex (Fig. 6.14).

In using the method the smallest possible droplet of contact liquid is placed in the center of the high-index glass of the refractometer. The curved surface of the stone is then placed on the droplet, the eyepiece removed, and the scale viewed from a distance of 10–16 in. If the spot (the contact of stone, liquid, and hemisphere) observed from this distance covers more than two or three scale divisions, it is too large. The stone should then be removed, the glass cleaned, and the stone replaced using as contact liquid only the small amount adhering to the stone. With the eye in a fixed position, the entire scale is scanned by turning the mirror. During this movement, the spot changes positions and appears to be superimposed successively on all scale positions. If the gem is tourmaline with ***R.I.*** (1.63) near the middle of the scale, the spot appears dark when superimposed on lower scale readings. When superimposed on higher readings, higher than 1.63, the spot has a light center. The line on the scale bisecting the spot when it is half light and half dark gives a reading for the refractive index of the stone. Frequently as the spot appears to move up and down the scale, the change from light to dark is very rapid and it may never appear to be half and half. In the example given, the spot may change from completely dark at 1.62 to competely light at 1.64. The refractive index can then be interpreted as approximately 1.63.

It is somewhat more difficult to obtain the refractive index by the spot method using a refractometer with a fixed eyepiece. However, it can be done in a manner similar to that described by observing the spot through the eyepiece from a distance of 14–16 in. The eye must now be shifted up and down. As this is done, the spot changes its position relative to the scale and appears dark in the lower numbered portions and light in the higher numbered portions. The position at which the spot is half light and half dark, or the point at which it suddenly changes from light to dark, can be observed as described above. However, with the eye at a distant point, the scale cannot be read. The position must be mentally recorded while the eye is moved toward the eyepiece to bring the scale into focus to make the refractive index reading.

* *Gems & Gemology*, Vol. VI, No. 2, Summer 1948.

Immersion Method

When colorless quartz (***R.I.*** ~ 1.55) is immersed in water (***R.I.*** 1.33), it has high relief; that is, the gem is clearly visible because of refraction and reflection of light as it moves from one medium to the other. The same stone immersed in a liquid of ***R.I.*** 1.55 becomes nearly invisible for now light moves from liquid to quartz with a minimum of refraction or reflection. However, a colorless sapphire (***R.I.*** near 1.77) in this liquid would stand out in sharp relief.

The mineralogist routinely determines refractive index by the *immersion method*. Using liquids with refractive indices between 1.41 and 1.78, calibrated at intervals of 0.01, he makes successive immersions until a match of refractive index of mineral and liquid is reached. Because this method requires crushing the mineral and viewing the small particles with a special high-power microscope, it is little used in gem identification.

However, the gemologist does use a modified immersion method to tell whether the refractive index of a gemstone falls within a specified range. Suggested for the purpose are the readily available liquids given in Table 6.1.

In making the test the stone is immersed in a liquid contained in a shallow (Petri) dish. The dish is held over a white sheet of paper on supports that allow a black card to be passed beneath it. The stone should be viewed from a point directly above while the card is slowly passed below it. If the edge of the card, seen both through the stone and liquid alone appears as a straight line, the index of refrac-

Table 6.1
IMMERSION LIQUIDS[a]

Liquid	R.I.	Liquid	R.I.
1. Water	1.33	5. α-Monobromonaphthalene	1.65
2. Amyl alcohol	1.42	6. Methylene iodide	1.74
3. Mineral oil	1.48	7. Methylene iodide + sulfur	1.78
4. Bromoform	1.60		

[a] Numbers 4 and 6 are the most commonly used heavy liquids.

tion of stone and liquid are the same. If the stone has a lower refractive index, the edge of the card moves into it ahead of the edge seen through the adjacent liquid. For a stone with higher refractive index, the reverse is true and the dark edge moves first into the liquid and then into the stone (Fig. 6.17).

Direct Measurement

This method of refractive index determination is sometimes called the Duc de Chaulnes method after the man who proposed it. Measurements made with it are only approximate but are helpful when the refractive index of a stone lies beyond the range of the refractometer (Fig. 6.18).

A microscope is necessary to make the three measurements required for the determination by this method (see Chapter 7). The stone, held in position

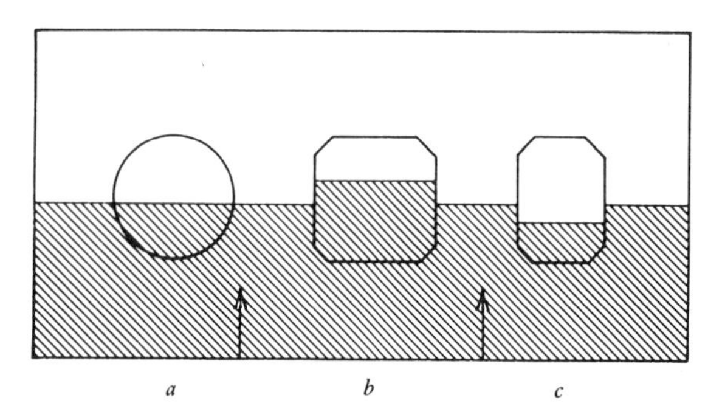

FIG. 6.17. Gemstones in immersion liquid with black card passed beneath. (*a*) **R.I.** of stone and liquid the same. (*b*) **R.I.** of stone less than that of liquid. (*c*) **R.I.** of stone greater than that of liquid.

by wax, is placed on a glass slide with the culet down, touching the slide, and the table parallel to the slide. The following measurements are necessary. (1) The microscope is focused on the glass slide and the reading shown on the fine adjustment focusing drum recorded. (2) The slide is moved to bring the stone into the field and the microscope tube racked upward until the table is in sharp focus. The difference in the reading in this position and the first reading obtained is the actual thickness of the stone, T. (3) The microscope is focused downward through the gem table until the culet or the point at the back of the stone is in focus. The difference in the reading on the focusing drum in this position and the second reading obtained is the apparent thickness of the stone, t. The refractive index of the stone is ***R.I.*** $= T/t$, that is, the actual thickness divided by the apparent thickness. It can be shown that $T/t = \tan i/\tan r$, not $\sin i/\sin r$ as in Snell's law. However, for small angles the ratio of their tangents approximates the ratio of their sines; therefore

$$\textbf{\textit{R.I.}} = \frac{\sin i}{\sin r} \approx \frac{\tan i}{\tan r} = \frac{T}{t}$$

The focusing knob on many microscopes is not calibrated to give the distance the tube is raised or lowered. However, with a little ingenuity, a centimeter scale can be attached to the movable housing of the microscope and a vernier fixed to the adjacent stationary support (Fig. 6.19). With it the vertical movement of the microscope can be determined within 0.01 mm.

Isotropic vs. Anisotropic

According to the manner in which they affect light passing through them, all transparent materials can be divided into two groups, isotropic and anisotropic. The isotropic includes gases, liquids, glass, amorphous substances such as opal, and all crystals of the isometric crystal system. They have a single refractive index, for light moves through them in all directions with the same velocity. In the anisotropic group, which includes all crystals except the isometric, the velocity of light varies with crystallographic direction resulting in a range in refractive

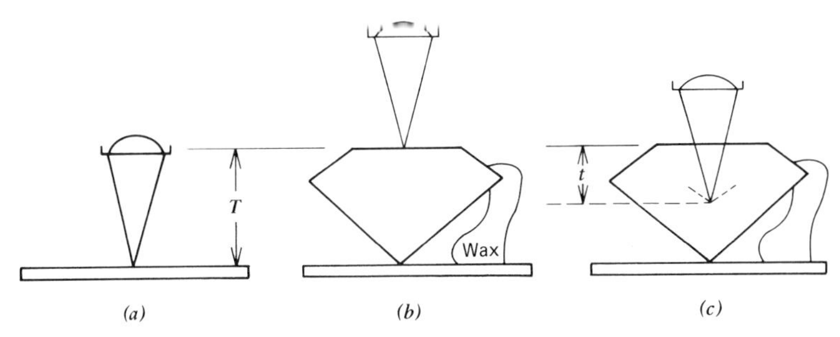

FIG. 6.18. Three readings for direct measurement of refractive index. (*a*) Focus on glass slide. (*b*) Focus on table, the large face at top of gemstone. (*c*) Focus through gemstone on culet, the small face at the bottom.

index. Except for special crystal directions, light entering an anisotropic crystal is broken into two polarized rays vibrating in mutually perpendicular planes and traveling with different velocities. Thus for a random orientation, an anisotropic crystal has two indices of refraction, one associated with each ray. This is known as *double refraction*.

Polarized Light

According to the wave theory (see above discussion of the nature of light under *Color*), light vibrates in all directions perpendicular to its direction of propagation. If the light is constrained to vibrate in a single plane, it is *plane polarized*. A source of polarized light is essential in gemstone examination for with it many characterizing crystallographic properties can be determined.

It was mentioned above that light passing through an anisotropic crystal is separated into two polarized rays vibrating in mutually perpendicular planes. If one of the rays can be eliminated, the other will pass out of the crystal as plane polarized light. The first efficient polarizer called the *Nicol prism* was based on this principle, that is, on double refraction. The prism was made from optically clear calcite, *Iceland spar*, with such strong double refraction that each ray (one designated as *O*, the other as *E*) produces a separate image when an object is viewed through a cleavage piece (Fig. 6.20 and Fig. 13.49). In making the prism, an elongated cleavage fragment was sawed at a specified angle and the two parts rejoined by cementing with Canada balsam (***R.I.*** 1.537). One of the two polarized rays into which the light is separated on entering the prism (***R.I.*** 1.66) is totally reflected when it strikes the Canada balsam (*O* ray, Fig. 6.21). The other ray with refractive index essentially that of the Canada balsam passes through the prism to emerge as plane polarized light (*E* ray, Fig. 6.21).

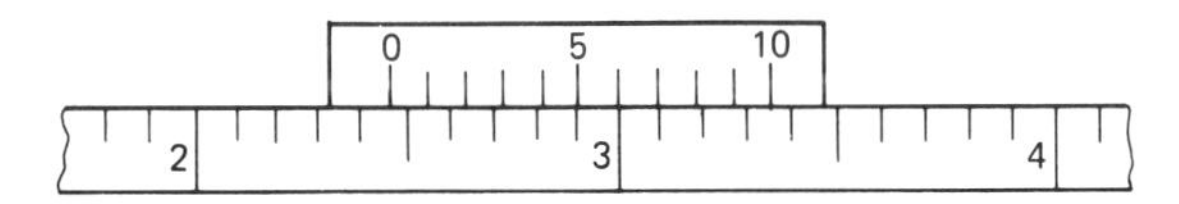

FIG. 6.19. Centimeter scale and vernier. Reading 2.46 cm.

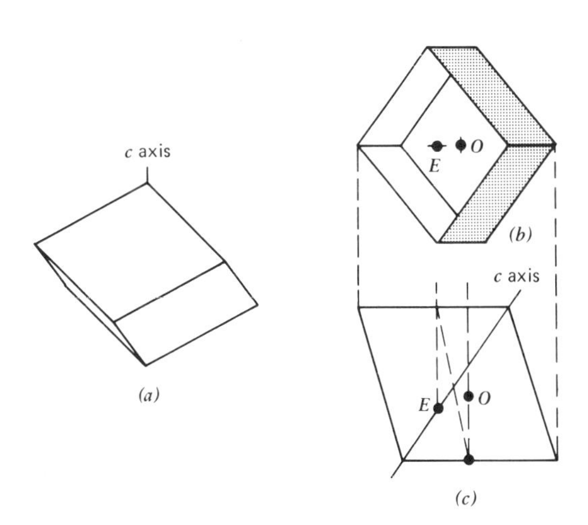

FIG. 6.20. Double refraction of calcite. (*a*) Calcite rhombohedron. (*b*) Double image with vibration directions indicated as seen through parallel faces. (*c*) Section showing paths of rays and apparent depth of images.

A second means of obtaining polarized light is by absorption. The two rays produced by double refraction may be absorbed differentially. If the absorption of one is nearly complete and the other is absorbed very little or not at all, the light passing through the crystal will be plane polarized. Among gem minerals, tourmaline offers the best example of this phenomenon. When light is passed through some tourmaline crystals at right angles to the c crystal axis, the ray vibrating at right angles to c is almost completely absorbed and the emerging light is polarized with vibrations parallel to c (Fig. 6.22*a*).

The most common means of producing polarized light today is with sheet polarizers such as Polaroid which utilize this property of differential absorption. In making them, crystals aligned on an acetate base absorb most of the light in one vibration direction but transmit nearly all wavelengths with little absorption in the other direction as plane polarized light (Fig. 6.22*b*). We shall see that polarizing plates made from such material are the basis of important instruments used by the gemologist.

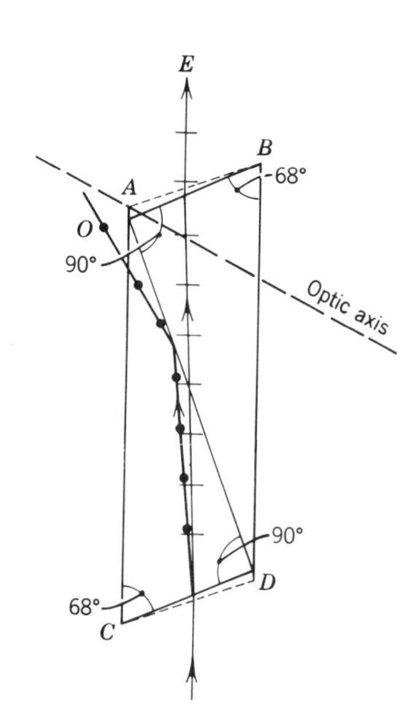

FIG. 6.21. Nicol prism. The ends of the initial cleavage rhombohedron of calcite, *AB* and *CD*, are ground to make angles of 68° with the prism edge and 90° with the sawed surface.

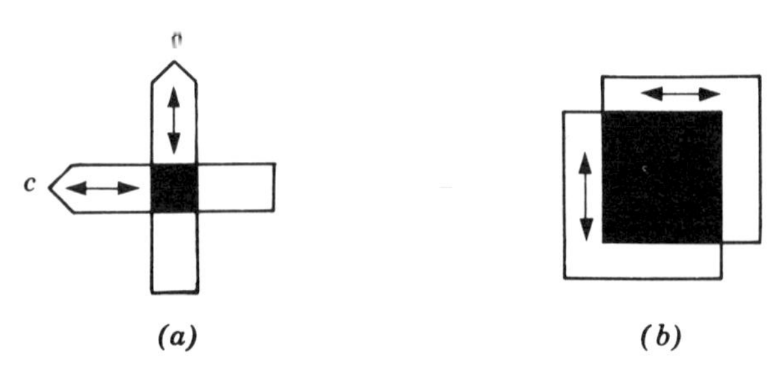

FIG. 6.22. Polarized light by absorption. (*a*) Tourmaline crystals. (*b*) Polaroid sheets. The arrows indicate directions of maximum transmission; directions of maximum absorption are at right angles.

A third method of obtaining polarized light is by *reflection*. Light striking a smooth nonmetallic surface is partially polarized with the vibration direction parallel to the reflecting surface (Fig. 6.23). Although unsatisfactory as a light source, the reflected beam is useful in orienting a polarizing plate in which the directions of absorption and transmission are unknown. The reflected light, viewed through the plate, is much brighter when the direction of transmission of the plate is parallel to the reflecting surface (Fig. 6.23*a*) than when turned 90° (Fig. 6.23*b*).

The Polariscope

A simple instrument yet one of the most important to the gemologist is the *polariscope*. It consists of two polarizing plates, jointly called the *polars*, one of which is supported 2–3 in. above the other. The conventional orientation of the plates is such that the lower one, the polarizer, transmits light vibrating N–S (front–back) and the upper plate (the analyzer) transmits light vibrating E–W. When held over a light source with the plates in these orientations, a minimum of light passes through the analyzer and the polars are said to be *crossed*. Such an instrument can be purchased with a built-in light source (Fig. 6.24), but one can be easily made in a home workshop.

Isotropic Substances

We have seen that light moves in all directions through isotropic materials with a uniform velocity;

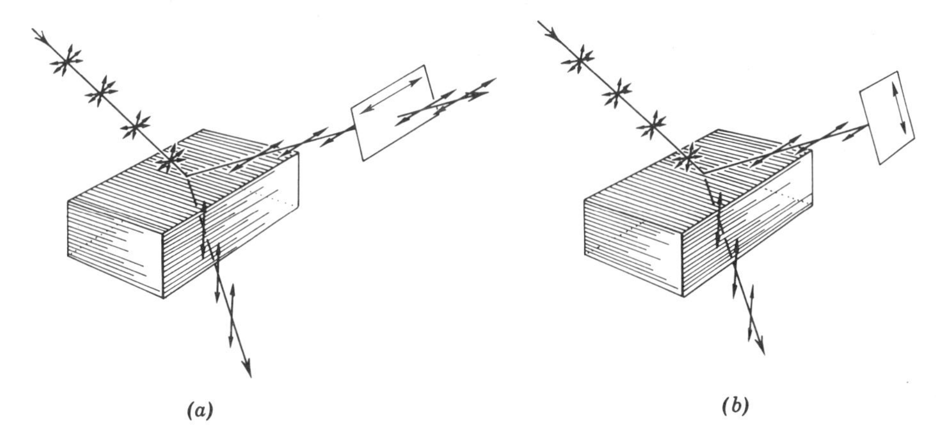

FIG. 6.23. Polarized light by reflection and refraction.

there is no double refraction and only one index of refraction. When an attempt is made to view a stone cut from glass or an isometric crystal through the polariscope, it appears dark. The light from the polarizer vibrating N–S passes through the stone to be eliminated by the analyzer which permits light to pass only if vibrating E–W (Fig. 6.25). One can thus quickly tell if a stone is isotropic for were it anisotropic, and in a random position, light would be seen to pass through it.

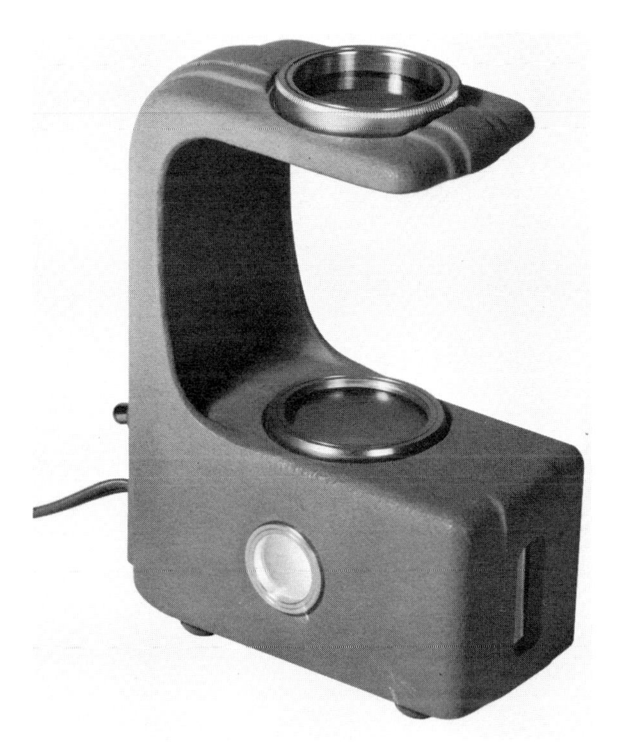

FIG. 6.24. Polariscope (GIA). Courtesy of Gemological Institute of America.

Uniaxial Crystals

For optical considerations anisotropic crystals are divided into two groups, *uniaxial* and *biaxial*. Uniaxial crystals are either tetragonal or hexagonal. In

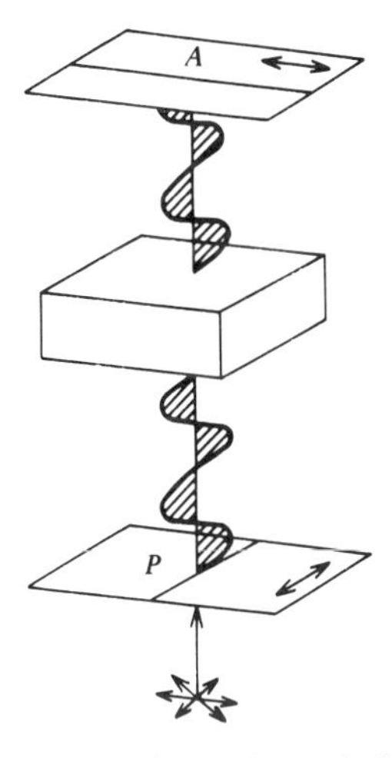

FIG. 6.25. Polarized light from the polarizer, *P*, passes through an isotropic substance in any orientation to be eliminated by the analyzer, *A*, and the substance always appears dark between crossed polars.

these crystal systems the *c* axis is unique; that is, it is of different length from the *a* axes, which are of equal length and lie in a plane at right angles to *c*. The optical symmetry agrees with the crystal symmetry, for light moving parallel to *c* also has a unique character; it suffers no double refraction. It moves through the crystal vibrating in all directions in the plane of the *a* axes (basal plane) and the crystal thus appears dark when viewed through a polariscope. For this reason the *c* axis is called the *optic axis* and since hexagonal and tetragonal crystals have only one such axis, they are called *uniaxial*.

When light moves through a uniaxial crystal in any direction other than parallel to *c*, it is broken into two rays traveling with different velocities. One, no matter its direction, has a constant velocity and always vibrates in the basal plane. It is called the *ordinary ray* (*O* ray). The other, the *extraordinary ray* (*E* ray), vibrates at right angles to the *O* ray in a plane that includes the optic axis. The maximum difference in velocity is obtained when light moves perpendicular to the optic axis with the *E* ray vibrating parallel to *c*. Since the indices of refraction are the reciprocals of the velocity, this situation also yields the maximum difference in refractive indices. The index associated with the ordinary ray is called ω (omega); that associated with the extraordinary ray is called ϵ (epsilon). The difference in the values of these two refractive indices is the *birefringence*. The *optic sign* of a uniaxial crystal is designated as *positive* if $\epsilon > \omega$ and *negative* if $\omega > \epsilon$.

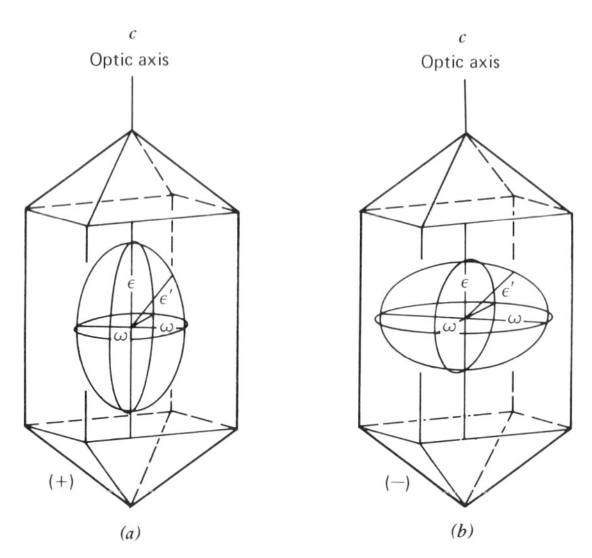

FIG. 6.26. The uniaxial indicatrix and orientation in tetragonal crystals. (*a*) Positive crystal. (*b*) Negative crystal.

Uniaxial Indicatrix

This is a geometrical figure helpful in relating the values of the refractive indices to their directions of vibration and to the direction of propagation of light through the crystal. For positive crystals the indicatrix is a prolate spheroid of revolution; for negative crystals it is an oblate spheroid of revolution (Fig. 6.26). Radial lines from their centers represent vibration directions, and the length of the lines are proportional to the refractive indices.

A section through the center of the indicatrix perpendicular to the axis of revolution is a circle (the circular section) the radius of which is proportional to ω. Thus light moving parallel to the axis vibrates in the circular section and has ω (***R.I.*** of *O* ray) as its only index of refraction. For any other direction of incident light there is double refraction and the *O* and *E* rays vibrate in a plane at right angles to the direction. A central section cut through the indicatrix parallel to this plane is an ellipse with major and minor semiaxes representing the refractive indices. Other sections through the indicatrix, parallel to planes of vibration of different directions of incident light, all have one semiaxis the same. This constant semiaxis, equal to the radius of the circular section, represents the refractive index of the *O* ray, ω. The other semiaxes vary in length between the radius of the circular section and the semiaxis of the indicatrix parallel to *c*. These are vibration directions of the *E* ray and the refractive indices, proportional to their lengths, are all designated ϵ'. Only in a vertical section through the indicatrix is ϵ itself a semiaxis of the ellipse (Fig. 6.26).

Estimation of Birefringence

The foregoing discussion is of practical as well as theoretical concern for the gemologist. Although the indicatrix may have a completely random orientation in reference to the facets, uniaxial gemstones are commonly cut with the table (the large top facet) either at right angles to the optic axis or parallel to it. If it is at right angles and one looks through the table, there is no double refraction and the back facets appear sharp. However, if the optic axis is parallel to the table, the light reaching the eye through the table is from both the *O* and *E* rays with maximum double refraction. If the birefringence is high, two images are seen; that is, there is *doubling of the back facets*. A stone should be examined from several different angles to make sure it is being viewed in the direction to give maximum birefringence. *Doubling* can be seen in some gems, for example zircon, with the unaided eye, but it is best observed with a lens or microscope. Doubling of the back facets indicates the stone is anisotropic and the amount the two images are separated gives an estimate of the birefringence.

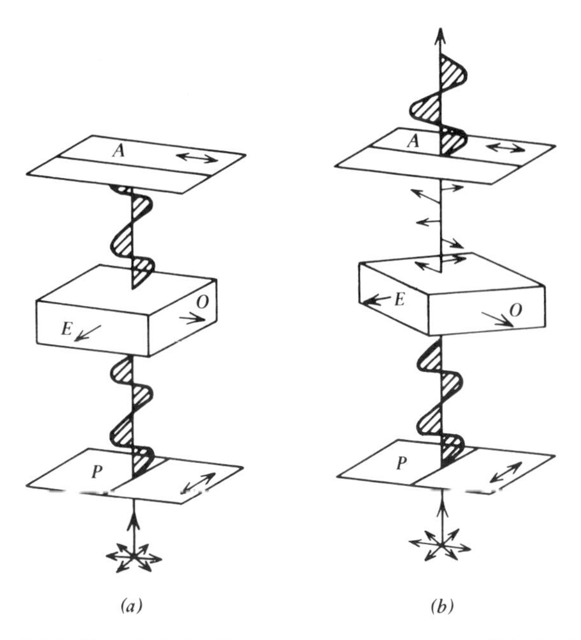

FIG. 6.27. (*a*) Extinction in a uniaxial crystal. Polarized light from the polarizer, *P*, vibrating N–S passes through the crystal as the *E* ray to be completely eliminated by the analyzer, *A*. A 90° rotation would bring the crystal to the other extinction position and the polarized light would pass through as the *O* ray. (*b*) Interference. With the crystal not at extinction, the polarized light entering it is broken into *O* and *E* rays vibrating at 90° to each other. In the analyzer each ray is resolved into a N–S and an E–W component but only the E–W components emerge. Since they travel through the crystal at different velocities, they are out of phase and interfere with the elimination of certain wavelengths.

Extinction

We have seen that isotropic substances remain dark in all positions between cross polars and that uniaxial crystals also remain dark when light moves parallel to the optic axis. There is another special condition in which uniaxial crystals present a dark field between crossed polars. This is when the vibration direction of the polarizer coincides with one of the vibration directions of the crystal (Fig. 6.27*a*). The light then passes through the crystal as the *O* or the *E* ray, to be completely eliminated by the analyzer, and the crystal is said to be at *extinction*. As the crystal is rotated from the *extinction* position, it becomes progressively brighter, reaching a maximum at 45°. At a 90° rotation it is again at extinction; there are thus four extinction positions in a rotation of 360°. The alternating positions of extinction and maximum transmission do not serve to distinguish uniaxial from biaxial crystals; they are shown by both. Their presence merely indicates that the crystal is anisotropic.

Interference

When a gemstone on the polariscope is not at extinction, white light striking it from the polarizer is separated into *O* and *E* rays vibrating in perpendicular planes. When they encounter the analyzer, each ray is again resolved into two components, one of which vibrates parallel to the direction of absorption, the other parallel to the direction of transmission. Because of the different velocities with which the two rays pass through the stone, they emerge out of phase; one is ahead of the other (Fig. 6.27*b*). Since they are constrained to vibrate in the same plane of the analyzer, they interfere with the elimination of certain wavelengths. This results in *interference colors* which are the complementary colors of the eliminated wavelengths and depend on the thickness of the crystal (gemstone), its orientation, and birefringence.

Gemstones have a great thickness in terms of the wavelengths of light; and although there is interference, the wavelengths transmitted are so closely spaced throughout the spectrum that the stone at the position of maximum brightness (45° position) appears to be illuminated by white light. But small crystal fragments in this position may show a variety of colors. When light passes through gemstones nearly parallel to an optic axis, they too may show interference colors arranged in bands concentric or parallel to thin edges.

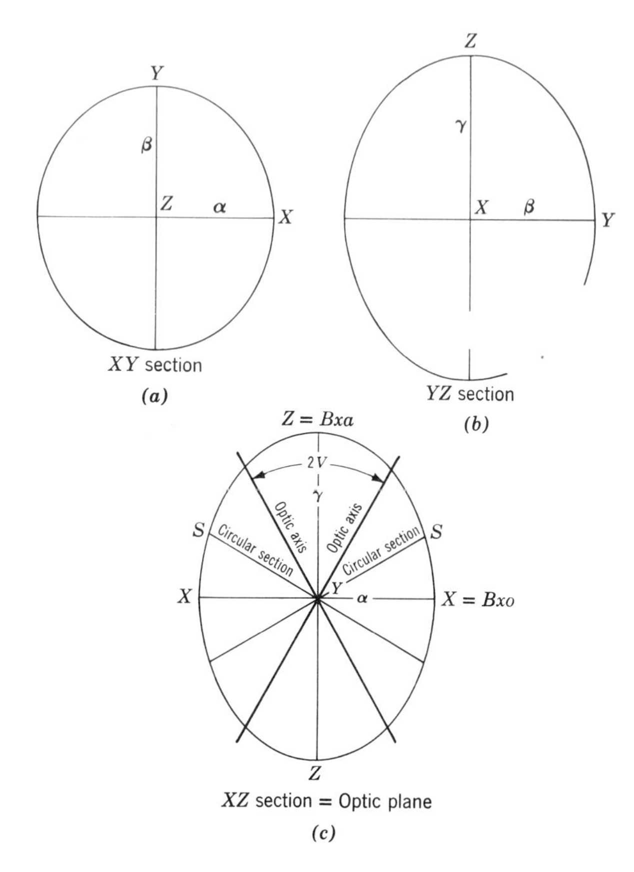

FIG. 6.29. Principal sections through the biaxial indicatrix of a positive crystal.

Biaxial Crystals

Crystals belonging to the orthorhombic, monoclinic, and triclinic crystal systems are called biaxial. They have two optic axes, directions along which light moves with no double refraction; in uniaxial crystals there is only one.

Biaxial Indicatrix

In biaxial crystals there are three principal refractive indices. The relation of the indices and their vibration directions can be best visualized with a *biaxial indicatrix* (Fig. 6.28). This is a triaxial ellipsoid so constructed that the three mutually perpendicular semiaxes coincide with the vibration directions of the indices and with their lengths proportional to the indices.

The three principal indices of refraction are designated as α (alpha), β (beta), and γ (gamma) and in Figure 6.28 are represented respectively by *OX*, *OY*, and *OZ*. α is the least, γ the greatest, and β intermediate; the birefringence is given as $\gamma - \alpha$. Biaxial crystals are defined as positive if β is closer to α than to γ and negative if β is closer to γ than to α.

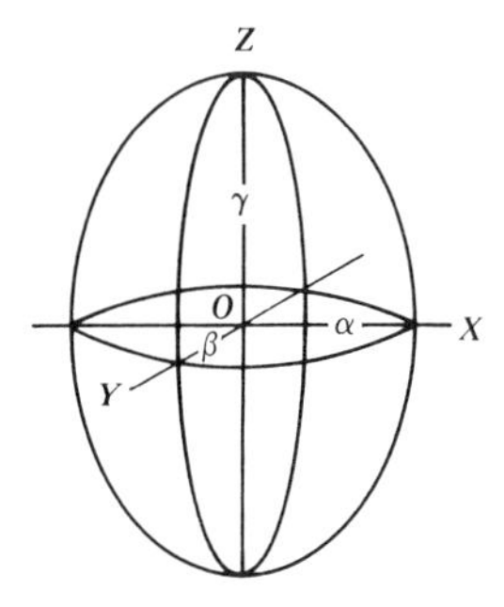

FIG. 6.28. Biaxial indicatrix. The principal optical directions *OX*, *OY*, *OZ* are at right angles and proportional respectively to the refractive indices α, β, and γ.

With two exceptions every section through the center of the indicatrix is an ellipse whose semimajor and semiminor axes represent the vibration directions and their lengths the indices of refraction when incident light is normal to the section. The principal sections of the indicatrix, *XY*, *YZ*, and *XZ*, are shown in Figure 6.29. Of greatest interest is the *XZ* section with γ its semimajor axis and α its semiminor axis. Between these extremes, there must be points on the ellipse whose radii are proportional to the intermediate index, β. In Figure 6.29*c* these are indicated by S. Sections through the indicatrix that include the Y (β) axis and these points are the two exceptions mentioned above; they are circular. All radii of these *circular sections* are equal to β. The optic axes are perpendicular to the circular sections and light traveling along them is not doubly refracted but behaves as light moving along the optic axis of a uniaxial crystal.

The XZ ($\alpha\gamma$) plane in which the optic axes lie is called the *optic plane* and the Y (β) direction is the *optic normal*. The acute angle between the optic axes is the *optic angle*, symbolized as $2V$. If γ bisects the optic angle, γ is said to be the *acute bisectrix* (Bxa); α is then the *obtuse bisectrix* (Bxo) and the crystal is positive (Fig. 6.29c). With α the Bxa and γ the Bxo, the crystal is negative. The closer β approaches α or γ, the smaller $2V$. When β lies halfway between α and γ, $2V = 90°$.

Distinction Between Uniaxial and Biaxial Gems

A doubly refracting gem frequently can be identified on the basis of specific gravity and average index of refraction. However, in some cases these measurements do not give unambiguous answers and determination can be made with more certainty if it is known that the stone is uniaxial or biaxial and is optically positive or negative. There are two ways of obtaining this information; one is using the refractive indices, the other is with interference figures.

Refractive Indices

Isotropic gemstones, because of their single refractive index, show a single shadow edge on the refractometer regardless of orientation. Uniaxial stones will in general show two shadow edges, one resulting from the O ray, the other from the E ray. The edges are clearly seen only if a monochromatic light source, such as sodium light, is used. If the stone is turned on the hemisphere, one edge remains stationary but the other edge moves. The constant edge gives the index of the O ray, ω; the movable edge, resulting from the E ray, gives various values of ϵ'. The reading for true ϵ can be taken only when the two edges are farthest apart and the difference between it and ω is the birefringence. If the readings from the movable edge are greater than ω, the stone is optically positive; if they are less than ω, the stone is negative (Fig. 6.30). If the facet, usually the table, in contact with the hemisphere has a random crystallographic orientation, the situation is as described above and ϵ' never equals ω.

However, very commonly uniaxial gems are cut so that the table is either parallel to or perpendicular to the optic axis. Special mention should be made of these two orientations. If the table is parallel to the c (optic) axis, a single shadow edge (ω) will be seen when the c axis is parallel to the general direction in which light moves through the refractometer, that is, parallel to the length of the instrument (Fig. 6.31*a*). As the stone is turned on the refractometer, a second shadow edge is seen giving the true value of ϵ after a rotation of 90° (Fig. 6.31*b*). If the table is at right angles to the optic axis, both edges remain stationary as the stone is rotated and give the maximum birefringence. Under these conditions it is impossible to tell which is the variable index. To determine this and thus obtain the optic sign another facet must be placed in contact with the hemisphere, or a polarizing filter may be used as described below.

As light leaves an anisotropic stone (uniaxial or biaxial) on the refractometer it is polarized in two directions vibrating at right angles to each other. In

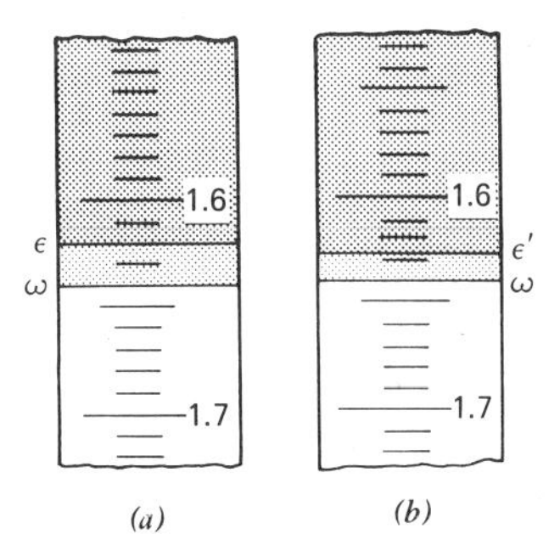

FIG. 6.30. Shadow edges on refractometer scale obtained with a tourmaline gem. (*a*) Readings of 1.620 and 1.640 give the maximum birefringence of 0.020. (*b*) Readings of 1.627 and 1.640 obtained after the gem is turned show that the higher index is constant and that the gem is negative.

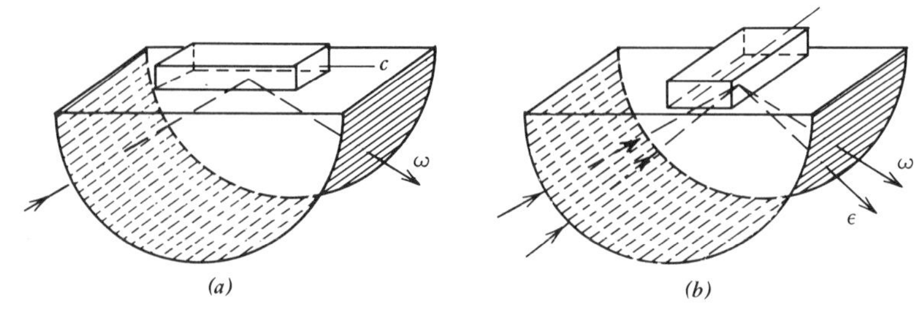

FIG. 6.31. (*a*) A single refractive index (ω) is measured when the *c* axis of a uniaxial crystal is parallel to the long dimension of the refractometer. (*b*) On a 90° rotation, ω and ϵ can both be measured giving the maximum birefringence.

uniaxial crystals one vibration direction is of the *O* ray, the other of the *E* ray. By placing a polarizing cap (or Polaroid sheet) over the eyepiece of the refractometer, one of the rays and its resulting shadow edge can be eliminated and in effect sharpen the remaining edge. This is particularly helpful when white light is used. When the facet in contact with the hemisphere is randomly oriented, the polarizing cap should be rotated until one of the shadow edges disappears. Turning the cap 90° will eliminate the other edge. In the event that the facet being tested is perpendicular to the optic axis and both shadow edges remain constant on rotation of the stone, the polarizing filter can be used to determine which edge results from ω and which from ϵ. In this orientation ϵ is vibrating parallel to the long dimension of the refractometer (i.e., parallel to a line joining the window and the eyepiece) and ω vibrates at right angles to this direction. If the polarizing cap is placed over the eyepiece with its transmission direction parallel to the vibration direction of ω, the shadow edge of ϵ is eliminated (Fig. 6.32*b*). Turning the cap 90° will eliminate the ω shadow edge (Fig. 6.32*c*). It is thus possible to tell which index is the greater and determine the optic sign.

Biaxial stones in general show two shadow edges on the refractometer. However, they differ from uniaxial for, when the stone is rotated on the hemisphere, both edges move. The minimum index indicated by the lower edge is α; the maximum index obtained from the higher edge is γ. The difference between the readings is the birefringence. To determine the optic sign, that is, (+) or (−), it is necessary to know whether the intermediate index, β, is nearer α or γ. When β is nearer to α than to γ, the crystal is (+); when β is nearer to γ the crystal is (−). γ' is used to denote indices between β and γ and α' for indices between β and α. Which of the above two situations obtains can be determined by rotating the stone on the hemisphere of the refractometer and noting the movement of the high (γ') and low (α') shadow edges. The value of β will equal the lowest reading of γ' and the highest reading of α'. If β is less than the midpoint between α and γ, the stone is positive; if β is greater than the midpoint, the stone is negative.

For example, consider a brazilianite gem with the facet in contact with the hemisphere not perpendicular to a principal optical direction (*X*, *Y*, or *Z*). The stone is rotated until the two shadow edges

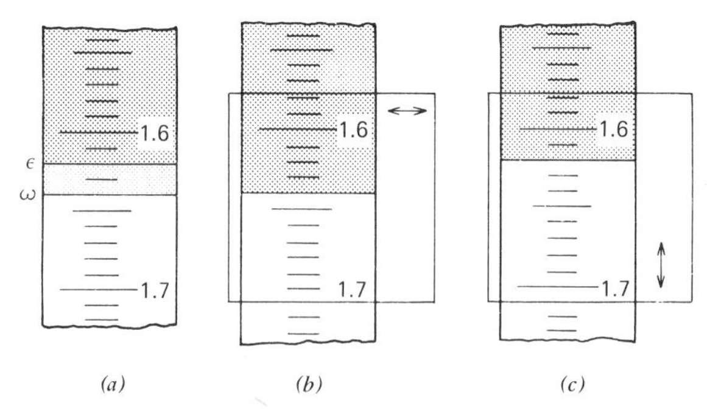

FIG. 6.32. (*a*) Two constant shadow edges resulting from tourmaline with facet on refractometer perpendicular to the optic axis. (*b*) Shadow edge of ϵ eliminated with polarizing sheet (arrow indicates direction of transmission). (*c*) Shadow edge of ω eliminated with polarizing sheet. With $\epsilon < \omega$ the optic sign is negative.

are farthest apart. The readings are high edge, 1.621 and low edge, 1.602. These are respectively the values for γ and α. Their difference (1.621 − 1.602 = 0.019) is the birefringence and the midpoint between them is 1.6115. On rotation from this position, the high edge moves lower, giving values for γ' that reach a low point of 1.609, the value of β. The highest value reached by the lower edge (α') is also 1.609. Since β is less than 1.6115 (the midpoint), it is closer to α than to γ and the gem is positive. A gem in which γ' does not pass the midpoint would be negative.

It is difficult for many to understand why α and γ can both be determined when a randomly oriented facet is in contact with the refractometer. Consider Figure 6.33*a* as representing such a case in which the stone is not shown but the *X, Y,* and *Z* directions of its indicatrix are indicated. None of these directions is parallel or perpendicular to the flat surface of the refractometer. In this random orientation there are two critical angles, one for α' and one for γ'. The two rays are totally reflected and emerge through the hemicylinder. We have learned that light moving through any anisotropic crystal is broken into two polarized rays vibrating at right angles to each other and at right angles to the direction of propagation. In using the refractometer, we must think of light entering the crystal and moving in the direction *OC* that lies in the plane in contact with the hemicylinder. The two rays into

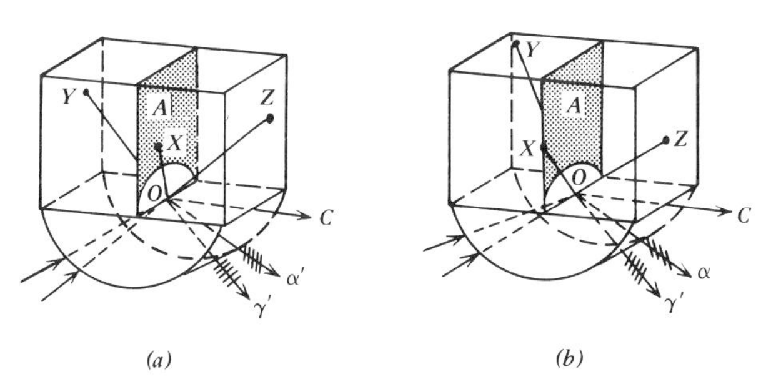

FIG. 6.33. Schematic representation showing the relation of the principal optical directions (*X, Y, Z*) to a randomly cut facet. (*a*) The shadow edges on the refractometer will give values for an α' and a γ'. (*b*) The same cut as in (*a*) rotated on the facet until one of the principal optical directions (*OX*) lies in plane *A*, at right angles to *OC*. The shadow edge readings now give values for a γ' and true α.

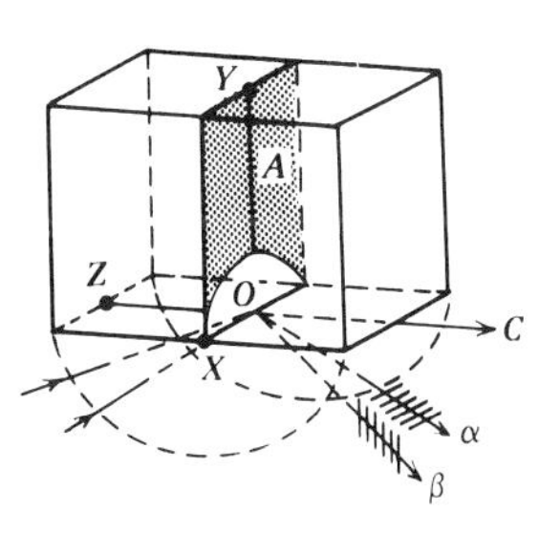

FIG. 6.34. Schematic representation of the principal optical directions (*X, Y, Z*) when one of them (here *OY*) is perpendicular to the facet in contact with the refractometer. As the stone is rotated, *OY* remains unchanged and a reading for β is obtained in all positions. The figure also shows the position in which the lowest reading, α, is obtained.

which it is broken must, therefore, vibrate in plane *A* (at right angles to *OC*). On emerging, the rays maintain their vibration directions, and give values for α' and γ'.

As the gemstone is rotated on the glass hemicylinder of the refractometer, the orientation of the indicatrix changes. When a position is reached such that *OX* lies in plane *A* (Fig. 6.33*b*), one of the totally reflected rays (vibrating parallel to *OX*) will give the value of true α. When the stone is turned so that *OZ* lies in plane *A*, the higher shadow edge, seen on the refractometer, will indicate the value of true γ.

If the facet in contact with the hemicylinder is perpendicular to one of the principal optical directions (*X, Y,* or *Z*), the other two principal optical directions must lie in the plane of the facet. If the direction perpendicular to the facet is *OY* (Fig. 6.34), the shadow edge indicating the refractive index vibrating in that direction (in this case β), remains stationary as the stone is turned. The lowest reading of the movable edge is α, as in Figure 6.34; the highest reading is γ. At one point in turning the stone from the α to the γ position, the movable edge coincides with the stationary β reading and there is only one shadow edge seen. If β is closer to α than to γ, the crystal is positive; but if β is closer to γ than to α, the crystal is negative.

It should be emphasized that the difference between the highest and lowest readings of refractive index of any anisotropic gem obtained from any facet is its full birefringence. Birefringence is an important determinative property for, although two stones may have similar average refractive indices, they may show a marked difference in birefringence.

Interference Figures

Interference figures are routinely used by mineralogists to distinguish between uniaxial and biaxial crystals and to determine the optic sign. They are usually obtained on crushed mineral fragments using a highpower polarizing microscope. Since gemologists rarely wish to crush gems and only a few have such a microscope available, a discussion here of this method is impractical. However, it is possible to obtain interference figures on gemstones and one relatively simple device to accomplish this is described in the Appendix.

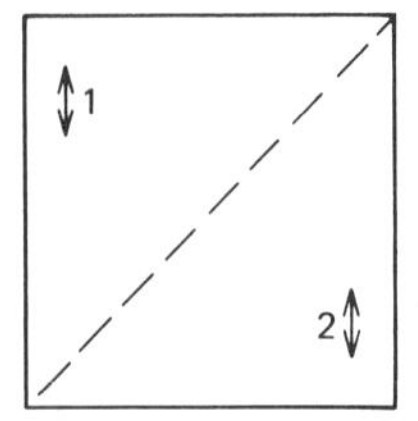

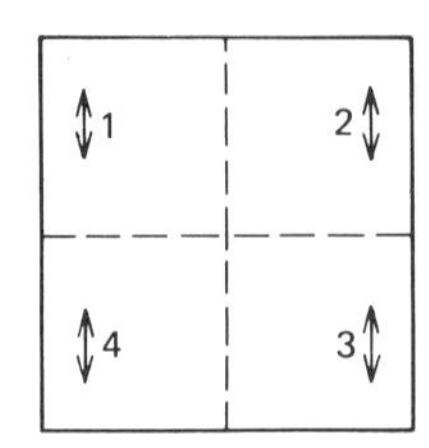

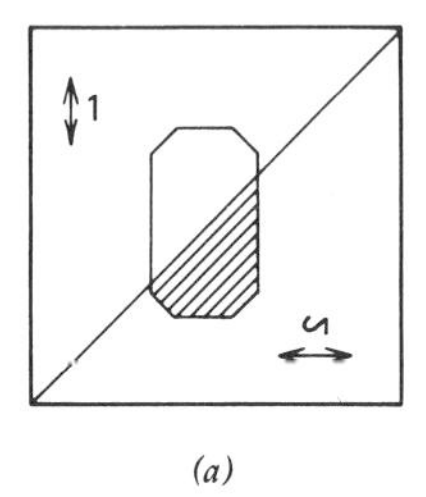

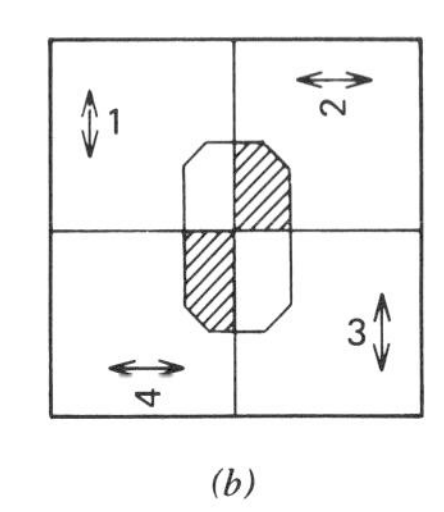

(*a*) (*b*)

FIG. 6.35. Square Polaroid sheet cut to observe pleochroism. (*a*) Sheet is cut on diagonal and portion 2 turned over and joined with 1. The transmission directions are then at right angles. (*b*) The sheet is cut into four small squares with 2 and 4 rotated 90° as shown.

Pleochroism and the Dichroscope

Light as it passes through anisotropic crystals may be absorbed differently in different vibration directions. The absorption in one direction may be nearly complete and at right angles negligible as in the tourmaline mentioned in the discussion of polarized light (q.v.). More commonly, however, different wavelengths are absorbed in one direction than in the other. The resulting color variation is called *pleochroism* (many colors).

Uniaxial gemstones with only two rays, *O* and *E*, can show only two pleochroic colors and are said to be *dichroic*. The colors vary as do the refractive indices. Light moving parallel to the optic axis shows only one color, that of the *O* ray. The color from the *E* ray becomes more pronounced, the greater the departure of incident light from the optic axis, and reaches a maximum at 90°. In some gems the colors resulting from the two rays are quite different but frequently they are merely shades of the same color differing only in intensity. The latter is true of ruby in which a deeper red results from the *O* ray than the *E* ray. This is expressed as absorption = $O > E$ or $\omega > \epsilon$. Absorption is independent of other properties and is considered, as are refractive indices, a fundamental optical property of colored crystals.

Biaxial stones with three principal directions of vibration may exhibit three different pleochroic colors, one resulting from vibrations parallel to each direction. The phenomenon is sometimes called *trichroism* but is usually referred to by the more gen-

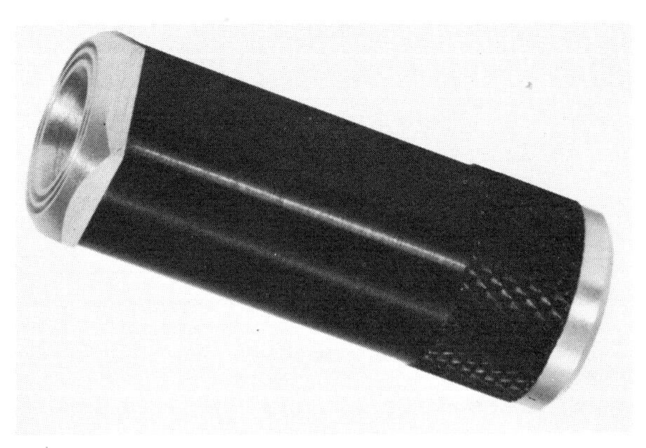

FIG. 6.36. Dichroscope (GIA). Courtesy of the Gemological Institute of America.

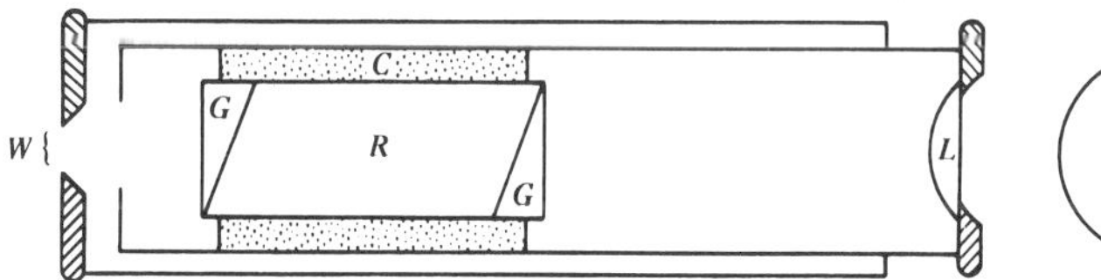

FIG. 6.37. Dichroscope, diagrammatic section. *R* is a calcite cleavage rhombohedron held by a cork setting, *C*, in a metal tube. Glass prisms, *G*, are cemented to the calcite to aid in light transmission. To the right of the drawing is a double image of the window, *W*, as seen through the lens, *L*.

eral term, pleochroism. Probably the most pronounced pleochroism in a biaxial stone is to be seen in gem zoisite, tanzanite. Light vibrating in the three principal directions yields deep red, sapphire blue, and yellow (or green) (see Plate VI).

Detection of Pleochroism

Since pleochroic colors result from rays vibrating at right angles to each other, it is possible to isolate one by eliminating the other. This can be done with the polariscope by rotating the stone between crossed polars to an extinction position. The upper polar should then be removed and the color of the stone noted in plane polarized light. If the stone is pleochroic, a different color or color intensity will be observed on rotating it 90°. Subtle color differences can be better seen if observed at the same time. This can be done by using a Polaroid sheet that is cut and rejoined so the directions of transmission are at right angles to each other as shown in Figure 6.35. The reassembled sheet is held over a light source and the gemstone is placed on the line of join (Fig. 6.35*a*) or at the center point (Fig. 6.35*b*). Both pleochroic colors will be transmitted side by side making comparison easy.

The Dichroscope

Although the pleochroic colors are seen simultaneously using a split Polaroid sheet, they originate from different parts of the stone. It is somewhat more satisfying to compare directly colors originating from the same portion of the stone. This can be done with a small ingenious instrument, the *dichroscope*. This instrument is a metal tube with a square or rectangular opening at one end and a lens at the other end (Fig. 6.36). Within the tube is mounted an elongated cleavage piece of optical calcite (Iceland spar) (Fig. 6.37). Because of the strong double refraction of calcite, a double image of the square aperture is seen through the dichroscope. If a pleochroic stone held over a bright light source is viewed, the two images have different colors. Although both images originate from the same portion of the stone, their vibration directions are at right angles to each other.

It should be remembered that pleochroism is lacking for light passing parallel to an optic axis, either uniaxial or biaxial, and is a maximum for light moving perpendicular thereto. It is therefore important in any method for detecting pleochroism that light be passed through the stone in several directions and that in each direction the stone be turned to achieve the maximum effect.

References and Suggested Reading

Anderson, B. W., 1971. *Gem Testing,* 8th ed. Van Nostrand Reinhold Co., New York. Chapters 2, 3, and 4.

Wahlstrom, E. E., 1969. *Optical Crystallography*, 4th ed. John Wiley & Sons, Inc., New York. 489 pp.

Webster, R., 1975. *Gems,* 3rd ed. Butterworth & Co. Ltd., London. Chapters 29, 30, and 31.

Wood, E. A., 1964. *Crystals and Light.* D. Van Nostrand Co., Inc., Princeton, N.J., 160 pp.

7 THE MICROSCOPE

The basic function of any microscope is to yield an enlarged image of an object to enable one to see details too small to be seen, or seen clearly, with the unaided eye. In gemology the most important use of the microscope is to inspect a gemstone for both internal and surface imperfections. A microscope may also be used to determine the refractive indices and other optical properties of a gem, but these are usually more conveniently measured by use of specialized instruments described in the preceding chapter.

Although most people do not regard it as such, an ordinary reading glass is a microscope. So also are the jeweler's loupe and the hand lens, for they all produce an enlarged image of an object. These magnifiers using a single lens are called *simple microscopes* in contrast to *compound microscopes* that employ two or more lens systems. The compound microscope is commonly designated merely as "microscope," the usage that is followed in this book.

Hand Lens

For the student of gemology the hand lens is the most important single piece of equipment. For, before turning to more elaborate devices, inspection with this simple magnifier may suffice to identify a gem.

Hand lenses of various magnifications and of comparable quality are available from several manufacturers. The highest-quality lenses, those recommended for the gemologist, are known as Hastings triplets or triple aplanatic lenses (Figs. 7.1c and 7.2). They are made of three elements; two external lenses of flint glass are cemented to a double convex lens of crown glass. This combination of glasses gives to the triplet a flat field and eliminates color fringes (chromatic aberration). Magnification should not be confused with quality, for a Hastings triplet with a magnification of 5 times or 20 times (written 5× and 20×) are of comparable quality. The higher the magnification, the closer the lens must be to the gem and the smaller the field of

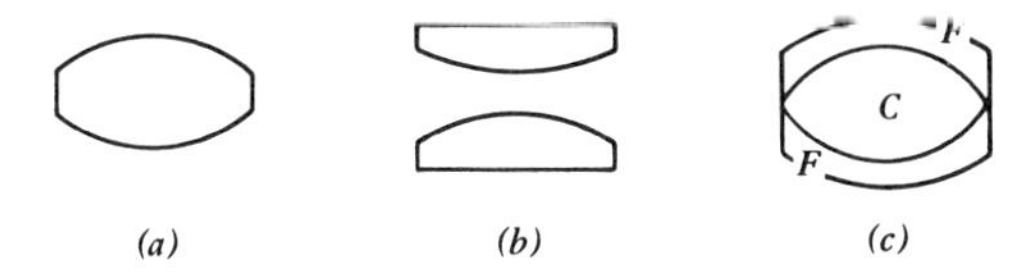

FIG. 7.1. (*a*) The least expensive magnifier is a double convex lens, acceptable for magnification up to 3×. (*b*) A better type is the doublet, consisting of two plano-convex lenses. (*c*) A still better type, known as a Hastings triplet or triple aplanat, consists of two external flint glass (*F*) lenses enclosing and cemented to a single double-convex crown glass lens (*C*).

view. All lenses should be held as close to the eye as possible. A magnification of 10× is most useful, for it gives an adequate field of view and is in focus about 1 in. from the gemstone.

Though extremely useful because of its small size and portability, a hand lens has serious shortcomings in its ability to reveal the internal features of a gemstone. Observations are usually made in improper lighting with the gem held in one hand and the lens in the other. For a thorough and critical examination of a gem it is necessary to use a compound microscope, particularly one especially adapted to gemological work.

Compound Microscope

A compound microscope consists fundamentally of two lenses, the objective and the ocular (eyepiece). The function of the objective lens, at the lower end of the microscope tube, is to produce a sharp image; the ocular enlarges this image. The total magnification of the microscope is the product of the magnification of the objective and the ocular. Thus a 10× objective and an 8× ocular give a magnification of 80 times. The familiar microscope used in biology and medicine is of this type. However, it is not suitable for gemological work because even the lowest magnification is usually too high, and the short working distance (distance between the objective and an object placed on the microscope stage) does not permit the examination of large objects. Another disadvantage is the inverted image, that is, features existing at the right of the object are seen at the left side of the field.

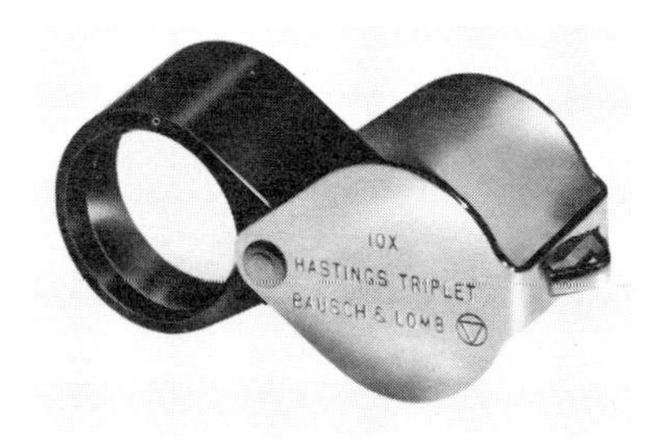

FIG. 7.2. A 10× triple aplanatic hand lens. Courtesy of the Bausch and Lomb Company, Rochester, New York.

The petrographic or polarizing microscope is a modification of the ordinary microscope equipped so that objects may be viewed both in plane polarized light and between crossed polars. The polarizing microscope is an important research tool in mineralogy and petrology for determining refractive index, optic sign, and crystal orientation but suffers from the same drawbacks as the biological type. It does, however, have limited use in gemology, particularly in examination of small stones or fragments of rough material.

The microscopes described above are traditionally the monocular type, that is, a tube with an objective at the bottom and an ocular at the top. However, they may be equipped with a binocular system which permits the use of both eyes simultaneously to make prolonged use less tiring. Since viewing is still done through a single objective, there is no three-dimensional or stereoscopic effect.

Low-power Binocular Microscope

Most useful in gemology is the low-power binocular microscope, also called a stereoscopic widefield microscope (Fig. 7.3). Several manufacturers each make a number of models of these instruments but the usual type has two separate matched objectives with prisms in front of each eyepiece to reinvert the image. These instruments provide true stereoscopic vision as well as the comfort of binocular viewing. Pairs of objectives may be mounted on a rotating nosepiece to permit rapid change in magnification, or the "zoom" principle may be used, permitting continuous change in magnification (Fig. 7.3). Magnification from 10× to 60× is readily obtained with these instruments, the range most useful in gemology.

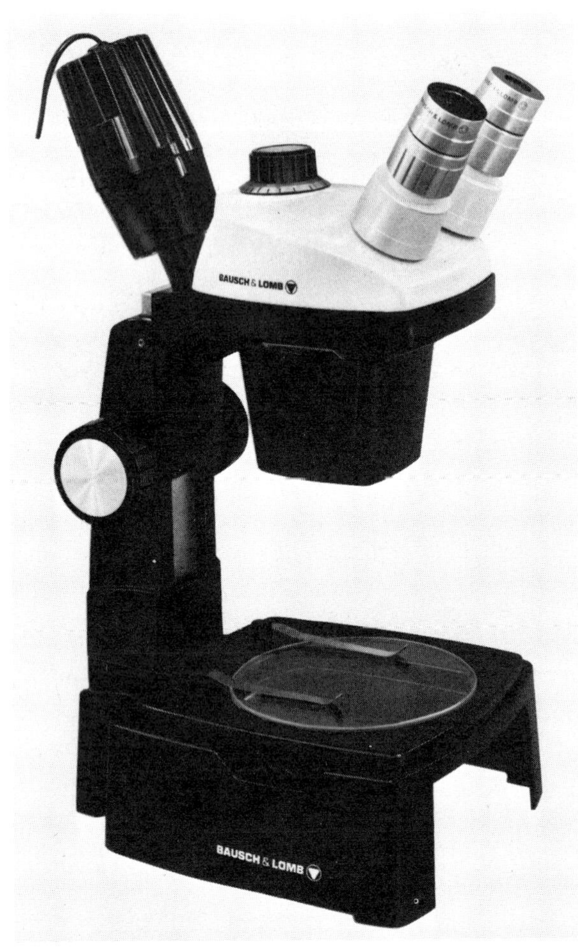

FIG. 7.3. Stereoscopic wide-field microscope manufactured by Bausch and Lomb. A "zoom" feature permits continuous change in magnification. The most useful range is 10–45×. Courtesy of the Bausch and Lomb Company, Rochester, New York.

Gemstone Examination with a Microscope

Most microscopes used for gemological work permit use of both incident and transmitted light. Examination of opaque gemstones, and most of those that are mounted, must be made with the light source above the microscope stage and the external features are observed by light reflected from the surface. The interior of a stone is best seen in transmitted light, that is, in light that passes through it from a substage illuminator.

Because of the external and internal reflections from the facets, viewing of the internal features of a gem is difficult even in transmitted light. Ideal conditions for internal examination require that light pass through the gem as through a transparent plate with parallel sides. As shown in Figure 7.4*a*, this condition does not exist with a gemstone placed table down on the microscope stage except for the very small area of the culet, if this facet is present. Elsewhere total reflection prevents one from viewing the interior of the gem. More light passes through the same stone resting on a pavilion facet (Fig. 7.4*b*), but still a large portion of the interior cannot be viewed. If one holds the stone in a pair of tweezers and looks at it from several directions, a better examination can be made but always with risk that important internal features have been missed. The problem can be minimized by the use of either an immersion cell or dark field illumination.

Immersion Cell

A simple means of overcoming the difficulties pointed out above is the immersion cell in which the gem is totally immersed in a liquid. Ideal conditions exist in the cell when the refractive indices of liquid and gem are the same and there is no reflection or refraction. The paths of parallel light rays coming from a substage illuminator in this situation are shown in Figure 7.5*a*. In practice it is not

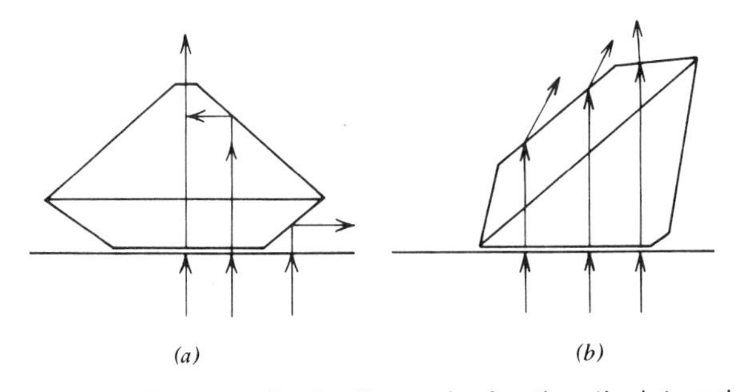

FIG. 7.4. Because of reflection and refraction, the internal features of a gemstone are viewed with difficulty using ordinary transmitted light. (*a*) With table down the only light passing directly through is in the small area of the culet. (*b*) More light passes through the gemstone resting on a pavilion facet.

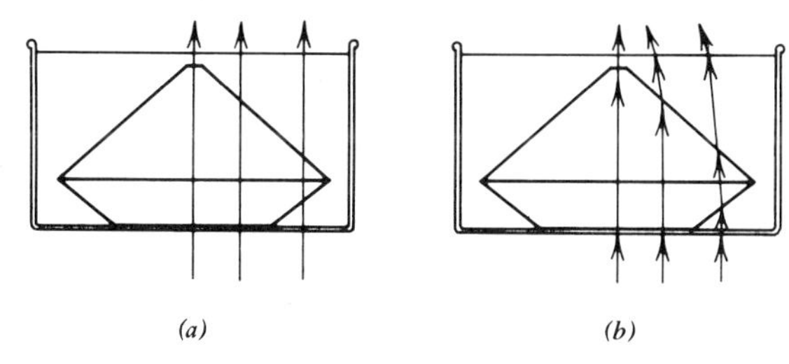

FIG. 7.5. (*a*) Gem immersed in liquid of the same refractive index. (*b*) Gem immersed in a liquid having a slightly lower refractive index than that of the gem.

necessary that the refractive index of the gem and liquid exactly match, although the closer they are to one another the better. The paths of light rays when the refractive indices are not the same are shown in Figure 7.5*b*.

The same liquids listed in Table 6.1, used for determining the relative refractive index, can also be used in the immersion cell.

The best immersion cells are made from a glass cylinder cemented at one end to a plate of optical glass. Less satisfactory is any shallow, flat-bottom dish such as a small beaker or the Petri dishes used in bacteria culture. The drawback of vessels of these latter types is that their bottoms usually contain swirl marks and bubbles which the inexperienced worker may interpret as being within the gem.

The principal disadvantages of an immersion cell are the handling of liquids, some of which have unpleasant odors and may be toxic, and the accompanying problem of cleaning both immersion cell and gemstone. Also, an immersion cell cannot be used for most mounted gems.

Dark Field Illumination

The problems of reflection that are minimized by use of an immersion cell may also be largely overcome by use of a lighting technique known as dark field illumination. In this method the gem is illuminated with a hollow cone of light, no part of which passes directly into the objective lens of the microscope. The gemstone lies at the apex of this cone where it deflects and reflects some of the light into the objective (see Fig. 7.6). The result is that

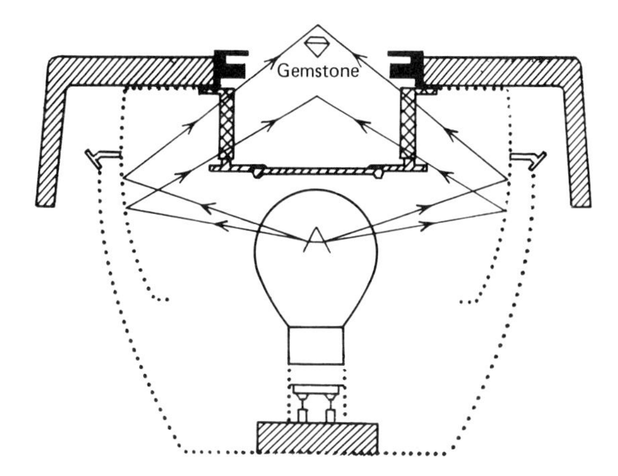

FIG. 7.6. Diagram of a dark field illuminator for use with a gemological microscope. Courtesy of the Gemological Institute of America.

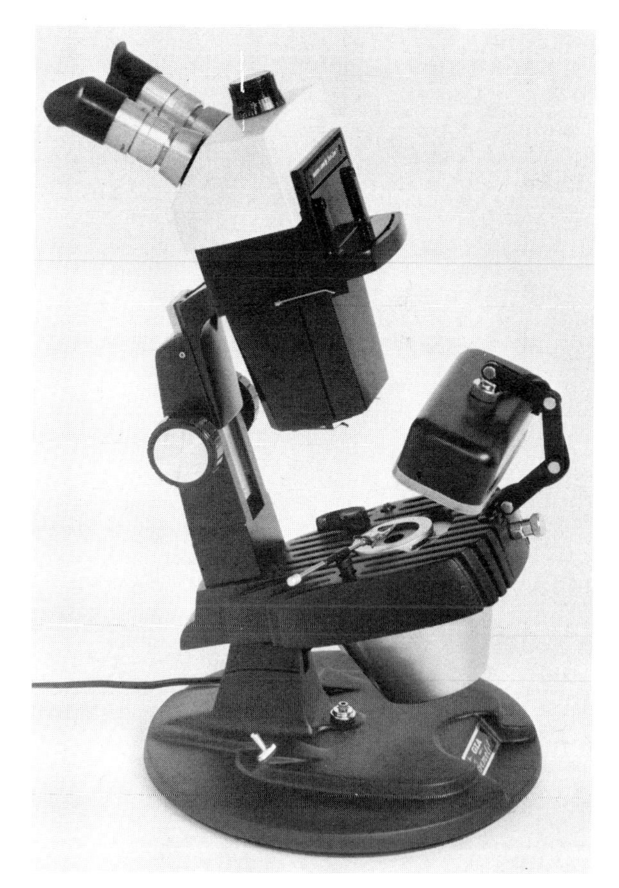

FIG. 7.7. The Gemolite, a stereoscopic wide-field microscope mounted on a dark field illuminator base. Courtesy of the Gemological Institute of America.

objects appear bright against a dark background. The best dark field illuminator is a parabolic reflector, and it is this principle that has been used in development of specialized gemological microscopes.

Gemological Microscopes

The Diamondscope* and Gemolite† are low-power binocular microscopes mounted on a dark field illuminator base, with an additional light provided for examination of the gem by incident light. Magnification of 10× through 45× can be obtained with the zoom feature. The stone holder can hold either loose or mounted gems (Fig. 7.7).

A more versatile instrument than those permanently attached to a dark field illuminator is a standard wide-field binocular microscope with stand, shown in Figure 7.8. Although not truly dark field, a satisfactory lighting is obtained by a small, circular, fluorescent "ring light," the stone being held in tweezers inside the ring and viewed against a dark background. The ring light can be raised to examine the stone in incident light or can be swung to one side to leave the stage unencumbered. The microscope can also be equipped with lower and upper polarizers and used as a dichroscope and polariscope, as well as with the crystal orienter (see Appendix). For use in this manner, transmitted light is provided by a mirror below the stage. This instrument has the distinct advantage that with an incident light source it can be used for examining such items as large pieces of jewelry, objets d'art, gem rough, and mineral specimens.

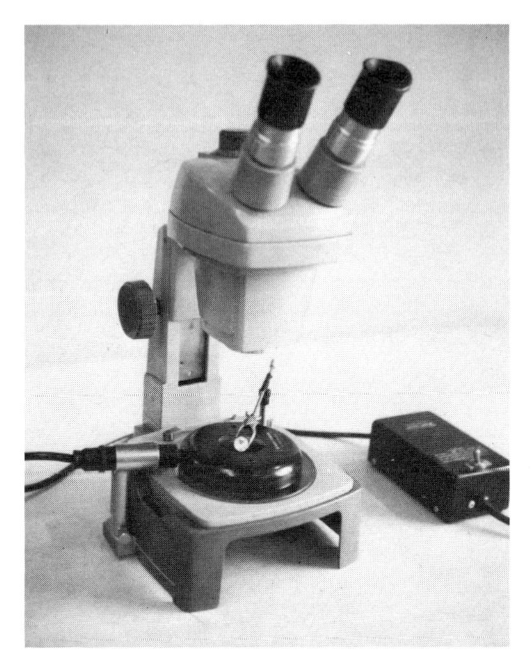

FIG. 7.8. A standard stereoscopic wide-field binocular microscope using a fluorescent "ring light" to illuminate gem against a dark background.

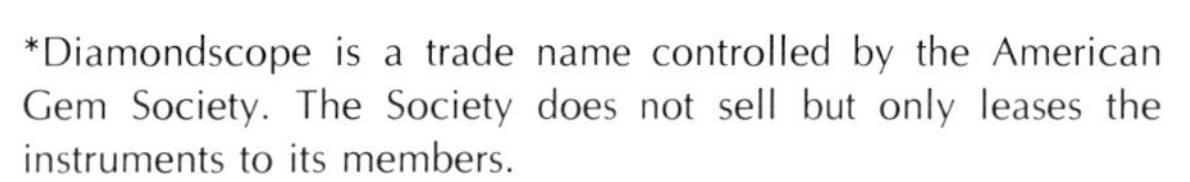

*Diamondscope is a trade name controlled by the American Gem Society. The Society does not sell but only leases the instruments to its members.

†The Gemolite is manufactured and sold by the Gemological Institute of America.

References and Suggested Reading

Liddicoat, R. T., 1975. *Handbook of Gem Identification*, 10th ed. Gemological Institute of America, Santa Monica, California. Chapter 7.

Webster, R., 1975. *Gems*, 3rd ed. Butterworth and Co., Ltd., London. Chapter 33.

8 THE SPECTROSCOPE

Spectroscopy is a very powerful analytical method by means of which the kinds and amounts of atoms making up a substance can be determined. Carried to its fullest extent the method requires complex and costly equipment and highly trained operators. However, spectroscopy can also be as simple as looking at the light transmitted through certain gemstones with an inexpensive hand-held instrument, and it is this simple application of spectroscopy to gem testing that is described here.

Spectroscopy refers to the investigation of spectra, the phenomena observed when the radiations from a source such as the sun are separated into their constituent colors or wavelengths. The separation may be done with a prism (refraction) or with a grating (diffraction). Instruments designed for this purpose are called spectroscopes if used for direct visual observation of the spectra, and spectrographs if photography or other methods of recording the spectra are used. The rainbow, formed by diffraction of sunlight in raindrops, is a familiar example of a spectrum.

The composite nature of white light was first demonstrated by Isaac Newton in 1664 when he passed a beam of sunlight through a glass prism and produced a colored image of the sun that he called a spectrum. We now know that visible light occupies a very small portion of the electromagnetic spectrum which ranges from very short wavelength gamma rays to very long wavelength radio waves (see nature of light discussed under *color* in Chapter 6).

A gemologist's interest in the electromagnetic spectrum is limited to the visible portion, and to a lesser extent to the light waves too short (ultraviolet) and too long (infrared) to affect the eye.

Emission Spectroscopy

The spectrum of a substance can be produced in a variety of ways but the one most frequently used in chemical analysis is emission spectroscopy, whereby electrons of the atoms are excited to electronic energy levels above their normal state by

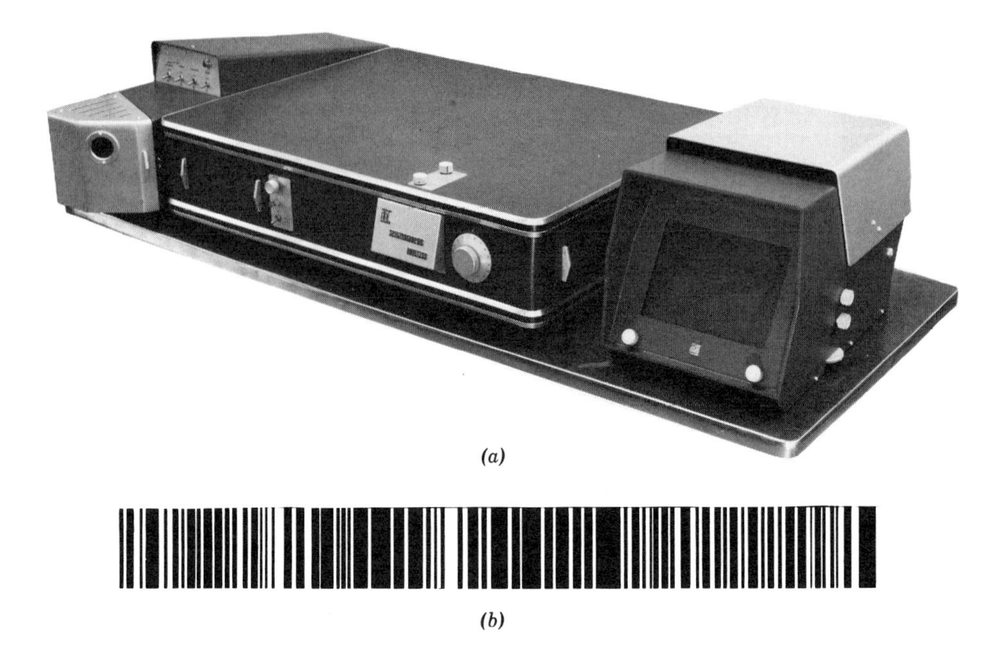

FIG. 8.1. (*a*) Spectrograph manufactured by Applied Research Laboratories. In using it the test sample is volatilized by an electric arc and the spectra recorded on a photographic film. (*b*) An emission spectrum. Courtesy of Applied Research Laboratories, Sunland, California.

means of the very high temperature of an electric arc. In a specimen so excited atoms emit light of characteristic wavelengths which are separated into a spectrum for study by visual, photographic, or electronic methods. Each kind of atom (chemical element) possesses its unique and characteristic collection of spectral wavelengths, whose recognition in the spectrum of the unknown substance provides unambiguous evidence of the presence of that element. An emission spectrum is a "bright line" spectrum consisting of bright lines on a dark background (Fig. 8.1*b*).

Since emission spectroscopy is a destructive test it has little use in gemology other than for identification of gem rough, which can usually be accomplished by simpler methods. However, emission spectroscopy has been refined to the point where very minute quantities of material can be analyzed, and a skillful spectroscopist could make a chemical analysis of a trace of powder scraped from the girdle of a cut gemstone.

Absorption Spectroscopy

It is also possible to study the spectrum of light transmitted through a substance. The result is a *continuous spectrum* of all wavelengths, but often, as illustrated in Figure 8.2, the continuous spectrum is crossed by dark bands or finer lines where certain wavelengths are completely obliterated. This is the *absorption spectrum,* also known as a "dark line" or reversed spectrum. The bands and lines occupy positions identical to those seen in an emission spectrum. The distribution of these bands or lines yields information that can greatly aid in the identification of certain gems, and of course the method of absorption spectroscopy is especially useful in gemology since it is nondestructive.

Absorption lines in the sun's spectrum were discovered by Wollaston in 1802. However, they are usually called Fraunhofer lines after the German physicist Fraunhofer, who in 1814 mapped several hundred of these lines. The wavelengths of the principal Fraunhofer lines have been precisely meas-

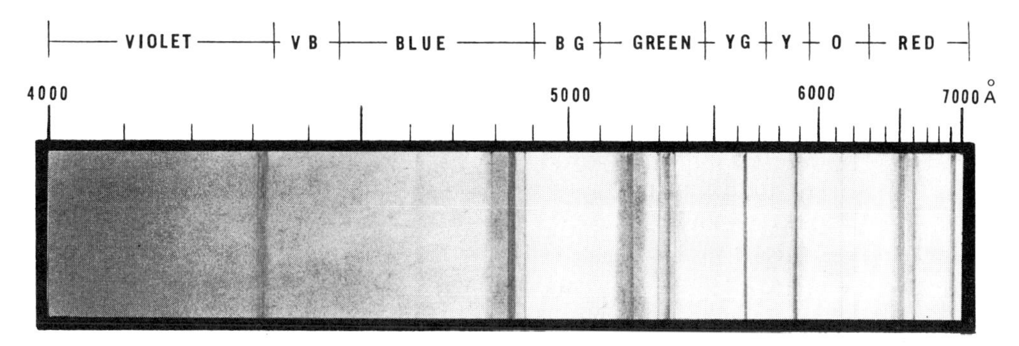

FIG. 8.2. The absorption spectrum of zircon from Burma (ordinary ray). Seen through the spectroscope the background would be a continuous spectrum from violet to red. Most of the absorption lines are here due to a trace amount of uranium. Courtesy of the Gemological Institute of America.

ured and have been assigned letters to identify them as given in Table 8.1. These are used as standards against which other lines are measured and instruments calibrated.

The two yellow lines (cause by sodium) lie so close together that they cannot be separated by an ordinary spectroscope such as is used in gem testing and so may be regarded as a single line. The lines B in the red and G in the violet are near the limits of easy vision of the spectrum and they have been selected as the standard interval on which to base the color dispersion of gemstones (see *Dispersion* in Chapter 6).

Absorption spectroscopy in gemology had its beginnings in 1866 when Arthur Church called attention to absorption bands in zircon. Sorby in 1869 thought that the absorption bands in zircon were caused by a new element which he called jargonium, but it was later found that they are caused by the presence of a trace of uranium.

Absorption spectroscopy may also be used in gemology to precisely define the color of a gem (or colors if it is pleochroic), by a detailed analysis of the continuous spectrum. This can be done by measuring the degree of transmistion of each wavelength of light using a complex instrument called a spectrophotometer. This method can produce valuable information about gemstones, but is seldom used because the instrument needed is not readily available.

Table 8.1
FRAUNHOFER LINES

Line	Color	Wavelength (Å)
A	Dark red	7593.8
B	Red	6867.2
C	Red-orange	6562.8
D	Yellow	5895.9
		5890.0
E	Green	5269.6
F	Blue	4861.4
G	Indigo	4307.9
H	Violet	3968.5
K	Violet	3933.7

Spectroscopes

Absorption spectroscopy in the study of gems may be carried out using a simple direct vision spectroscope. The one most frequently used is the prism type, in which the light is passed through an adjustable slit, and then through a train of glass prisms (Fig. 8.3). Direct vision spectroscopes operating on the diffraction grating principle are also available. A grating instrument is better for observation of the blue end of the spectrum but does not transmit as much light as a prism spectroscope.

Several instruments are available in which the spectroscope is permanently mounted on a stand,

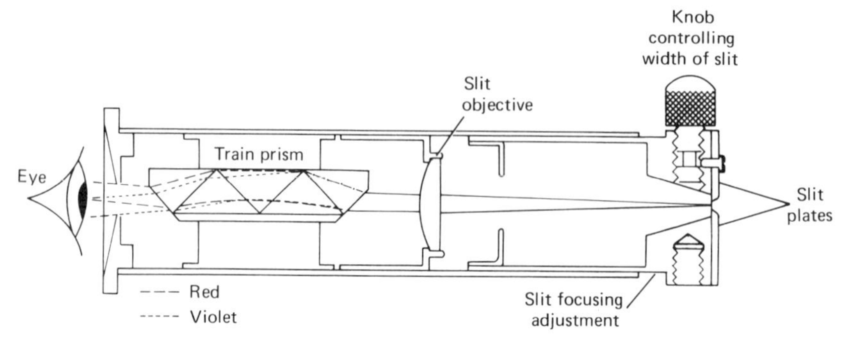

FIG. 8.3. A direct vision prism spectroscope of the type manufactured by R. and J. Beck, London.

or sometimes a microscope, with a light source and a holder for maneuvering the gem. One of these is shown in Figure 8.4.

Some gemologists prefer to first observe the gem under a low-power monocular microscope and then hand-hold the spectroscope over the microscope tube after first removing the eyepiece. Whichever method is used, a high-intensity light source is a prime requirement. In some instances, especially for dark-colored gems, better results are obtained by using reflected rather than transmitted light.

Some of the spectroscopes used in gem testing

FIG. 8.4. The GIA spectroscope unit, showing spectroscope, stone holder, and light source. Courtesy of the Gemological Institute of America.

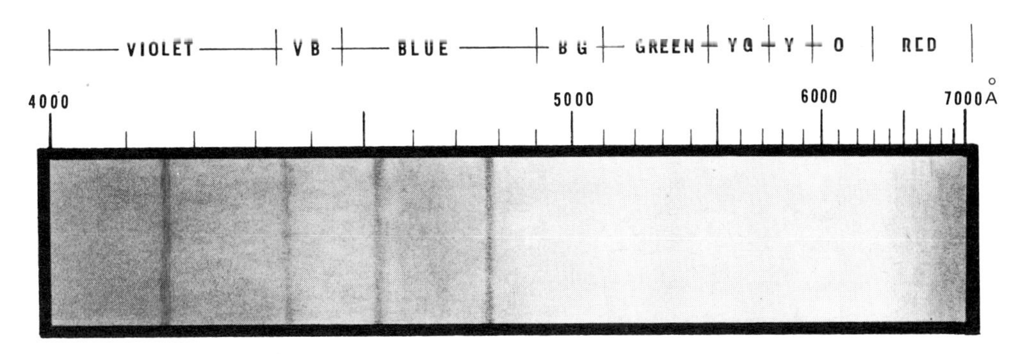

FIG. 8.5. Absorption spectra of a pale yellow diamond. Courtesy of the Gemological Institute of America.

incorporate wavelength scales so that the actual wavelength of the absorption band or line can be measured in angstrom units. This is the ideal method, for then identification of the lines can be made by reference to wavelength tables. In actual practice the spectrum produced by these small direct vision spectroscopes is so condensed that accurate wavelength measurement is not practical. To the experienced practitioner of the "art" of spectroscopy the visual arrangement of the bands and lines is sufficient for identification.

The absorption spectrum produced by a particular gemstone depends on its chemical composition. For example, peridot, in which ferrous iron is a major element, has three prominent absorption bands due to iron. Frequently the absorption spectrum of a gemstone is characteristic of the minor element present as the coloring agent. Among the best examples are the allochromatic gems whose color is due to small amounts of chromium, such as ruby, emerald, and alexandrite. Other elements whose presence gives characteristic absorption spectra are manganese, cobalt, vanadium, copper, selenium, uranium, and certain rare earth elements. In some instances, diamond in particular, an absorption spectrum is caused not by trace elements but by characteristics of the crystal structure.

Not all gem species yield absorption spectra that are distinctive enough to be useful in gem identification. In the descriptions of individual gems, the absorption spectrum is given only for those for which it is strong and characteristic.

References and Suggested Reading

Anderson, B. W., 1971. *Gem Testing*, 8th ed. Butterworth & Co., Ltd., London. Chapter 8.

Liddicoat, R. T., 1975. *Handbook of Gem Identification*, 10th ed. Gemological Institute of America, Santa Monica, Calif. Chapter 8.

Webster, R., 1975. *Gems*, 3rd ed. Butterworth and Co., Ltd., London. Chapter 32.

9 INCLUSIONS IN GEMSTONES

Inclusions in gemstones, here defined as internal imperfections other than fracture or cleavage, are important for several reasons. They are one of the factors that determine the quality and value of a gem. They are an essential feature of phenomenal gems, such as stars and cat's eyes. They may yield much information of scientific value regarding the origin of gemstones and aid in the determination of the provenance of certain gems. For some gem materials the species of an unknown stone may be determined from inclusions alone, without need for further tests. A "map" of the inclusions in a gem may allow later identification, as for example, in the case of theft. Lastly, and most important, inclusions may provide a means of establishing whether a gem is of natural origin, or whether it is synthetic or imitation.

A microscope, particularly one equipped with both light and dark field illumination, is essential to the study of gemstone inclusions.

See Plate III for illustrations of various types of inclusions.

Classification

The world leader in the study of gemstone inclusions, and the foremost proponent of their use in identification and in the determination of the genesis and provenance of gems, is Edward J. Gübelin. The classification proposed by Dr. Gübelin, based on mode of origin, is as follows.

1. *Protogenetic* (preexisting). Inclusions that were present before the gemstone crystal formed, and which were enveloped by the growing host crystal, are termed protogenetic, from the Greek "proto," meaning first in time. Examples are bits of preexisting rock, such as in garnet formed during metamorphism, and earlier formed crystals of a related or even the same species, such as diamond in diamond.

Protogenetic inclusions cannot always be recognized as such, for the same mineral pair (inclusion plus host) may result from another process. Protogenetic inclusions are frequently rounded or corroded, as a result of having been partially dissolved by the host. However, they are also frequently sharp and angular.

2. *Syngenetic* (contemporary). Inclusions that were formed at the same time as the host crystal, from the same melt or liquid, are termed syngenetic, from the Greek meaning simultaneously formed. In geology the term is applied to a mineral deposit formed contemporaneously with the enclosing rocks. Examples are rutile in quartz, chrome diopside and olivine in diamond, and byssolite fibers in demantoid garnet. Syngenetic inclusions are generally sharp and angular.

Syngenetic liquid inclusions are also common. These are cavities formed during crystal growth that are filled with liquid, liquid plus gas bubbles, or even liquid, gas, and crystals. These liquid inclusions are of particular scientific interest for they represent bits of the fluid from which the crystals were growing. Their study can lead to important information about the chemistry of the fluids, and the temperature and pressure conditions during crystal growth. Other syngenetic features frequently observed in gemstone crystals are twinning and color banding.

3. *Epigenetic* (subsequent). Inclusions that were formed after the growth of the crystal are termed epigenetic, a word derived from the Greek, meaning "after the origin." In geology the term is applied to a mineral deposit of later origin than the enclosing rocks. Epigenetic inclusions may be caused by chemical alteration, crystallization from enclosed liquid inclusions, exsolution, penetration of material into cracks and fissures, and structural damage by radioactive emmanations. Examples are rutile needles responsible for asterism in corundum, formed by exsolution, and dendritic crystals of manganese oxide in moss agate.

It is often not possible to determine whether an inclusion observed in a gemstone is protogenetic, syngenetic, or epigenetic. Thus this classification, although of scientific significance, is not a practical one. For descriptive purposes it is more convenient to divide gemstone inclusions as follows:

1. Solid inclusions.
2. Cavities filled with liquid or gas or both.
3. Cracks and fissures filled with gas, liquid, or solid.
4. Growth phenomena, including twinning, color zoning, and swirl marks.

Solid Inclusions

Solid inclusions may be the same mineral as the host, as diamond in diamond, or a mineral different from the host, as rutile in corundum. Some of the most frequently seen inclusion–host pairs seen in gemstones are the following:

Inclusion	Host
Apatite	Garnet, corundum, spinel
Calcite	Spinel, corundum, emerald
Chrome diopside	Diamond
Chromite	Emerald, serpentine
Diamond	Diamond
Garnet	Corundum, diamond
Hematite	Topaz
Mica	Quartz, corundum, emerald
Olivine	Diamond
Pyrite	Sapphire, emerald
Quartz	Aquamarine, topaz, emerald
Rutile	Corundum, quartz
Spinel	Ruby
Zircon	Garnet, corundum

Liquid and Gas Inclusions

Internal cavities in crystals may be filled with liquid or gas, or both. Sometimes solid matter is also present in the form of small crystals of a mineral different from the host crystal. When both liquid and gas are present the inclusion is called *two-phase* (Fig. 9.1).

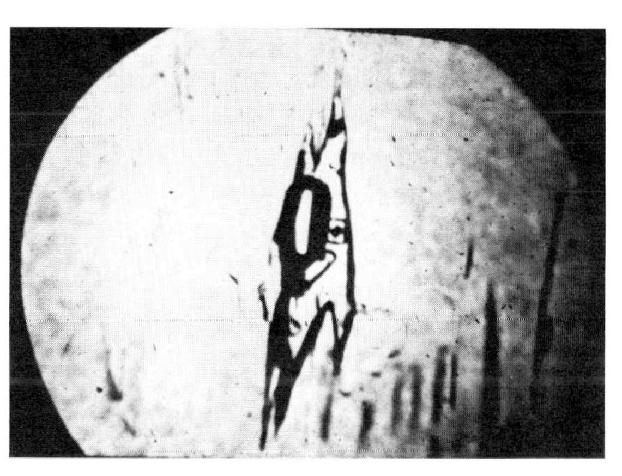

FIG. 9.1. Two-phase (liquid and gas) inclusion in beryl.

When solids are also present they are referred to as *three-phase* inclusions. Some primary liquid inclusions are in cavities having a shape imposed by the regular atomic structure of the host crystal, called negative crystals. Many liquid and gas inclusions have served to partially fill and heal cleavage cracks and fractures and are known as *feathers*. Important in the identification of synthetic and imitation gems are gas bubbles trapped in the material during manufacture.

Growth Phenomena

Though not inclusions in the strict sense, as are the features just described, growth phenomena of several types are important internal features of gemstones and are often diagnostic of mode of origin. Growth pehnomena are generally caused by interruptions in the steady growth rate of a crystal owing to changes in temperature, or to changes in the chemistry of the solutions in which the crystals are growing. The principal growth phenomena are twinning, color zoning, swirl marks, and curved growth lines as in synthetic corundum.

References and Suggested Reading

Liddicoat, R. T., 1975. *Handbook of Gem Identification*, 10th ed. Gemological Institute of America, Santa Monica, Calif.

Gübelin, E. J., 1953. *Inclusions as a Means of Gemstone Identification*. Gemological Institute of America, Santa Monica, Calif.

Gübelin, E. J., 1972. Inclusion: Clues to Gem Growth. In *Treasures from Land, Sea, and Lab. Jeweler's Circular Keystone*, October 1972/Part II, pp. 40–43. The Chilton Co., Philadelphia.

Gübelin, E. J., 1974. *Internal World of Gemstones*. ABC Edition, Zürich.

Webster, R., 1975. *Gems*, 3rd ed. Butterworth and Co.,Ltd., London. Chapter 34.

10 METHODS OF GEM SYNTHESIS

Introduction

Many of the most important gems including diamond, emerald, ruby, sapphire, and opal have been manufactured by man. These synthetic gems are not imitations but possess the same chemical, physical, and optical properties as the naturally occurring minerals. One of the tasks of the gemologist, and frequently a difficult one, is to distinguish natural gems from synthetic ones. There are also man-made materials that have no known natural counterpart that may be used as imitations of natural gems.

In the earliest recorded attempts to synthesize gems, natural stones were buried in the ground in the hope that they would reproduce or grow larger. Later, alchemists attempted to imitate the processes of nature, and throughout the nineteenth century chemists experimented with crystal growth. Microscopic crystals of ruby were reportedly made in 1837 by Marc Gaudin by fusing alum to which chromium sulfate was added to give the red color. In 1877 Edmund Frémy and Charles Feil published on their success in producing ruby crystals by fusing aluminum oxide and lead oxide, a method now known as flux growth. But none of these crystals was clear or of useful size. Auguste Verneuil was Frémy's assistant from 1887 to 1890, and during this period they described production (by flux growth) of ruby crystals weighing up to 1/3 carat which were used experimentally in jewelry.

In 1902 Verneuil published a description of the flame fusion method of making synthetic ruby, a method he probably perfected in 1892. This date marks the beginning of gem synthesis on a commercial scale, and Verneuil's technique, somewhat modified, is today the most important in the manufacture of gem materials.

The science of crystal growth has advanced rapidly in recent years, and research in using crystals for industrial purposes such as transistors and lasers has led to many important advances in gemstone synthesis.

There are three major methods of crystal synthesis: (1) growth from a vapor, (2) growth from a melt, and (3) growth from a solution. It is not only very difficult to grow large crystals by the vapor growth technique, but most of the crystals thus grown are of nongem materials. A possible exception is silicon carbide, which can be grown by sublimation, that is, from a vapor; but the crystals are rarely transparent and usually in plates, too thin to be cut as gems. The important methods of gem synthesis are thus variations of growth from either a melt or a solution, and it is these that are described in the following section. Not included are some methods used for growing crystals of nongem material for research and industry.

Melt Growth

In all the melt growth techniques, the source material, that is, the material melted, has essentially the chemical composition of the desired crystalline compound. The major differences in methods lie in the manner in which the material is melted and in which it is caused to solidify to yield single crystals.

Verneuil Technique (flame fusion)

This method as first developed on a commercial scale (about 1902) was so successful that by 1907 production of synthetic ruby had reached 5 million carats annually. Synthetic sapphire and spinel are also produced in large quantities by this method, as well as strontium titanate, rutile, and other crystals for research and industry.

The Verneuil method uses an inverted oxygen–hydrogen torch. Powdered material of the chemical composition of the desired crystal (for ruby, aluminum oxide plus coloring agent) is fed from a hopper above the torch and melts as it passes through the intensely hot flame. The molten drops of aluminum oxide fall onto a ceramic rod that is slowly rotated and lowered, and a carrot-shaped "boule" forms consisting of a single crystal of corundum. The boule at first grows upward into a mushroom shape, and then the rate of feed and the lowering of the boule are controlled to produce for gem use a boule about 20 mm in diameter, 65–70 mm long, and weighing 400–500 carats. A diagram of a Verneuil furnace is shown in Figure 10.1, and boules of synthetic corundum are illustrated in Plate XII.

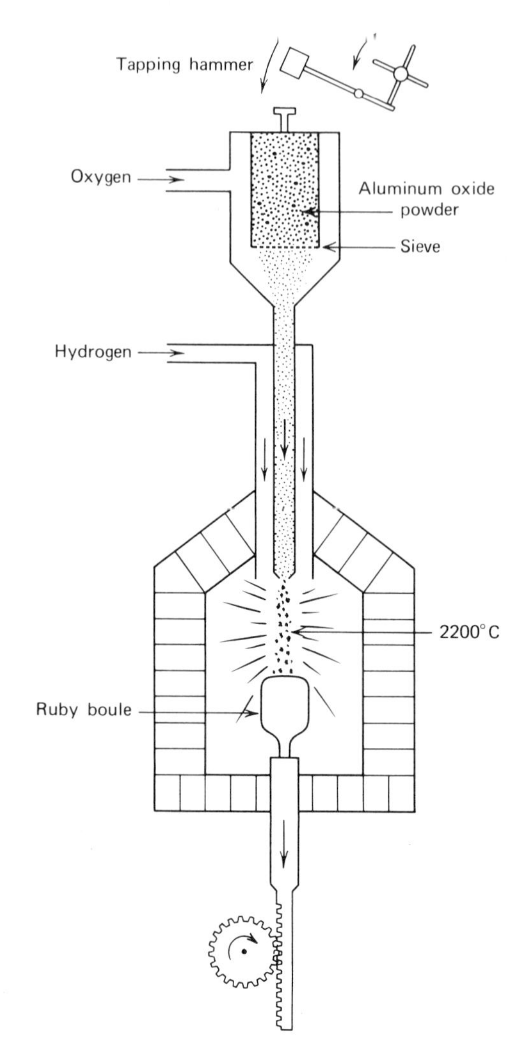

FIG. 10.1. Diagram of a Verneuil furnace. Finely divided aluminum oxide powder passes from the hopper through a sieve with the aid of a tapper and then through the flame fed by oxygen and hydrogen. As the boule crystallizes, it is slowly lowered in the furnace.

On cooling, boules are normally strained and, when tapped, split lengthwise into two nearly equal parts. Thus, originally, it was not the boule but the half boule that determined the maximum size of the cut stone. However, after many years of working with half boules, it was learned that strain can be relieved by annealing, that is, keeping the boule for a few hours at a temperature just below its melting point and then slowly cooling.

Improvements in the Verneuil method made since 1960, prompted by the need for large ruby crystals for lasers, now allow production of rods up to 2.5 cm in diameter and 30 cm long, as well as large disk-shaped boules 5 cm thick and 15 cm or more in diameter.

The Czochralski Process

In this technique, often called crystal pulling, a seed crystal on the end of a rod is placed so that it touches the surface of a melt held in a crucible. The rod is then slowly rotated and withdrawn. As it is withdrawn, material continually crystallizes on the end of the rod (Fig. 10.2). The pulling rate may vary from 1 mm/hour to 10 cm/hour and the rotation from 10 to 100 rpm depending on the type of crystal being grown. For growing corundum an iridium crucible must be used, capable of withstanding a temperature of 2100°C. The method is especially useful for growing rod-shaped crystals. Ruby crystals up to 5 × 40 cm in size are grown using this technique.

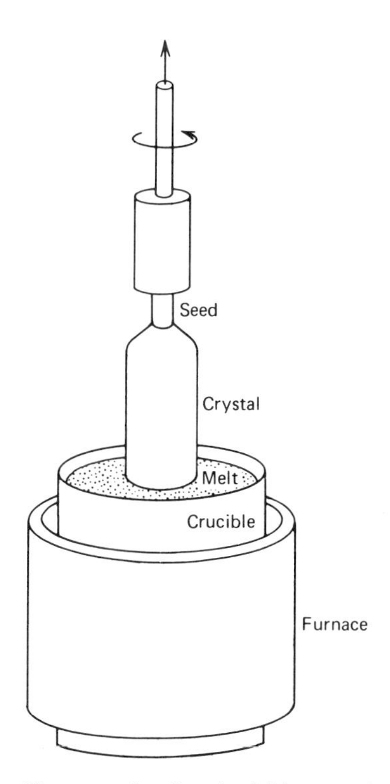

FIG. 10.2. Diagram of a Czochralski apparatus. The crucible is filled with powder and melted in the furnace. A rod with a seed crystal attached to the lower end is lowered until it touches the melt and is rotated and slowly withdrawn, "pulling" the crystal from the melt.

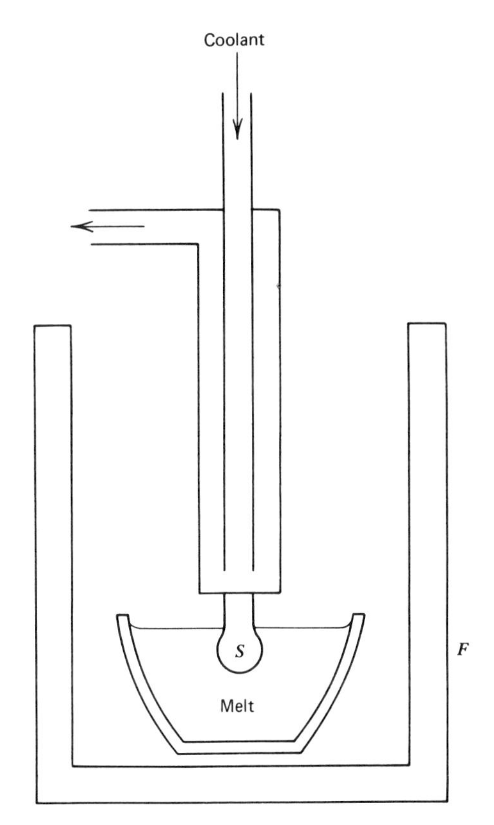

FIG. 10.3. Diagram of a Kyropoulos apparatus. A furnace (*F*) contains crucible filled with melt. A seed rod (*S*) is held in the molten material and cooled by circulating liquid. The melt is then allowed to cool very slowly and the entire contents of the crucible solidify into a single crystal.

A variation of the Czochralski process is the Kyropoulos technique. Here, instead of the crystal being pulled, a seed crystal is held at the top of the melt and crystallization is induced by slowly lowering the temperature of the crucible (Fig. 10.3). The molten material crystallizes on the seed crystal, forming large single crystals eventually conforming to the shape of the crucible.

Another variant is the so-called heat exchanger method by which large single crystals of sapphire are grown in vacuum. Because crystals grow from the bottom to the top of the crucible, turbulence of the melt is suppressed resulting in high crystal per-

FIG. 10.4. Boule with "windows" and coring cut from crystals grown by the heat exchanger method. Boule 20 cm in diameter, 12 cm high. "Windows" small, 10 cm in diameter; large, 25 cm in diameter. Coring 10 cm in diameter, 20 cm high. Courtesy of Crystal Systems, Inc., Salem, Massachusetts.

fection. Large single crystal boules, such as illustrated in Figure 10.4, can be grown. Although only colorless sapphire is at present grown by this method, colored varieties could possibly be manufactured by adding the appropriate coloring agents.

"Skull Melting"

This is a melt growth method used to crystallize materials that melt at such high temperatures, or are so chemically reactive, that conventional crucibles cannot be used. The "skull," a water-cooled container, is filled with the material to be melted. The whole "skull" is not heated but, by the use of a radio frequency generator, only the central material is brought to extremely high temperatures and melted within a shell of its own powder. On cooling, crystals usually grow from the bottom upward. Cubic zirconia, with a melting temperature of about 2750℃, is a diamond substitute grown in crystals several centimeters long by this method (Nassau, 1977).

Growth from Solution

In growth from solution, a saturated solution is made of the material to be crystallized. On lowering of the temperature, or evaporation of the solvent, crystals form from the supersaturated solution. Such is the type of formation of sugar crystals that form at room temperature from a supersaturated water solution of sugar.

Unfortunately, materials crystallizing from solution at or near room temperature lack the attributes of gems. But if, as in nature, the temperature of the solvent can be made sufficiently high (a condition which may also require high pressure), many gem materials become soluble and crystallize from solution with lowering temperature. This is known as *hydrothermal growth* if water is the solvent.

Some materials, quite insoluble in high temperature water, are soluble in a flux (a solvent of high melting point). Several minerals have been synthesized from such solutions, a method called *flux growth*.

Hydrothermal Growth

For growth of gem material from hydrothermal solutions, specially designed apparatus is required to withstand the necessary high temperature and high pressure. This is a heavy-walled steel vessel called an autoclave or "bomb." In the preparation for hydrothermal growth, the source material, water, and seed crystals are placed in the bomb. The bomb is then lowered into a furnace and heated to a temperature of several hundred degrees centigrade. As the temperature increases, the expanding water with feed material in solution fills the bomb, and the pressure increases. The final pressure is determined by the amount of filling and the temperature. The method is a laboratory duplication of the way many gem quality crystals grow in nature, such as in pegmatites.

An apparatus for growing quartz crystals hydrothermally is shown in Figure 10.5. The feed material, small pieces of pure crystalline quartz, is placed in the lower portion of the pressure vessel and the seed crystals are held in racks in the upper portion. The "bomb" is then filled to the appropriate level with the "hydrothermal solution," water made alkaline to increase the solubility of the nutrient quartz. The furnace heaters that surround the autoclave are arranged to maintain a higher temperature at the bottom than in the upper portion, creating a temperature gradient that causes the feed material to dissolve in the higher temperature region

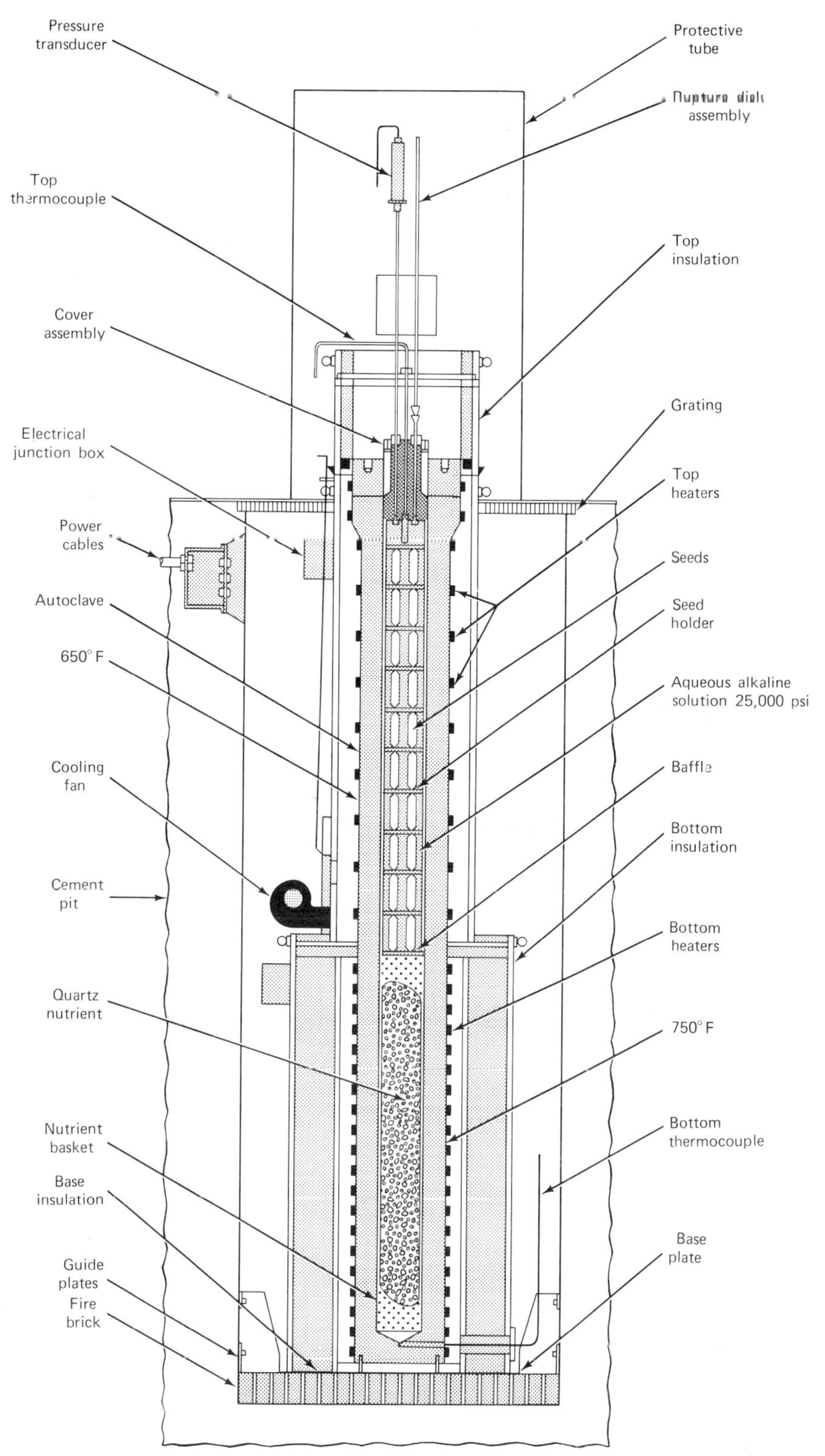

FIG. 10.5. Apparatus for hydrothermal growth of quartz crystals. On heating, the entire bomb becomes filled with silica-rich aqueous solution. Quartz dissolves in the hotter, lower region of the bomb and deposits on the seed plates in the cooler upper region. Courtesy of Western Electric Company, North Andover, Massachusetts.

and deposit on the seed crystals in the lower temperature region.

In addition to quartz, ruby and emerald have been grown hydrothermally, although most commercial production of the latter two is by other methods.

Flux Growth

In the flux growth technique, a powder of the desired material is mixed with a flux and the mixture brought to a molten state in an inert crucible (commonly platinum) (Fig. 10.6). After thorough mixing of the molten material, the crucible and its contents are allowed to cool very slowly. When a certain temperature is reached, different for each substance, the solubility is exceeded and crystals of the synthetic gem begin to grow in the flux and increase in size as the temperature is lowered further. When cool, the crystals are extracted from the solidified mass by dissolving the flux in an appropriate solvent.

A flux is a substance that, when molten, will dissolve a powdered material with which it is mixed that has a much higher melting point. The flux growth method is thus useful in growing crystals having high melting points. Some of the flux materials used are lead fluoride, boron oxide, lead oxide, lithium oxide, and molybdenum oxide. The melting must be carried out in vessels of platinum or iridium because of the high temperature required and because of the powerful solvent action of the molten flux.

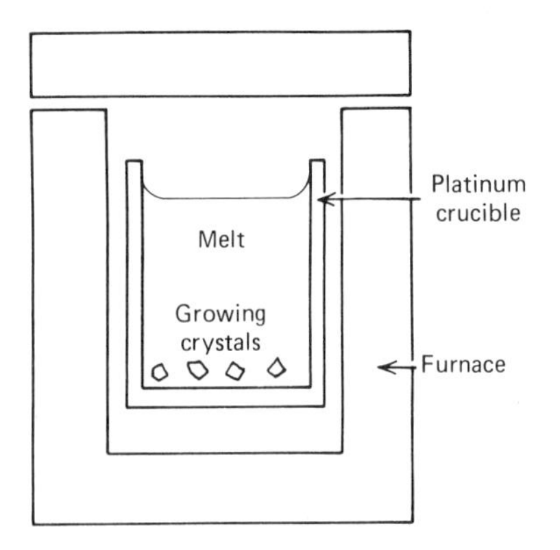

FIG. 10.6. Diagram of a flux growth apparatus. The desired material and flux are melted in the crucible and the entire mass slowly cooled. Crystals form in the mass and are removed by dissolving the flux.

The gem materials ruby and emerald are manufactured by the flux growth method.

Diamond Synthesis

The earliest claim of any importance in the synthesis of diamond was in 1878 by James B. Hannay, a chemist of Glasgow, Scotland. His method was to place in a sealed steel tube a mixture of bone oil and paraffin to which metallic lithium had been added. The mixture was heated in a furnace for about 14 hours. He conducted a series of about 80 experiments, and in three of these found small transparent particles he believed to be diamond. A dozen of these particles he sent to the Keeper of Minerals in the British Museum, M. H. Story-Maskelyne, who pronounced them diamonds. The particles were preserved and in 1943 were tested by X-ray diffraction. Although 11 of the 12 particles proved to be diamond they have since been shown to be natural, not synthetic.

The next scientist to claim to have made diamonds was Henri Moissan, Professor of Mineralogical Chemistry in Paris. In 1904 he announced that he had been successful in making diamonds by placing pure carbon and iron in a crucible, melting the mixture in an electric furnace, and then quickly cooling the molten mass. The crystals he was able to recover had a specific gravity of 3–3.5, and produced carbon dioxide when burned in oxygen. Moissan therefore concluded that the particles were diamond. But it was later learned that Moissan's assistant had added diamond chips to the melt.

Based on present-day knowledge of the graphite–diamond equilibrium curve (Fig. 10.7) neither Hannay or Moissan could have produced diamond, for their methods can not attain the required very high temperature.

Unsuccessful attempts by many investigators followed until February 15, 1955, the General Electric Company announced that their scientists had succeeded in making diamond in their laboratory in Schenectady, New York. The key to success by General Electric was the development of apparatus

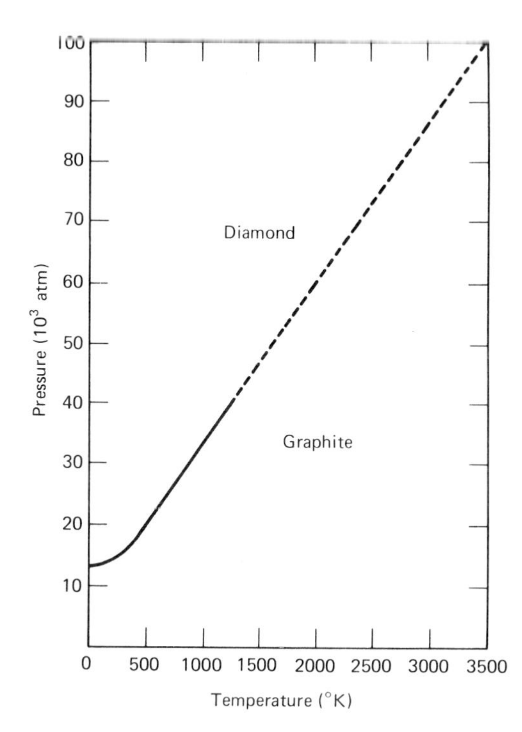

FIG. 10.7. The graphite–diamond equilibrium curve, after R. Berman and F. Simon. Solid line, calculated (up to 1200°K); dashed line, extrapolated. Temperature °K (Kelvin) = °C + 273. 10^3 atm = 14,700 lb/in.^{2} (psi) ≅ 1 kilobar (kb).

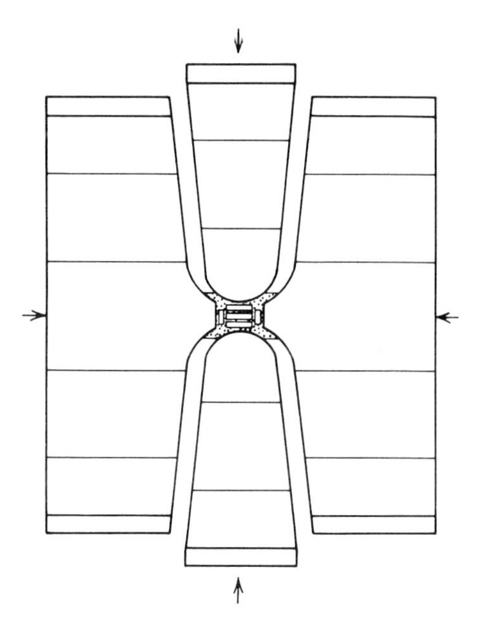

FIG. 10.8. Cross section of the pressure chamber used for synthesizing gem diamonds.

capable of creating and maintaining temperatures of more than 2500°C (5000°F) and pressures up to 1,500,000 psi.* By 1958 General Electric was producing synthetic diamond abrasives for sale at prices competitive with natural diamond. By late 1959 De Beers through its subsidiary Ultra High Pressure Units, Ltd. was manufacturing industrial grade diamond. Today over 50,000,000 carats of diamond abrasives are produced each year, manufactured in several countries, and synthetic diamonds have been produced in many research laboratories throughout the world.

Gem Diamond Synthesis

On May 28, 1970 General Electric announced that it had made clear, gem quality diamonds, some of them weighing as much as a carat (faceted up to half a carat) (see Fig. 13.8). For synthesis of gem quality diamonds the method used for making industrial grade diamond is modified so that the required temperature and pressure can be maintained for as much as several days. This allows the crystals to grow slowly, a requisite for nearly perfect crystals of large size.

The pressure chamber (Fig. 10.8) is a ring-shaped tungsten carbide piece supported by a series

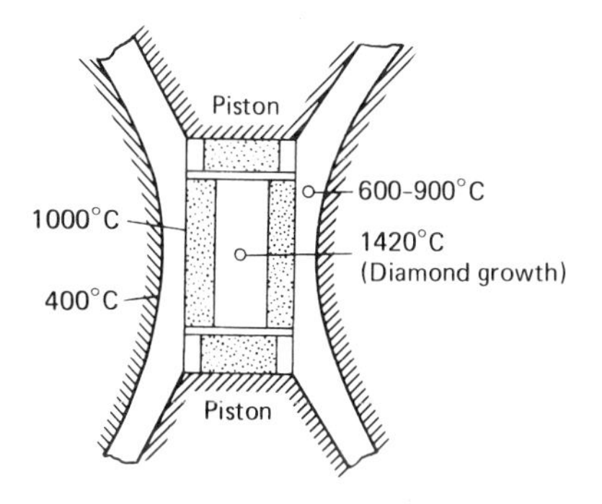

FIG. 10.9. Enlargement of the central shaded portion of Figure 10.8 showing the temperatures in the superpressure cell.

* Units of pressure may be given as pounds per square inch (psi), atmospheres (1 atm = 14.7 psi), or kilobars (1 kb = 987 atm or 14,500 psi).

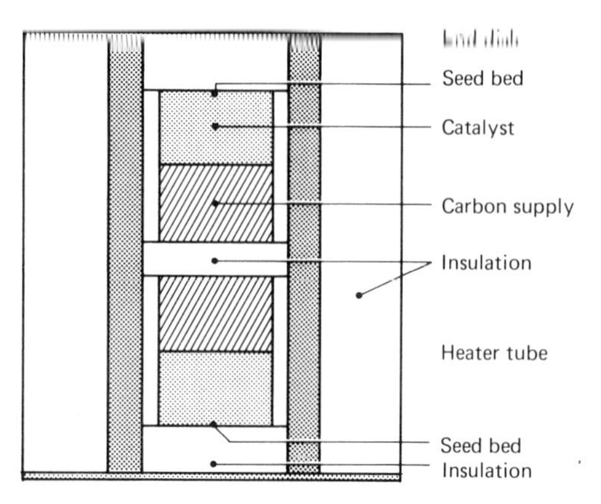

FIG. 10.10. Cross section of superpressure cell for growing gem diamonds showing the arrangement of the starting materials.

of tight steel belts. The pressure is generated by forcing together conical tungsten carbide pistons that wedge into a die, with a gasket of pyrophyllite. The reaction zone container is also made of pyrophyllite. The pressure applied to the reaction cell is 60,000 atm (approximately 900,000 psi). The reaction zone is inside a carbon tube heater, which is heated by passage of electric current from one piston, through the heater, and out through the opposite piston. The temperatures maintained in the superpressure cell are shown in Figure 10.9.

The arrangement of the pressure cell is shown in Figure 10.10. In the center is a source of carbon in the form of industrial grade synthetic diamond. On each side of this carbon source is a catalyst, such as iron or nickel, which becomes molten under operating conditions. At the ends is a diamond seed crystal. A temperature differential of about 10–15°C is maintained between the center and the ends of the cell. The carbon dissolves in the hotter part (center) of the cell and deposits on the diamond seed crystals in the cooler region. In a typical experiment the temperature of the hot part of the cell would be 1455°C (2650°F) and of the growing crystal 1425°C (2600°F), and these conditions are maintained for several days, at 60,000 atm pressure.

The G.E. diamonds have been made colorless, as well as in a variety of colors including various hues of blue and yellow.

References and Suggested Reading

Arem, J. E., 1973. *Man-Made Crystals.* Smithsonian Institution Press, Washington, D.C.

Holden, A. and P. Singer, 1960. *Crystals and Crystal Growing.* Doubleday and Co., Inc., Garden City, N.J.

Nassau, K., 1964. Growing Synthetic Crystals. *Lapidary Journal,* Vol. 18, pp. 42–45, 313–317, 386–389, 474–477, 588–595, 690–693.

Nassau, K., 1977. Cubic Zirconia, the Latest Diamond Imitation and Skull Melting. *Lapidary Journal,* Vol. 31, pp. 889–904, 922–926.

Nassau, K. and R. Crowningshield, 1969. The Synthesis of Ruby. *Lapidary Journal,* Vol. 23, pp. 114–119, 313–314, 334–338, 440–446.

Nassau, K. and J. Nassau, 1971. Dr. A. V. L. Verneuil and the Synthesis of Ruby and Sapphire. *Lapidary Journal,* Vol. 24, pp. 1284–1296, 1442–1447, 1524–1532.

Webster, R., 1975. *Gems,* 3rd ed. Butterworth and Co., Ltd., London.

Webster, R., 1970. Modern Synthetic Gemstones. *Journal of Gemmology,* Vol. 12, pp. 101–148.

11 IMITATION, COMPOSITE, AND TREATED GEMS

Introduction

From archaeological finds we know that man used mineral fragments and pebbles for personal adornment before the dawn of civilization. But with advancing culture, he learned to work them into more pleasing shapes. Such gem material, fashioned into beads, has been found in Egyptian tombs constructed more than 5000 years ago. But these were the tombs of the pharaohs and noblemen. Human nature has not changed in 5000 years and the less privileged people of that time also wanted to possess ornaments. As a result, imitations of the precious objects were made to satisfy that desire. Egyptian glass beads are known dating from about 2500 BC and the so-called faïence of Egypt was made much earlier. Because glass is the oldest material from which imitation gems were made, it seems appropriate that it be discussed first in this chapter.

Glass

Glasses can be thought of as liquids that are so stiff that under ordinary definitions they have the properties of a solid. The definition accepted by the American Society for Testing Materials (ASTM) is, "Glass is an inorganic product of fusion which has cooled to rigid conditions without crystallizing." The term "paste," from the Italian *pasta* meaning dough, has been applied to the glass of imitation stones because the ingredients are mixed wet to assure uniformity of the batch.

Types of Glass

Glass used in gem imitations are of two main types, crown glass and flint glass. Crown glasses, with essential ingredients silica, soda, and lime, are the more common and are used in making bottles, window glass and optical glass. They are also used in making molded imitation gems for costume jewelry. Flint glass composed of silica and soda but with lead oxide replacing the lime of crown glass is sometimes called *stross* after the Austrian, Joseph Stross, who is credited with its discovery. Stones cut from flint glasses more closely resemble the gems they imitate because the presence of lead gives them a higher refractive index and a greater dispersion. There are many other types of glasses such as silica glass made by fusing pure quartz and borosilicate glasses used in laboratory and cooking glassware, but these special glasses are rarely used in imitation gems. Natural glasses are described in Chapter 13.

Coloring of Glasses

To obtain colored glass a coloring agent, usually a metallic oxide, is mixed with the batch. The oxide and resulting color in general are manganese, purple; cobalt, blue; selenium, red; iron, yellow and green; copper, red, green, and blue; gold, red; chromium, green; and uranium, yellow-green. Some colorless glass imitations are colored by spraying the pavilion facets with a pigment of suitable color. The iridescent colors of many imitations result from vacuum sputtering of thin films such as are used to coat camera lenses. Many of the cheapest glass imitations are "foiled"; that is, their pavilion facets are covered with a thin film that acts as a mirror to enhance their brilliance.

Properties of Glass Imitations

If imitations are molded, as is frequently the case, they are readily identified because edges between facets are rounded and the facet surfaces are not flat. In some cases glass imitations are carefully cut making identification more difficult, but frequently only the crown facets are polished.

The physical properties of glasses that are useful for identification are as follows:

1. Hardness. The hardness is 5–6, less than the gems they imitate. A steel file can scratch the hardest glass and the softer lead glasses can be scratched by a knife blade or a needle point.

2. Index of Refraction The range of refractive index is normally from 1.48 to 1.70, although it can be as high as 1.95. Since glasses with refractive indices greater than 1.70 are very soft, they are seldom used as gem imitations.

3. Single Refraction. Glass is isotropic (singly refractive). Some glasses, because of strain, show anomalous double refraction, but never to the degree that would be confused with true double refraction.

4. Density. This property varies widely with chemical composition, and may range from 2.2 to 4.2. It is especially affected by lead oxide; the greater the amount of lead oxide, the higher the density.

5. Luster and Fracture. Glasses break with a conchoidal fracture and usually show a vitreous luster on the fracture surface; some high-lead glasses have an adamantine luster. These properties help distinguish opaque glasses imitating chalcedonic quartz (black onyx) from the natural mineral which has a dull luster on a fracture surface.

6. Heat Conductivity. When glass is held in the hand it feels warm, whereas a natural or synthetic crystalline material, because of its better heat conductivity, feels cool to the touch.

7. Air Bubbles and Flow Lines. The usual glass imitation has spherical or elongated bubbles and frequently shows flow lines or irregular color bands (see Plate III). These features, characteristic of viscous liquids, are inherited by the glass to which they cool.

Glass is often used to imitate beryl (emerald and aquamarine), quartz (rock crystal, citrine, and amethyst), and topaz. Glass imitations of sunstone and aventurine are called *goldstone* and *aventurine* glass. It must be remembered that because of wide compositional differences, glasses may have the same refractive index as any of the above gems. However, the gems are doubly refractive whereas glass is isotropic.

Plastic

Plastics, which have countless familiar uses in our modern civilization, are also made to simulate gem materials. They are formed into useable products by heating and/or molding. The first such material to come into use was celluloid, a cellulose plastic. It has now largely been replaced by a host of synthetic resin plastics. Among these are phenol–formaldehyde resins (bakelite), methyl methacrylate resins (plexiglass, lucite), polystyrene resins, and polyvinyl chloride resins.

The gemologist is concerned only with determining whether or not the imitation is plastic, not which of the multitude of plastics it is. Because of their low density (1.05–1.55), low refractive index (1.33–1.42), and low hardness, plastics are easily identified. Plastics are not only soft enough to be scratched by a knife but they are usually sectile. That is, they are not brittle and a scratch remains as a small furrow. It should be noted, however, that opaque plastics contain fillers that can greatly increase their density and hardness.

Probably the most widely used plastic is a clear form of bakelite made to imitate amber. The material has both a specific gravity and a refractive index higher than amber and is harder and tougher than amber. Plastic imitations of lapis lazuli and turquoise are frequently encountered.

Doublets, Triplets, and Foil Backs

Doublets and triplets are composite stones devised to overcome the inferior hardness of glass imitations or to produce one large stone from two smaller pieces. Doublets, as the name implies, are made by joining two pieces of material either with cement or by fusion (Fig. 11.1). Triplets (Fig. 11.2) consist of two layers of colorless material joined by a cement that imparts color to the stone. Foil backs (Fig. 11.3) are made by applying a mirrorlike back to the stone, either to enhance its brilliance or color or to produce a starlike effect. The term *assembled stones*

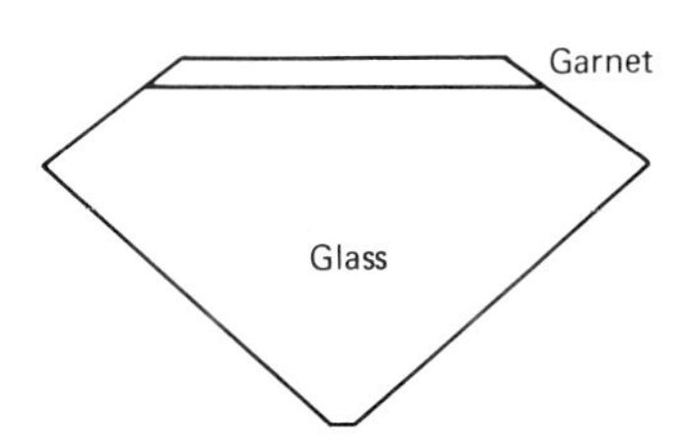

FIG. 11.1. A doublet consisting of a garnet top fused to a lower portion of glass.

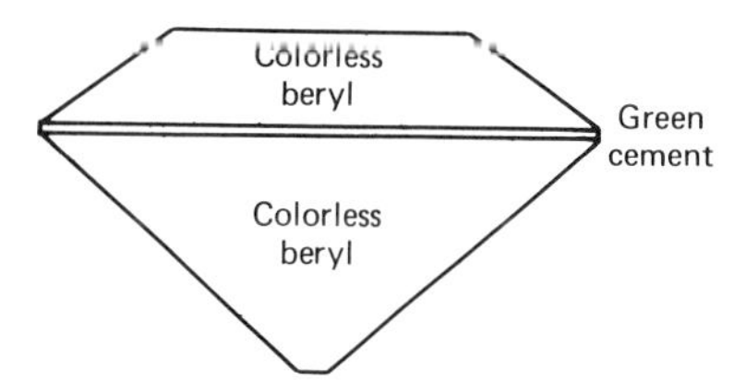

FIG. 11.2. A triplet consisting of two pieces of colorless beryl joined by a layer of green cement.

has been proposed by Robert M. Shipley to designate all three of these substitutes. Many types of assembled stones have been made. Some of the most frequently encountered are as follows*:

1. **Genuine Doublet or Triplet.** In these the top and bottom of the stone are of the same species. For example, in a diamond–diamond doublet the stone is cut from two thin slices joined by a colorless cement. A triplet made to simulate an emerald consists of two pieces of light-colored beryl joined by a layer of emerald-green cement. In a sense these stones can not be considered imitations, since they do not purport to be minerals other than they are. The deception is in one case the size and in the other the color.

2. **Semigenuine Doublet or Triplet.** Only one portion of the stone, usually the crown, is of the species it imitates. For example, an emerald imitation may be made up of a colorless beryl top and a colorless quartz pavilion, joined by a layer of emerald-green cement.

* Classification according to Richard T. Liddicoat, Jr., 1962. *Handbook of Gem Identification,* 6th ed., pp. 136–137.

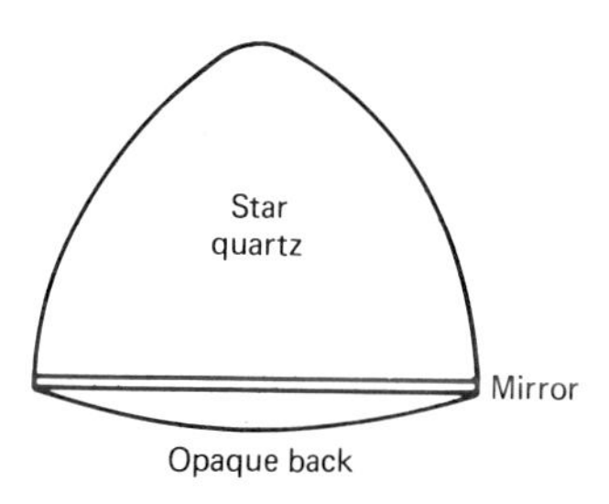

FIG. 11.3. A star quartz foil back made to imitate star sapphire.

3. False Doublet or Triplet. One or more parts of the stone is a natural material but the finished stone imitates a gem not used in its construction. Examples are a doublet with a garnet crown and red glass pavilion made to imitate ruby; a triplet made to imitate emerald consisting of quartz crown and pavilion joined by emerald-green cement; a false doublet made to imitate diamond with neither part a natural material, for example, a strontium titanate pavilion and a colorless synthetic sapphire crown.

Three of the most important doublets now seen in the jewelry trade are listed here.

Opal Doublet

These doublets are thin slices of precious opal cemented to a piece of either common opal or black glass. Sometimes these stones are difficult to distinguish from a genuine stone consisting of a thin seam of precious opal naturally occurring on opal matrix. This is particularly true when, as is frequent, the stone is in a bezel mounting that completely conceals the edge of the stone. A form of opal triplet has been made in which the opal doublet has a convex cap of clear quartz which protects the opal.

Star Stones

Star quartz cabochons backed by a mirror, or by blue glass and a mirror, are made to imitate star sapphires. Another variation is to take natural star sapphire with a good star but gray color and make a triplet with blue cement, or a doublet with a back of blue sapphire or synthetic sapphire. Still another star stone substitute consists of a cabochon of synthetic sapphire across the back of which have been ruled fine lines intersecting at 60° angles.

Garnet–glass Doublet

Perhaps the most frequently encountered composite stones are garnet–glass doublets, especially in older jewelry made before the use of synthetics became widespread. The purpose of the garnet top is impart hardness, not color. Frequently the garnet portion is slightly larger than the table, and is fused or cemented to colored glass, usually red or green.

Less Common Doublets and Triplets

In recent years so many composite stones made of various combinations of natural and synthetic materials have appeared on the market that it is impossible to describe them all. The following, not already mentioned, have been described in the gemological literature:

Doublet of natural yellow sapphire crown on a pavilion of either blue or red synthetic sapphire.
Opal triplets with synthetic spinel or corundum tops.
Doublets made with a diamond top on a base of quartz, synthetic sapphire, spinel, or glass.
Emerald triplet with crown and pavilion of synthetic, colorless spinel.
Doublets with colorless quartz crown and variously colored glass pavilions.

Detection of Doublets and Triplets

That a stone is a doublet or triplet can usually be determined by observation under the microscope at low magnification. The composite nature of the stone is more clearly seen by immersing it in a liquid. Immersion not only reduces surface reflections but frequently reveals that different parts have different refractive indices.

If a garnet–glass doublet of any color other than red is placed table down on a sheet of white paper, a red ring appears on the paper produced by the garnet top. Examination of the crown facets by reflected light will show the line where the two parts are joined, accentuated by the difference in luster between garnet and glass. It must be remembered that the garnet top will show genuine inclusions which can lead to an incorrect identification. The cement or fused layer usually contains spherical bubbles that lie in the plane joining the two parts.

Except for genuine doublets or triplets, refractive index readings taken on crown and pavilion facets will vary widely.

Liddicoat (1975) has pointed out that the cement in triplets usually fluoresces strongly under shortwave u.v. light, whereas the other parts of the stone do not.

Treated Stones

A wide range of gem materials now being used in the jewelry trade are treated in a number of different ways. In general these are natural gem materials and the treatment enhances their color to make them appear more valuable than they really are. The principal methods of treating are dyeing, heat treating, and irradiation.

Dyeing

The most obvious way to change the color of a gem is to stain it with a dye of the desired color. The process is restricted essentially to gem materials that are sufficiently porous to permit penetration of the solutions containing the pigmenting chemicals. Microcrystalline quartz, chalcedony (see Plate XI), which can be dyed almost any color is the outstanding example; however, turquoise, opal, coral, jade, and pearl also can be artificially tinted.

Heat Treating

The color of some gem minerals is enhanced or completely changed when stones are heated under the proper conditions. In many cases the "proper conditions" are a trade secret and the cause of the color change is poorly understood. In others a change in the oxidation state of the chromophores (see discussion under *Color* in Chapter 6) is believed to be the determining factor in color change.

Although heat treating may be carried out in the rough, most stones are cut before heating. The most common method is to embed the stones in sand and heat in an electric oven. But sometimes good results may be obtained with a single stone by heating it in a test tube over an open flame.

The gemstones most commonly heat treated are quartz, beryl, zircon, topaz, tourmaline, and zoisite (tanzanite). The color changes produced by heating are mentioned in the descriptions of the individual gems.

Irradiation

A color change has been produced in many gems by irradiation but in only a few is the change permanent. Exceptions are topaz and diamond. Colorless or brownish topaz may become a light to deep blue when exposed to gamma radiation. Permanent color changes are brought about in diamond by exposing it to various subatomic particles such as produced by radium, the cyclotron, cobalt 60, and the atomic pile. The colors resulting from such exposures are given in Chapter 13 (see DIAMOND under *Important Gemstones*).

References and Suggested Reading

Liddicoat, R. T., 1975. *Handbook of Gem Identification*, 10th ed. Gemological Institute of America, Santa Monica, Calif. Chapter 11.

Webster, R., 1975. *Gems*, 3rd ed. Butterworth and Co., Ltd., London. Chapter 18.

Webster, R., 1964. *Composite Stones—a Survey of Their Types and Their Properties. Journal of Gemmology*, Vol. 9, pp. 160–176.

12 CUTTING AND POLISHING OF GEMS

Introduction

Some natural gem crystals have a brilliance, symmetry, and beauty that cannot be improved upon by man. These marvels of the mineral kingdom should never reach the lapidary's wheel. However, the preponderance of gem rough must be shaped and polished to reveal its full beauty.

Minerals have been shaped by man since the beginning of the Old Stone Age a million years ago. This early working was not to give symmetry to objects of adornment but to make tools and weapons. As time passed, the artifacts were more skillfully shaped and during the Neolithic period at the end of the Stone Age, they were polished as well. It was probably about this time that man first polished the rough surfaces of hard minerals with the sole objective of enhancing their beauty.

Cut stones can be grouped into two basic types, cabochons with curved surfaces and faceted stones bounded by smooth planes. Cabochon was the earliest style, for the primitive lapidary merely polished a pebble already rounded by stream action. Later rounded surfaces were ground on angular fragments of gem material and polished. Because of the more advanced technology necessary to position symmetrically and polish flat facets, it was much later that faceted stones made their appearance. Today the type of cutting is determined largely by the nature of the material. Generally, transparent material is faceted, whereas translucent or opaque gems are cut as cabochons. But heavily flawed transparent gem material, as some emerald, is better cut cabochon than faceted. Sometimes the desired optical effect dictates the style, such as the required cabochon cut of star and cat's eye gems.

The purpose of cutting a gem is to enhance the beauty of the material. For opaque and translucent material this is accomplished by shaping and polishing to produce a high degree of surface reflection of light. For transparent gem material the brilliance and color are improved not only by the highly polished surface but, in the properly proportioned stone, by internally refracted and reflected light.

Gem cutting methods, to be covered later in this chapter, are not the same for all gems. Because of its superior hardness, the methods used to cut and polish diamond differ from those used for other gems. A person who cuts and polishes diamonds works only with diamond and is known as a *diamond cutter;* a *lapidary* cuts all other gemstones.

Cabochon Cut

There are three basic styles of cabochon cuts but the variations in them may be many. (1) The simple cabochon has a domed top and a flat base. The curvature of the top may vary greatly from one of very low profile to a very high dome. (2) The double cabochon has a domed top and bottom. The convexity of both top and bottom may be the same, but usually the top is higher and thus has a lesser radius of curvature than the lower part. (3) The hollow cabochon has a domed upper surface but the lower part is hollowed out, or concave. This type of cut is used for very deep colored material, such as deep red garnet, to increase light transmission. In other types of cabochon cuts the bottom may or may not be polished but in the hollow cabochon, the concave surface is polished to enable light to pass through. The outline of a cabochon cut gem as viewed from above may be circular, oval, rectangular, or other polygonal shapes, or may have a fancy shape such as a heart. Examples of the basic cabochon styles are given in Figure 12.1.

The cabochon cut is used for translucent gem materials such as turquoise, jade, chalcedony, and many other less common minerals. It is especially desirable for opaque gems to bring out pleasing textures, designs, or color contrasts. Gemstones with special optical effects such as asterism and chatoyancy are always cut as cabochons. Examples are star sapphire, cat's eye, moonstone, and opal.

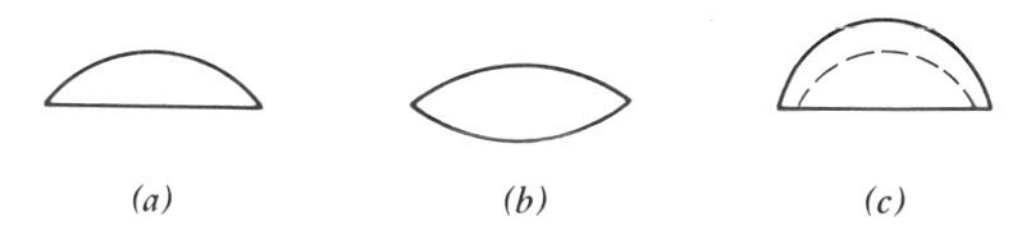

FIG. 12.1. Cross sections of cabochon stones. (*a*) Simple cabochon. (*b*) Double cabochon. (*c*) Hollow cabochon.

Faceted Cut

Compared to the cabochon cut the faceting of gemstones is a relatively recent art and probably had its beginning in the fourteenth century.

The degree of brilliance of a faceted gem depends on the optical properties of the material (primarily the index of refraction) and, equally important, on the way in which the facets are placed. In the early styles of facet cutting, sometimes merely polishing natural crystal faces, these factors were not considered. Eventually when the importance of refraction and internal (total) reflection became understood, styles of cutting evolved that made best use of the optical properties of the gem material. How these properties of refractive index, critical angle, and total reflection, discussed in Chapter 6, are used to improve the brilliance of a faceted gem is shown in Figure 12.2.

When light passing through a gemstone strikes a back facet at an angle greater than the critical angle, it is totally reflected. If the angles of the back facets of the gem are correct for a given refractive index, a large proportion of the light entering the top of the stone will be totally reflected twice and returned to the eye of the observer (Fig. 12.2*a*). If the gem is improperly proportioned, there is refraction (leakage) of light through the back of the stone (Figs. 12.2*b* and 12.2*c*), where the light ray strikes a back facet at an angle less than the critical angle. In the first case both brilliance and color are improved whereas in the second they are diminished. Color depends in part on the length of the light path and thus is deeper in the correctly cut stone than in one of improper proportions. In Figure 12.2 the critical angle for quartz, 40.5°, is used.

As refractive index rises, the critical angle becomes smaller and it is possible to decrease the inclination of the main crown and pavilion facets and still maintain brilliancy. Thus gemstones of high

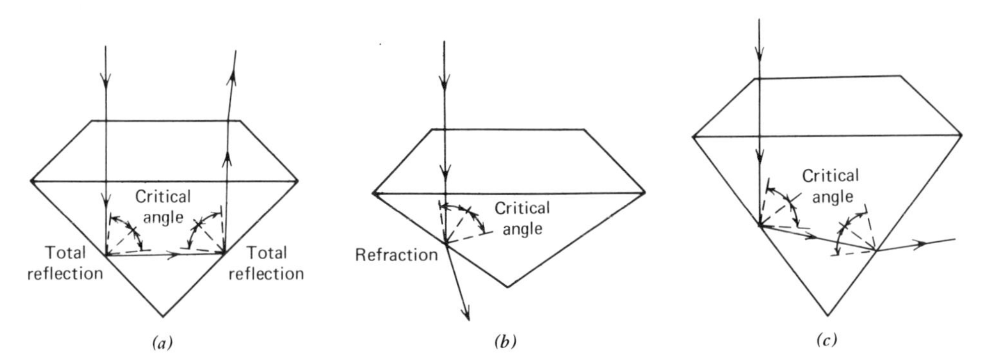

FIG. 12.2. (*a*) Total reflection of light passing through a properly proportioned gemstone. (*b*) Leakage of light through the back of a gemstone cut too shallow. (*c*) Leakage of light through the back of a gemstone cut too deep.

refractive index may be cut shallower in relation to their width. In diamond, with ***R.I.*** 2.42 and ***C.A.*** 24.4°, the recommended crown and pavilion angles are 34.5 and 41°, respectively, and it is important for brilliant cut diamonds that these angles be closely adhered to. For gems of lower refractive index it becomes less important that the theoretical angles be used, especially in colored stones, and there is lack of agreement among various investigators as to the optimum angles for various gem materials. Tables of angles for faceted gems may be found in Kraus and Slawson (1947) and Sinkankas (1962) (see references at end of chapter).

The paths taken by light rays in a gem having double refraction are shown in Figure 12.3. Only a few natural gemstones have strong double refraction (high birefringence) and ordinarily this is not of concern to the lapidary. However, in synthetic rutile the extreme anisotropism causes a "fuzzy" appearance which can be reduced by proper orientation. That is, the table should be perpendicular to the c crystallographic axis.

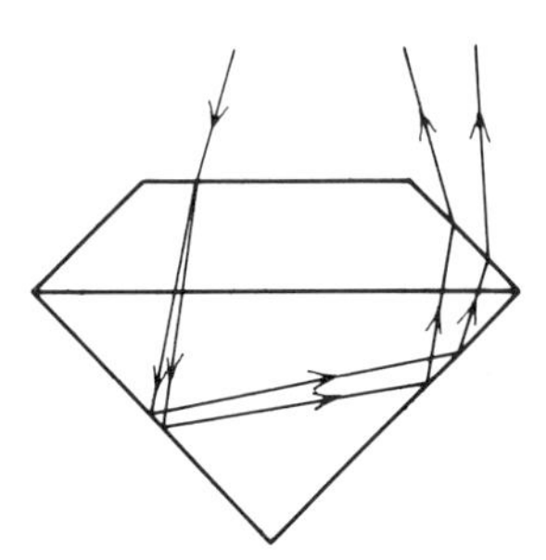

FIG. 12.3. Passage of light through a gemstone having strong double refraction (high birefringence), causing doubling of the back facets.

Dispersion and the resulting "fire" displayed by a faceted gemstone also depend to some extent on proportion, in particular the angle of the top facets. Diamond is the only important gemstone with strong dispersion and the necessity of adhering to proper proportions to accentuate its "fire" is discussed in Chapter 13 (see DIAMOND under *Important Gemstones*).

Today there are two basic styles of cutting faceted gems, the step or emerald cut and the brilliant cut as shown in Figure 12.4. All other cuts are derived from them, and almost infinite variations of both are possible. However, the earliest faceted stones, particularly diamond, did not conform to either of these basic designs.

Rose Cut

The first of the styles of cutting using flat facets was the rose cut which was probably developed before the sixteenth century. It consists of facets on the top only, the base being flat. Two of the many possible variations in the rose cut are shown in Figure 12.5. The rose cut was used for diamonds, and in Victorian jewelry the red garnets from Bohemia were often cut in this style.

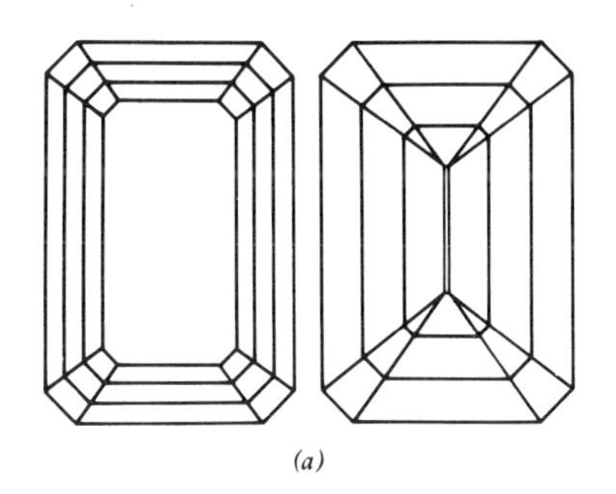

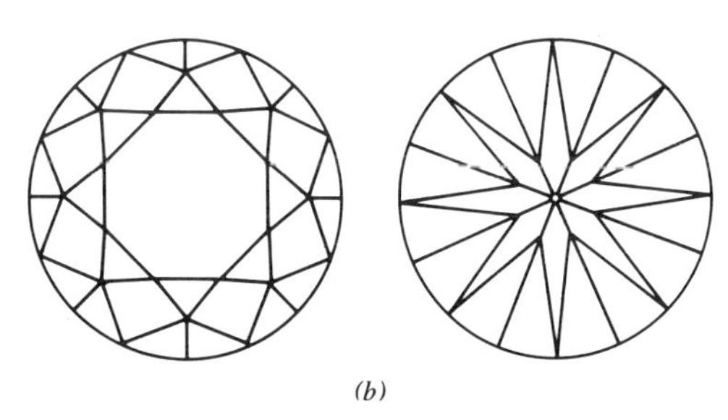

FIG. 12.4. The two basic styles of cutting faceted gems. (*a*) Step or emerald cut. (*b*) Brilliant cut. In both the top facets are shown on the left, the bottom facets on the right.

Table Cut

The next advance in the faceting of gems was the table cut, shown in Figure 12.6. It probably evolved because of the octahedral shape of many diamond crystals, and was much used for diamond.

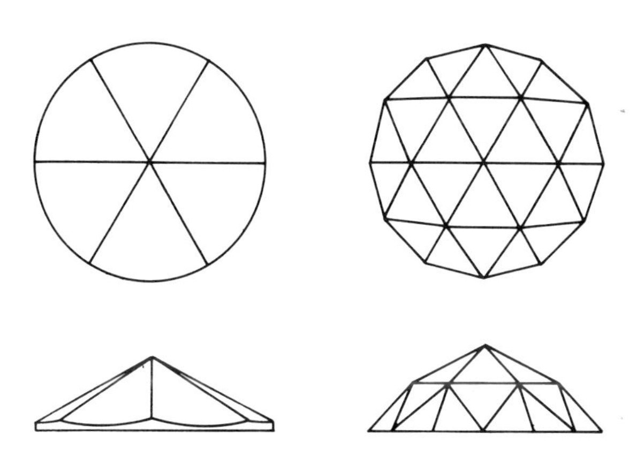

FIG. 12.5. The rose cut, the first of the styles of cutting using flat facets.

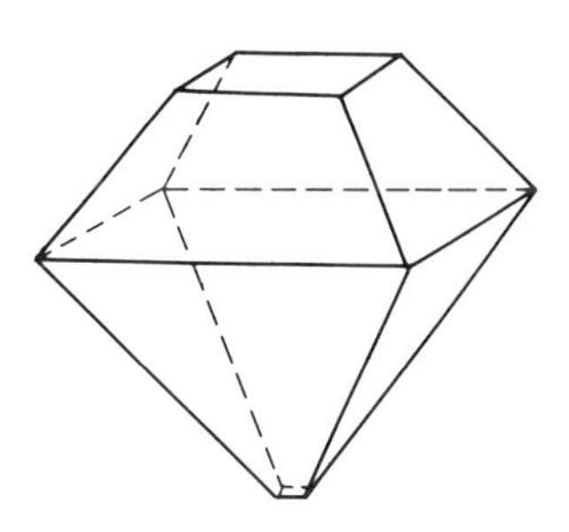

FIG. 12.6. The table cut, an early form of faceting used for diamonds.

Step or Emerald Cut

This style of cutting, also sometimes known as the *trap cut* or *cushion octagon,* probably evolved from the table cut. As shown in Figure 12.4*a*, it is characterized by a large table facet flanked by parallel rows of trapezoid-shaped facets. The pavilion has similar rows of parallel facets decreasing in steepness toward the basal facet, or culet. This style is often called emerald cut because it is much used for this gem.

The outline of step cut stones may be square, rectangular, triangular, kite shaped, keystone shaped, lozenge, and many others. Small step cut stones with a long rectangular shape are known as baguettes. Some of the possible variations in the step cut are shown in Figure 12.7.

Still another modification of the step cut is the scissors cut. Here the form is rectangular and each trapezoidal side facet is replaced by four triangular facets. Two forms of the scissors cut are shown in Figure 12.8.

Brilliant Cut

The cut most frequently used for gems having a round shape is the brilliant. In fact, this style has become so widely used for cutting diamond that the term "brilliant" has become synonymous with diamond. The brilliant cut is said to have been first used by a Venetian craftsman, Vicenzio Peruzzi, at the end of the seventeenth century. It is the style

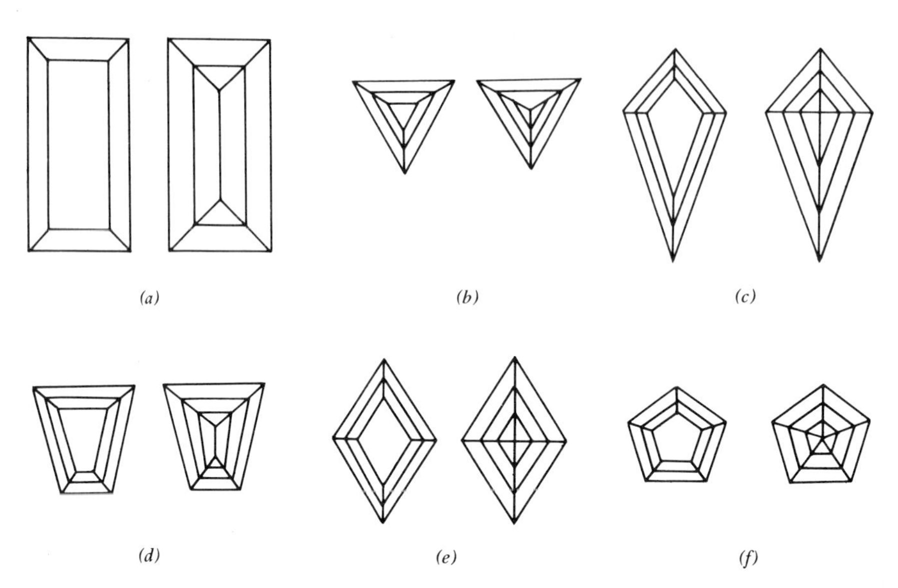

FIG. 12.7. Some of the many possible variations in the step cut. (*a*) Baguette cut. (*b*) Triangle cut. (*c*) Kite cut. (*d*) Keystone cut. (*e*) Lozenge cut. (*f*) Pentagon cut. In each pair left drawing front, right drawing back.

most suitable for cutting all gems to attain greatest brilliance, although the importance of close adherence to optimum angles and proportions is less for deeply colored material than for gems that are colorless or nearly so.

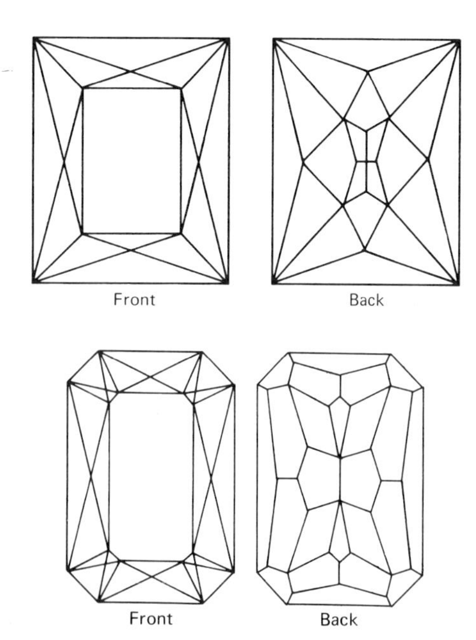

FIG. 12.8. The scissors cut, another modification of the step cut.

The standard brilliant cut (Fig. 12.4*b*) consists of 58 facets, 33 on the top or crown of the stone, and 25 on the base, or pavilion. Except for the table and culet all facets are either triangular or kite shaped.

Modifications of the standard round brilliant are the pear, oval, and marquise cuts, shown in Figure 12.9. These deviations from the round form give the cutter and designer greater latitude but note that in these other shapes not all the pavilion facets can be held to the proper angle; hence some light leakage results, reducing overall brilliance.

Sinkankas has pointed out that in gems of greater than 10 carats the 58 facets of the standard brilliant cut gives too simple an effect, and for gems of large size it is customary to add facets, still retaining the basic brilliant form (Fig. 12.10).

A forerunner of the modern brilliant cut, seen in older pieces of jewelry, was the old-mine cut. Old-mine cut stones are faceted in the brilliant style but have a small table, high crown, large culet, and their outline is more square than round. An example is given in Figure 12.11.

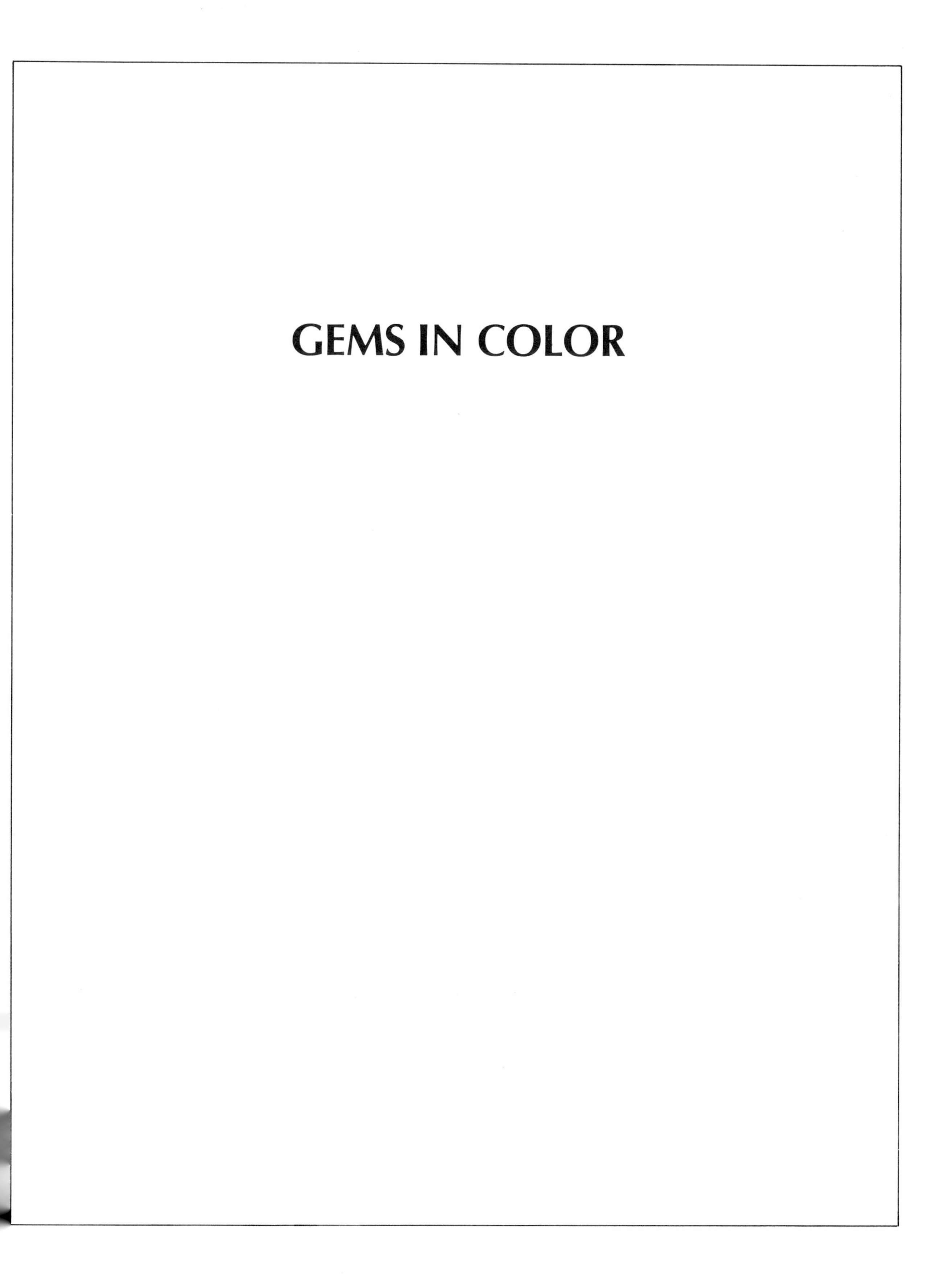

GEMS IN COLOR

PLATE I Mineral Carvings

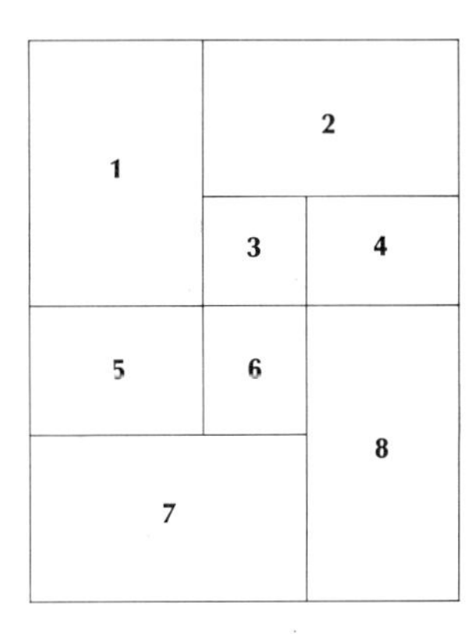

1 JADE (jadeite)

2 SODALITE
Sculptured bird.

3 BLUE SAPPHIRE
The Lincoln head is 6 cm high and weighs 1318 carats. Sculptor, Norman Maness. The Kazanjian Foundation, Beverly Hills, California.

4 CHALCEDONY
Chinese carving.

5 JADE (nephrite)

6 AGATE
Chinese snuff bottle.

7 CHALCEDONY
Carved bowls.

8 CORAL

Photographs 1, 2, 7, and 8 by Lee Boltin, courtesy of the Smithsonian Institution, Washington, D. C.

PLATE II Diamond, Ruby, Sapphire, and Zircon

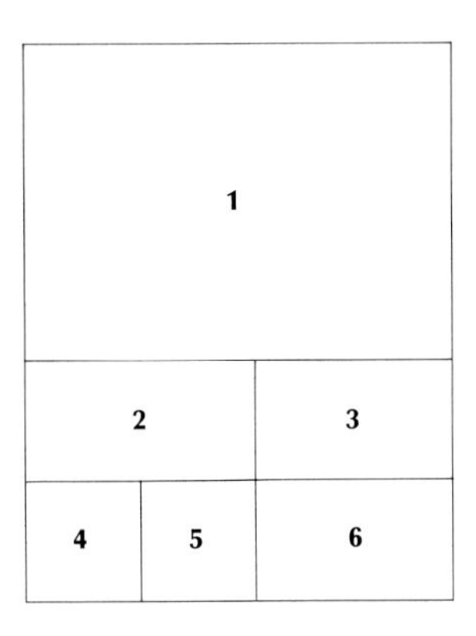

1 DIAMOND
Diamond sizes range from 4 carats (*center*) to 0.05 carat.

2 DIAMOND
A diamond in a kimberlite matrix, South Africa.

3 RUBY
A ruby in a marble matrix, Burma.

4 STAR RUBY
Rosser Reeves star ruby, 168 carats.

5 SAPPHIRE
Star of Asia sapphire, 330 carats.

6 ZIRCON

Photograph 1 is of the DeBeers Colored Diamond Collection, courtesy of N. W. Ayer ABH International.

Photographs 2, 4, 5, and 6 courtesy of the Smithsonian Institution, Washington, D.C.

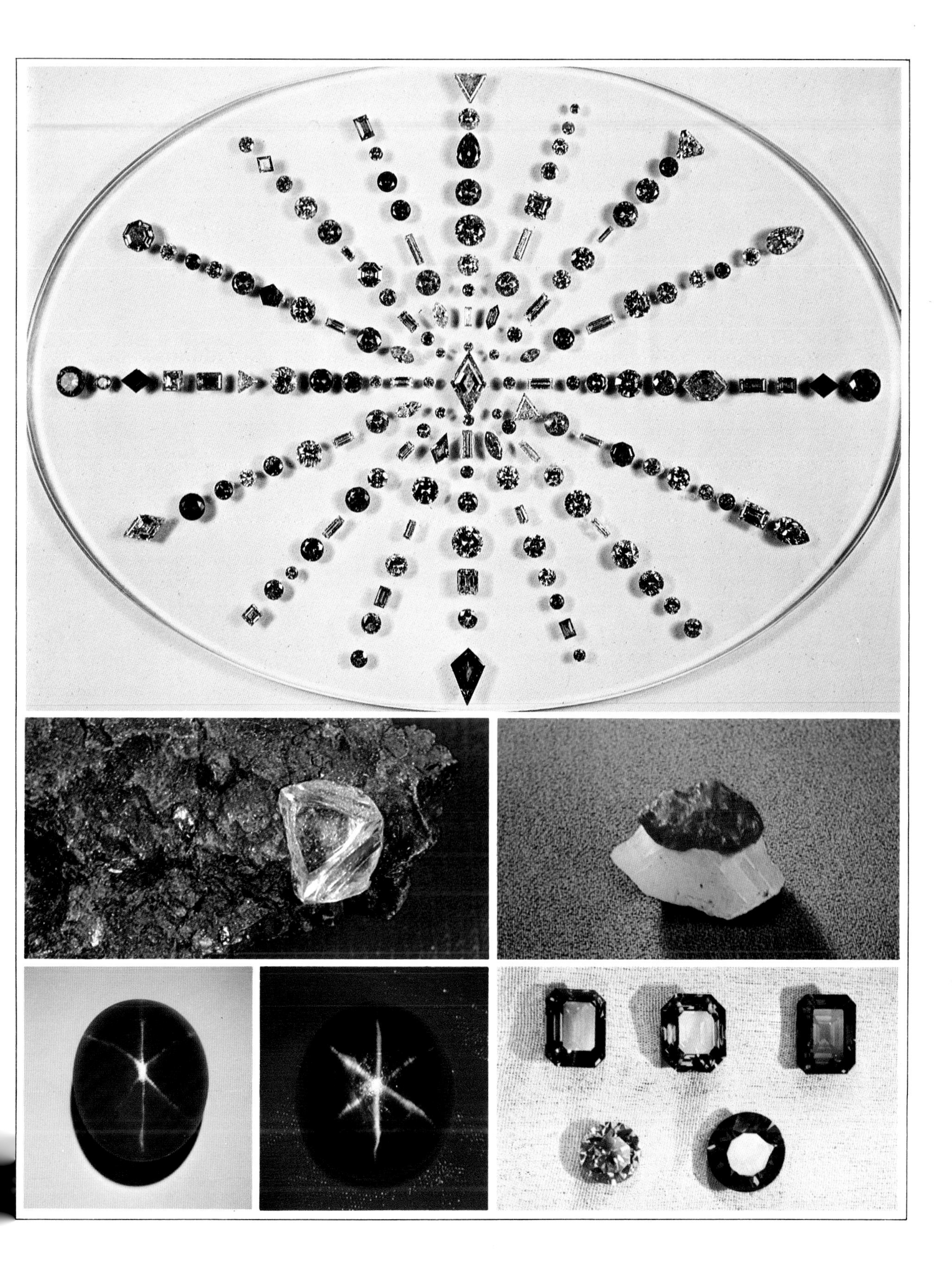

PLATE III Inclusions in Gemstones

1	2	3
4	5	6
7	8	9
10	11	12

1 **RUBY**
Flux inclusions in synthetic ruby.

2 **RUBY**
Bubbles in synthetic ruby.

3 **RUBY**
Hexagonal crystal inclusions in Chatham synthetic ruby.

4 **DIAMOND**
Pyrope garnet in diamond.

5 **DIAMOND**
Chrome diopside in diamond.

6 **DIAMOND**
Grossular garnet in diamond.

7 **CORUNDUM**
Fingerprint inclusions in natural corundum.

8 **CORUNDUM**
Color zoning in natural corundum.

9 **SAPPHIRE**
Curved striae in synthetic alexandrite-like sapphire.

10 **GLASS**
Elongated bubble in glass.

11 **GLASS**
Swirl lines in glass.

12 **EMERALD**
Flux inclusions in Gilson synthetic emerald.

Courtesy of the Gemological Institute of America.

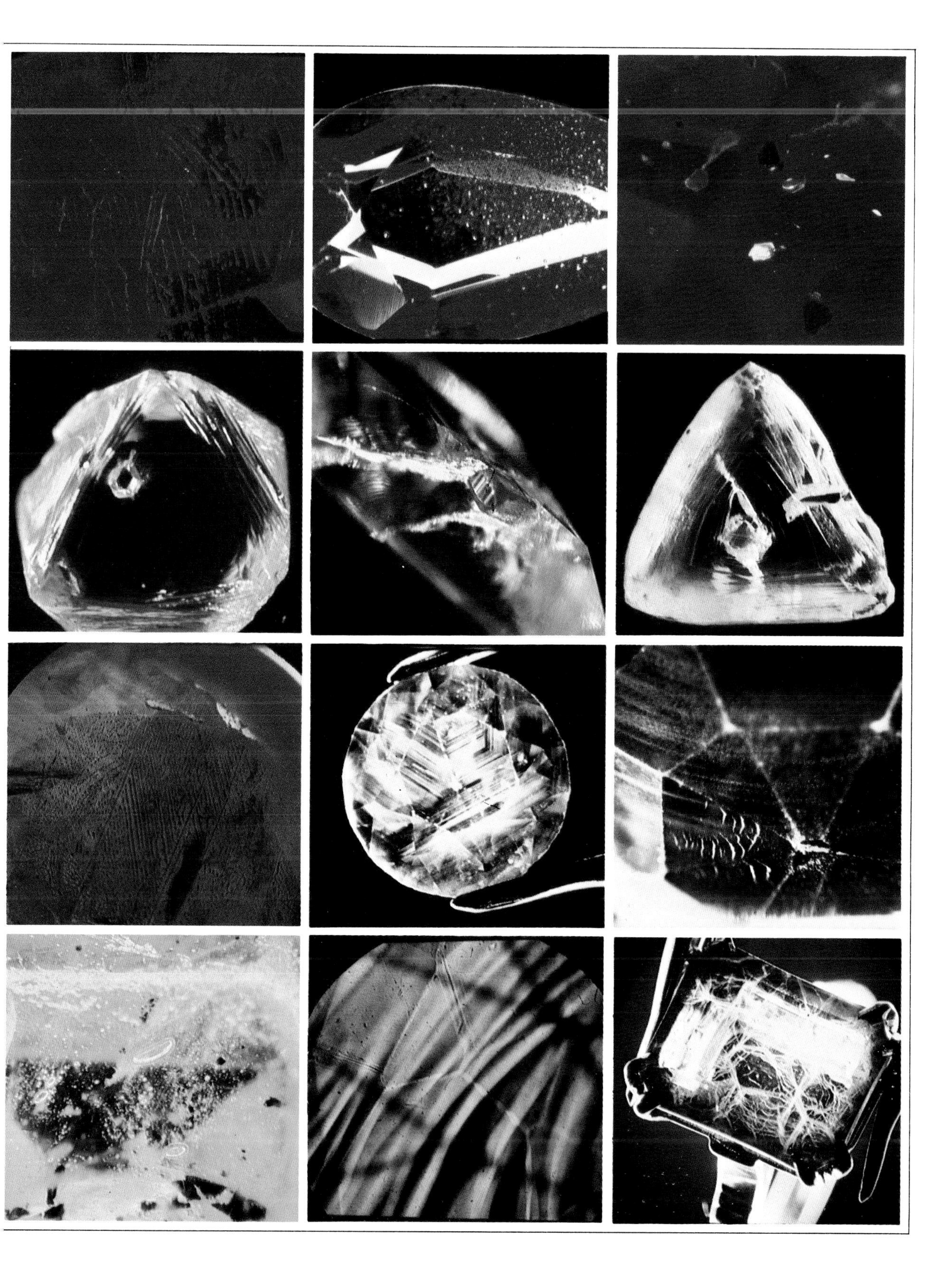

PLATE IV Sapphire

Courtesy of A. Ruppenthal, Idar-Oberstein, Germany.

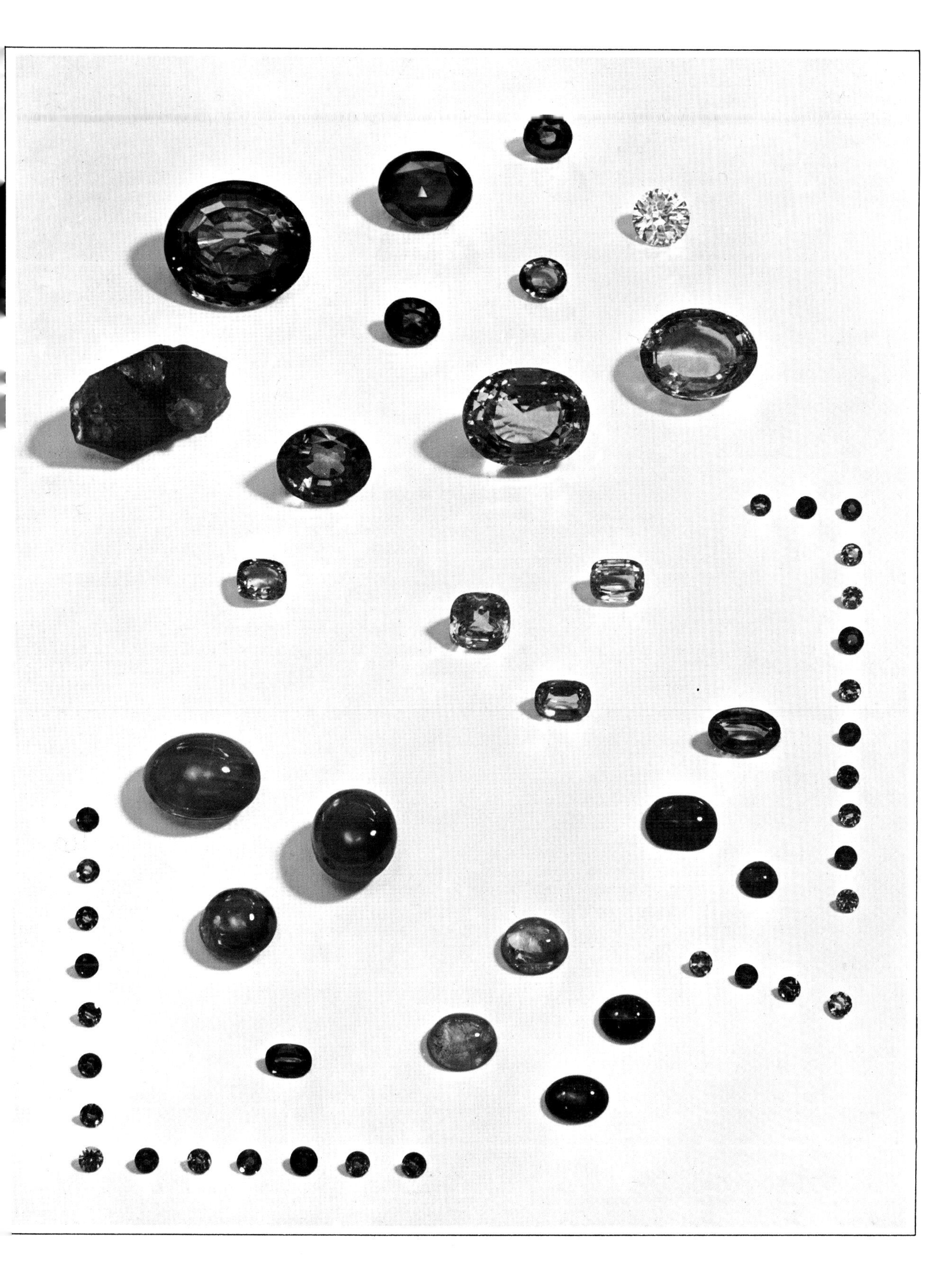

PLATE V Topaz, Peridot, and Spinel

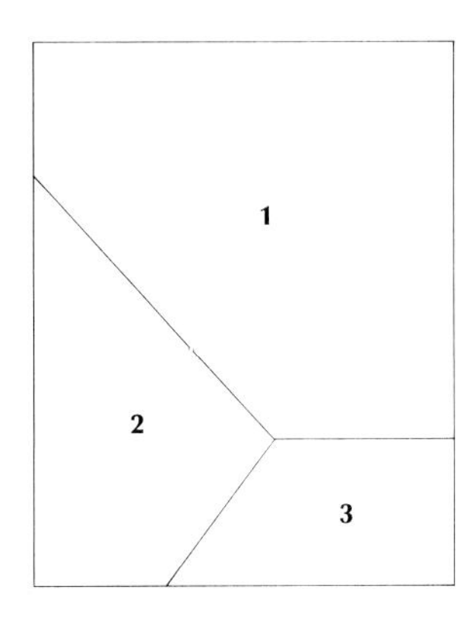

1 **TOPAZ**

2 **PERIDOT**

3 **SPINEL**

Courtesy of A. Ruppenthal, Idar-Oberstein, Germany.

PLATE VI Spodumene, Zoisite, and Iolite

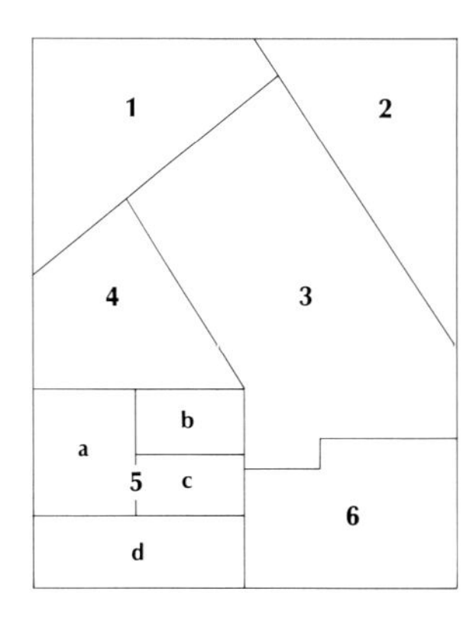

1 SPODUMENE CRYSTAL

2 KUNZITE CRYSTAL

3 KUNZITE

4 HIDDENITE

5 ZOISITE (tanzanite)

a. Crystal.
b. Without heat treatment.
c. Heat treated.
d. Pleochroism.

6 IOLITE

Spodumene (kunzite, hiddenite) and iolite courtesy of A. Ruppenthal, Idar-Oberstein, Germany.

PLATE VII Beryl

Courtesy of A. Ruppenthal, Idar-Oberstein, Germany.

PLATE VIII Tourmaline

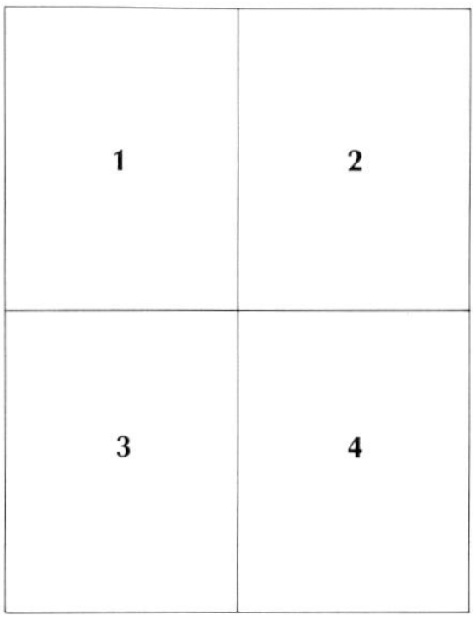

1	2
3	4

1 HAMLIN NECKLACE
This necklace is set with 70 tourmaline gems mined at Mount Mica, Maine, by Augustus C. Hamlin in the late 1800s. The largest green stone weighs 34.25 carats. The necklace is on exhibit at the Harvard University Museum.

2 TOURMALINE CRYSTALS
Gem tourmaline crystals on albite, Madagascar.

3 and 4 TOURMALINE GEMS
These cut stones illustrate the wide variety of colors in tourmaline gems.

Photograph 2 courtesy of the Smithsonian Institution, Washington, D.C. Photograph by Lee Boltin.

The photographs of the cut stones (3 and 4) appeared in the 1977 Spring issue of *Gems and Gemology* and are reproduced here by courtesy of the Gemological Institute of America.

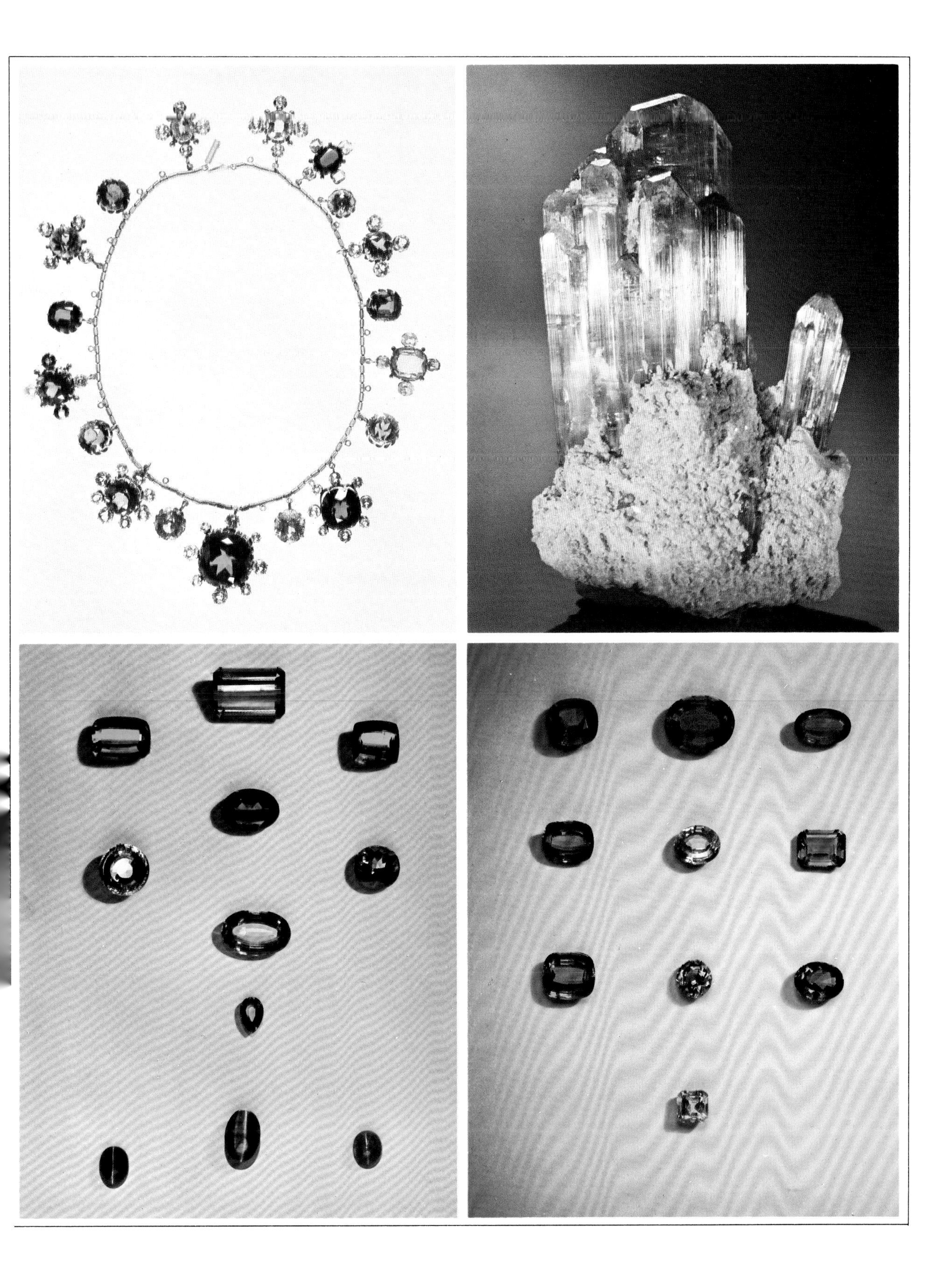

PLATE IX Uncommon gems

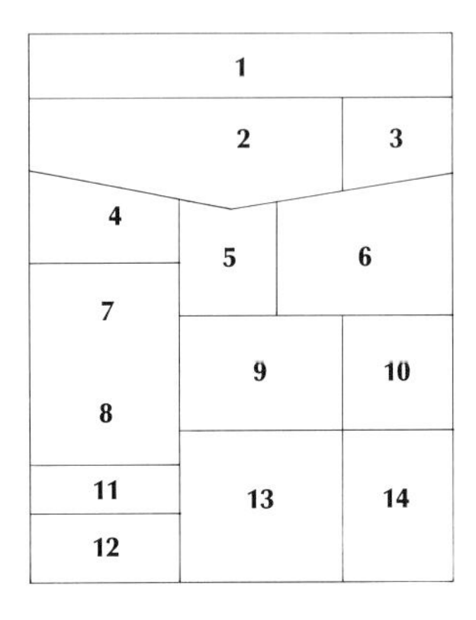

1 FLUORITE

2 SPHALERITE

3 PETALITE

4 AMBLYGONITE

5 SINHALITE

6 DIOPSIDE

7 DIOPTASE

8 APATITE

9 EUCLASE

10 ENSTATITE

11 BARITE

12 PHENAKITE

13 BRAZILIANITE

14 SCHEELITE

Courtesy of A. Ruppenthal, Idar-Oberstein, Germany.

PLATE X Opal

Courtesy of A. Ruppenthal, Idar-Oberstein, Germany.

PLATE XI Quartz, Garnet, and Sphene

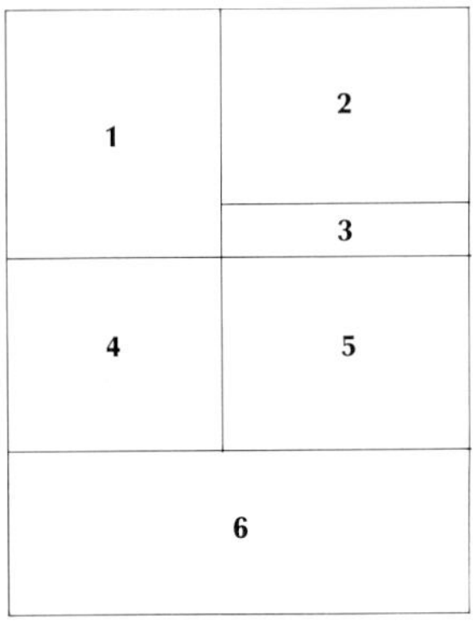

1	2
	3
4	5
6	

1 AMETHYST CRYSTALS

2 GARNET
Top center: Yellow-orange grossular. *Top left:* Almandine. *Bottom left and top right:* Green grossular. *Center:* Hessonite. *Bottom center:* Spessartine. *Bottom right:* Brown grossular.

3 SPHENE

4 QUARTZ
Top center and bottom left: Citrine. *Top right and left:* Amethyst. *Center:* Rutilated quartz. *Bottom right:* Maderia red quartz. *Bottom center:* Smoky quartz.

5 SYNTHETIC QUARTZ
Quartz has been synthesized in many colors some of which, for example, the blue and green, do not occur in nature. The elongated crystal (*upper right*) is 7 in. long.

6 DYED AGATE
A single slice of agate sectioned and reassembled after each section has been dyed a different color. The left end is the natural color.

Photographs 1 and 3 courtesy of the Smithsonian Institution, Washington, D.C. Photograph 1 by Lee Boltin.

The photographs of cut stones appeared in the 1977 Spring issue of *Gems and Gemology* and are reproduced here by courtesy of the Gemological Institute of America.

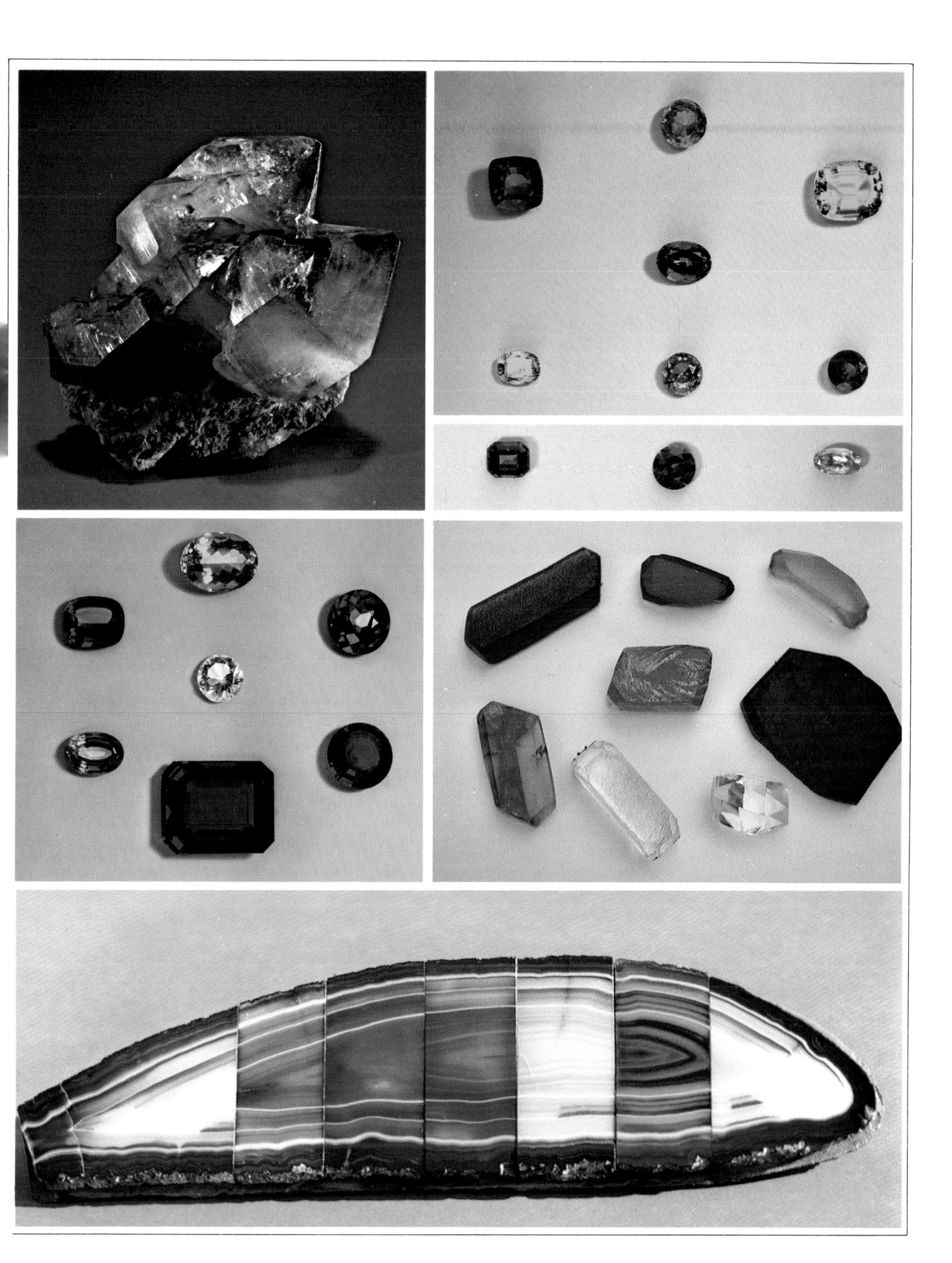

PLATE XII Synthetic, Treated, and Imitation Gems

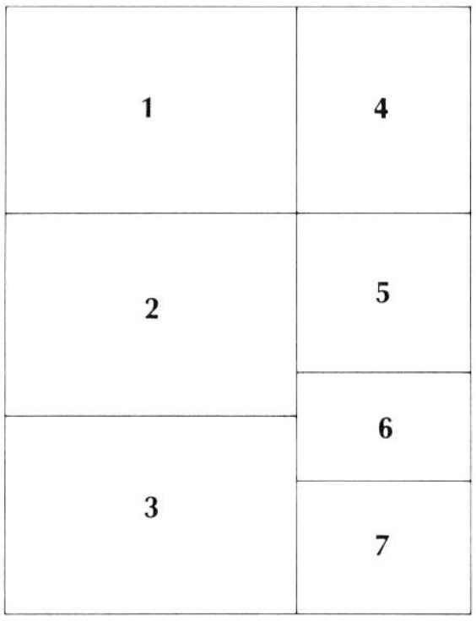

1	4
2	5
	6
3	7

1 CORUNDUM
Variously colored boules of synthetic corundum with gemstones cut from similar material.

2 TURQUOISE
Center: Natural turquoise. *Left, top and bottom:* Natural turquoise impregnated to improve color and porosity. *Left, center:* Turquoise apparently made by cementing turquoise powder (by Gilson). *Right:* Three simulated turquoises made of plastic.

3 COLOR CHANGE BY RADIATION
These natural gem materials were essentially colorless before irradiation. *Left:* Three topaz stones, blue and brown. *Center, top:* Blue-black pearl. *Center:* Greenish-yellow quartz. *Bottom, center:* Smoky quartz and amethyst (amethyst is synthetic). *Right:* Three beryls.

4 EMERALD
Cluster of Gilson flux-grown emeralds.

5 YAG
Faceted gemstones cut from flux-grown crystals.

6 RUBY
Flux-grown ruby and cut stone (X3).

7 OPAL (simulated)

Simulated opals made by J. L. Slocum. Courtesy J. L. Slocum, Rochester, Michigan.

Numbers 2 to 6 courtesy of K. Nassau, Bell Laboratories.

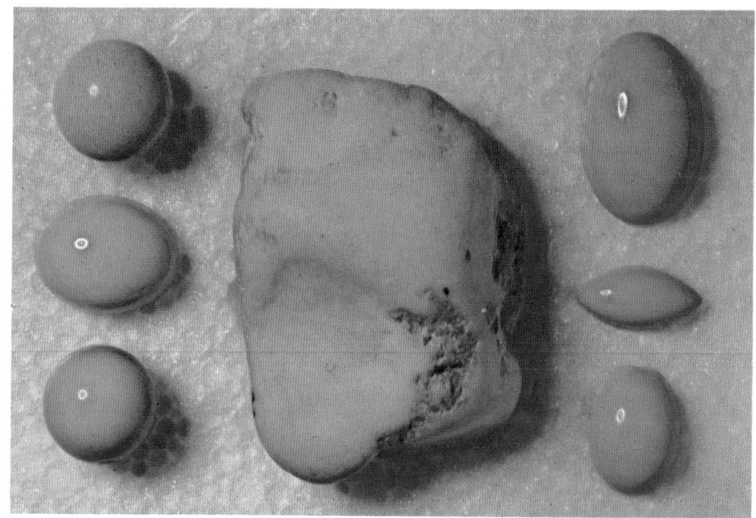

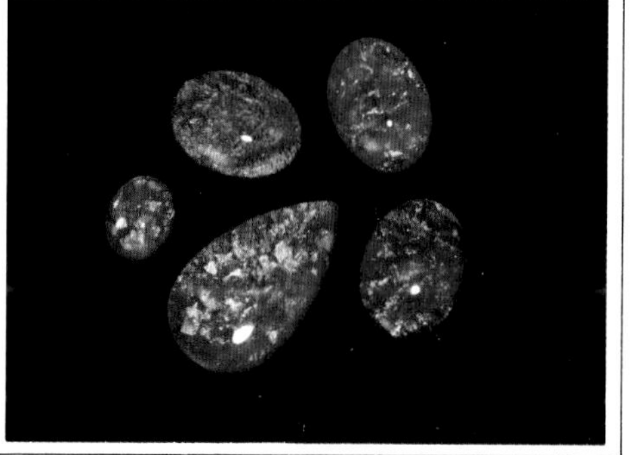

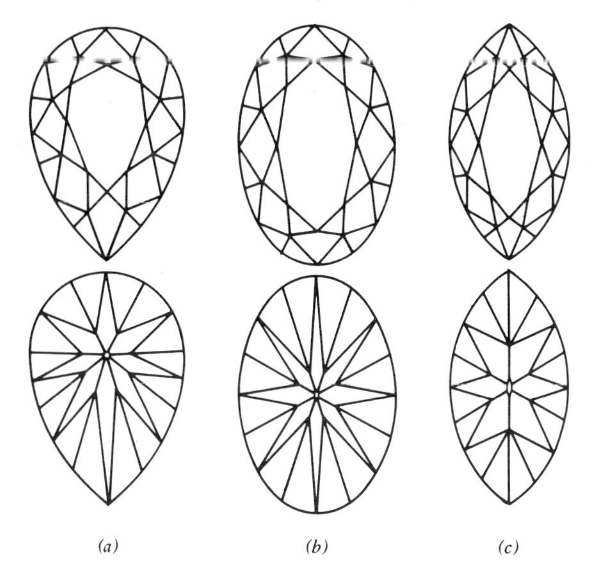

FIG. 12.9. Modifications of the round brilliant. (*a*) Pear-shaped. (*b*) Oval. (*c*) Marquise. Top drawing front, bottom back.

Following the old-mine cut was the European cut, a round stone like the modern brilliant, but with somewhat smaller table, higher crown, and larger culet (Fig. 12.12).

The antique cushion cut, also in the basic style of the brilliant, may vary widely in detail. It is often used for colored gems, especially those cut in the Far East (Fig. 12.13).

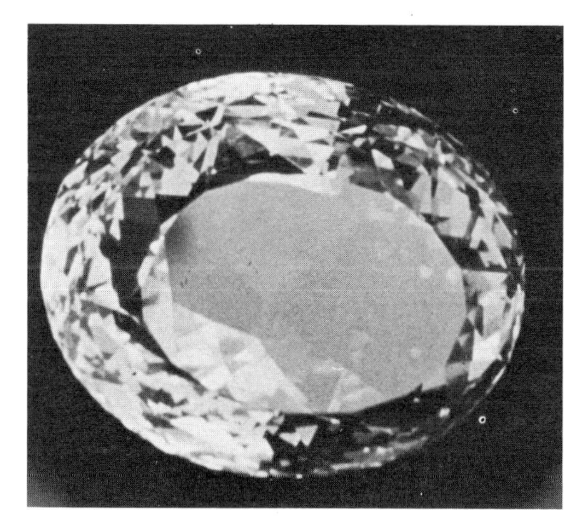

FIG. 12.10. A 578 carat oval brilliant beryl, with added facets, cut by John Sinkankas. Courtesy of the Smithsonian Institution, Washington, D.C.

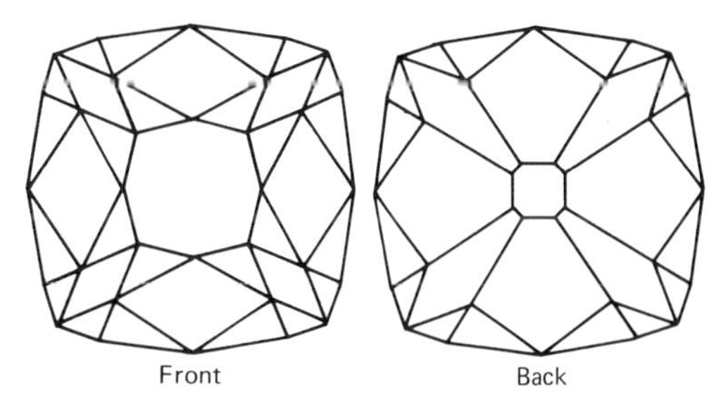

FIG. 12.11. The old-mine cut, a forerunner of the modern brilliant cut.

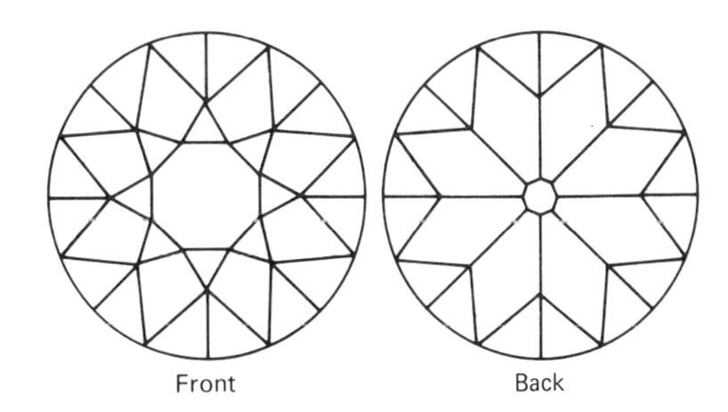

FIG. 12.12. The European cut, like the modern brilliant cut but with smaller table, higher crown, and larger culet.

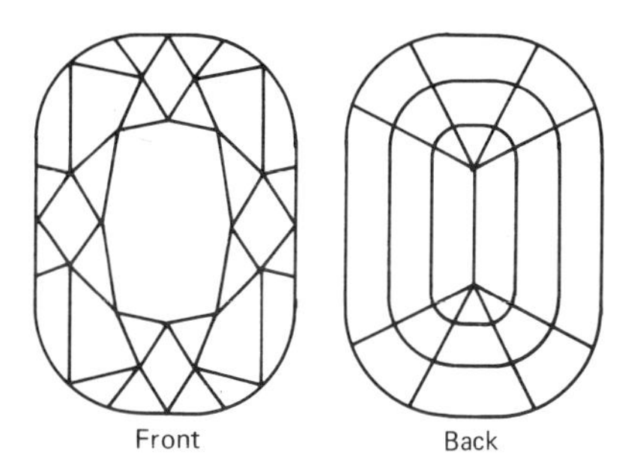

FIG. 12.13. An example of the antique cushion cut, which may vary widely in detail.

Carved Gems

The term *cameo* is used to designate stones on which a raised figure or design has been cut. Generally material is used which has two differently colored layers, in one of which the raised design is cut, the layer of the second color forming the background. An *intaglio* is a similar style in which the design is incised. A stone with a raised design in a concave depression is a *cuvette*. *Stone cameos* are those cut from onyx, a layered form of chalcedony (quartz), and cameos cut from much softer shell material are called *shell cameos*. Gemstones are also carved into countless other forms such as leaves, flowers, animals, and stylized designs and may range in size from very small to sculptured pieces of proportions limited only by the availability of suitable material. Some examples are shown in Plate I.

Methods of Cutting Gems Other Than Diamond

The methods used for cutting both diamond and other gems, though similar in most respects, differ in that in diamond cutting the facets are ground and polished in one operation, whereas for other gems grinding and polishing are two distinct steps. The steps are briefly described in this section. For more detailed information about methods and equipment the reader should consult the references at the end of the chapter.

Sawing

The rough material is first brought to the required size by sawing with diamond saws, circular blades with diamond powder locked into the rim. The saw blades may be very thin copper or bronze only a few inches in diameter, used for very valuable material, or heavy steel several feet in diameter for slabbing large pieces. The saw blades are generally run vertically. Feed may be by hand for small pieces, or the material may be clamped in a vise and fed mechanically. A water or oil coolant is used. Diamond saws of a wide variety of design and size are available; one is shown in Figure 12.14.

Grinding

Gemstones are given their rough shape by grinding with abrasive wheels. The most widely used wheels are made of silicon carbide, but may be of another synthetic abrasive, or even sandstone. Coarse grit wheels are used for the initial shaping and fine grit wheels for later slower removal of material. The grinding wheels are usually run on horizontal shafts, and must be supplied with water to cool the work and wash away the rock dust generated by the grinding. Grinding wheels are used to shape cabochons, preform faceted gems, shape carvings, and perform many other tasks in the lapidary shop.

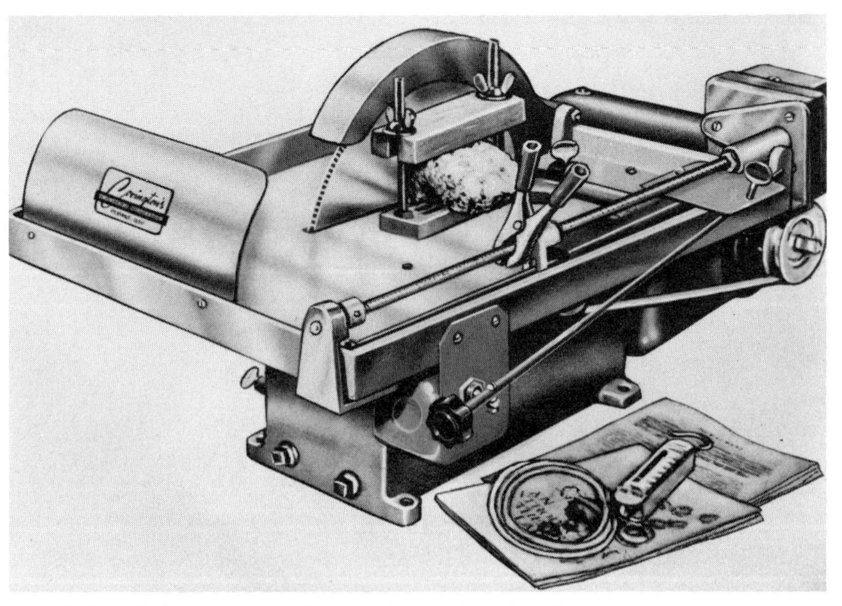

FIG. 12.14. A diamond saw. The sawing is done by diamond grit locked into the rim of a metal blade. Pictured here is a trim saw with 8 in. blade. Blades come in sizes from an inch or two to several feet. Courtesy of the Covington Engineering Corp., Redlands, California.

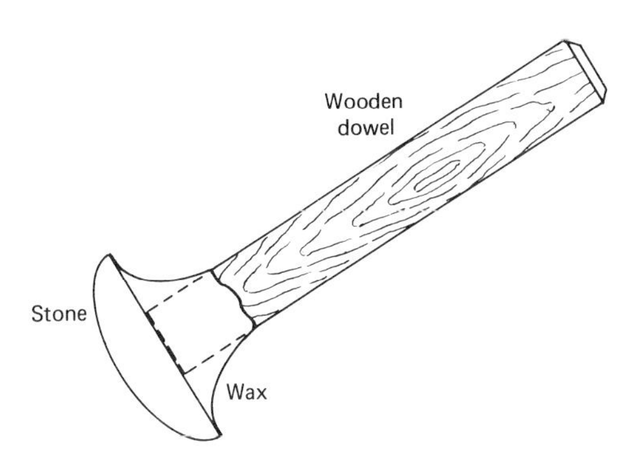

FIG. 12.15. A holder for gems being ground and polished is called a dop stick. The gem is fastened to the end of a dowel with dopping wax. Adapted from J. Sinkankas, 1962, *Gem Cutting*, D. Van Nostrand Co., Inc.

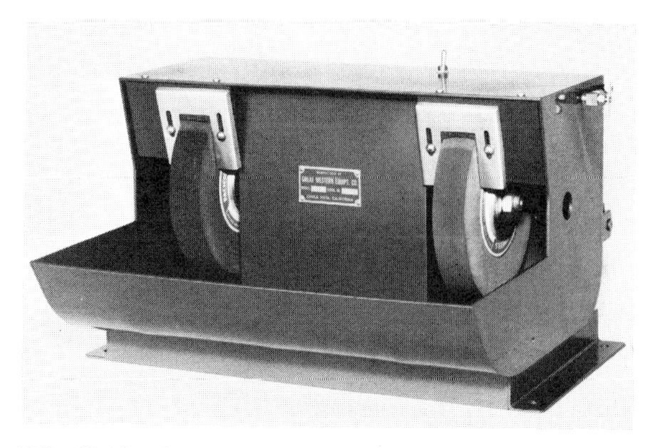

FIG. 12.16. Silicon carbide grinding wheels are used to rough out the shape of a gemstone. This process is called grinding.

In the grinding operation the piece may be hand-held, especially in the beginning phase of the shaping process. Later, in both cabochon cutting and preforming faceted gems, the piece is usually fastened to the end of a holder, a piece of hardwood dowel called a dop stick. The stone is fastened to the stick with dopping wax made of sealing wax and shellac (Fig. 12.15).

A typical grinding wheel such as may be found in both amateur and professional lapidary shops is shown in Figure 12.16. Figure 12.17 shows the large diameter sandstone wheels still used in the famous gem cutting center of Idar-Oberstein, West Germany.

Lapping and Sanding

The next step in the finishing of gemstones is lapping. This process consists of holding the stone against a horizontal metal lap to which abrasive grit

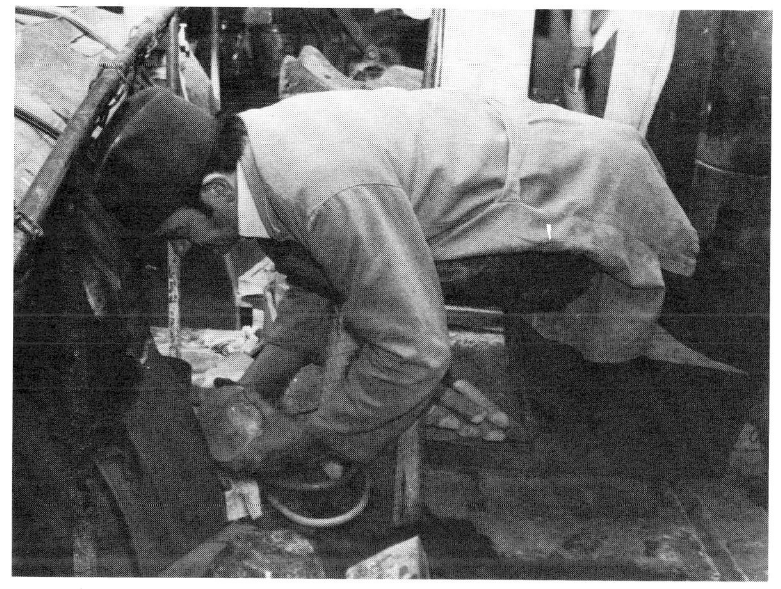

FIG. 12.17. In Idar-Oberstein, West Germany, some grinding is still done on large sandstone wheels. Here the cutter, lying in front of a sandstone wheel, is grinding a rock crystal sphere. The cutters lie on a wooden stool to be able to use both arms and legs to press large pieces against the wheel. All the forming is done by hand. Courtesy of Gerhard Becker, Idar-Oberstein, West Germany.

(usually silicon carbide or aluminum oxide, but sometimes diamond) is fed, with water as a lubricant. As in grinding, the operator begins with a coarse grit and proceeds to finer and finer grits, until the surface has a finely frosted finish and is ready for polishing. In the final stages of lapping a grit as fine as 1200 mesh may be used. The laps may be made of almost any metal, but the most commonly used are either mild steel or cast iron (Fig. 12.18).

Since the surface of a metal lap is essentially unyielding, lapping is used for flat surfaces. These may be the facets of a faceted gemstone, or large flat surfaces. If the gemstone has a curved surface such as a cabochon, it is prepared for polishing by grinding on either a curved or flexible surface. This process is sometimes called *sanding* because it is often done by means of sanding cloth, either wet or dry, covering a rotating disk or as a continuous belt, as in woodworking. It may also be done by applying abrasive with water to grooves in a wooden disk, to leather or canvas stretched over a yielding surface,

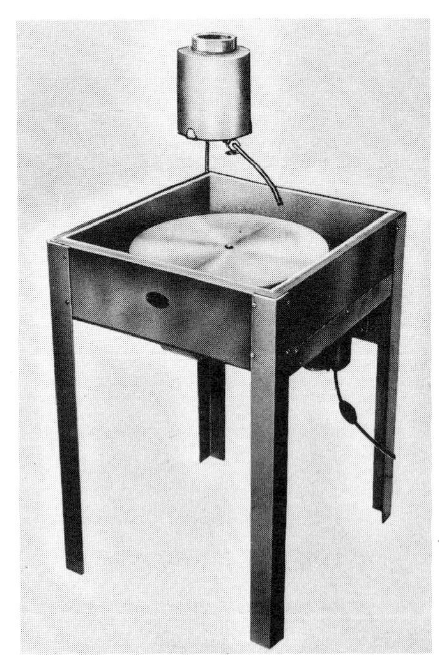

FIG. 12.18. Metal lap. Flat surfaces are prepared for polishing by holding the stone against the rotating lap to which abrasive grit is fed with water as a lubricant. Courtesy of the Covington Engineering Corp., Redlands, California.

FIG. 12.19. Belt sander used to prepare stones for polishing by grinding on a flexible surface. Courtesy of the Covington Engineering Corp., Redlands, California.

etc. Whatever method is used, the purpose is to remove the flat spots made by the grinding wheel and, by using progressively finer grits, prepare the surface for polishing. Figure 12.19 shows a belt sander.

Polishing

The final step in finishing a gemstone, polishing, is done by pressing the stone against a buff that has been charged with a polishing agent. Exactly what takes place during this process is not fully understood. Some believe it is just superfine grinding, until the scratches cannot be seen. G. Beilby proposed in the early 1900s that during polishing, a very thin layer (called the *Beilby layer*) has flowed over the scratches. Later work has indicated that no

Beilby layer forms on material of very high melting point, such as diamond, whereas on others it may form and immediately recrystallize, or it may form and remain amorphous.

Oxides of various metals are the most frequently used polishing agents. These may be iron oxide (rouge), chromium oxide, tin oxide, silicon oxide (tripoli), and others. Also used is diamond, in the size range 3200–6400 mesh.

The polisher used depends on the type of gemstone. Cabochons, flats, and spheres are polished on polishing buffs made of felt, leather, wood, or cloth. Faceted stones are polished on flat surfaces (laps), less yielding under pressure, made of metal (such as copper or tin), plastic, or wood.

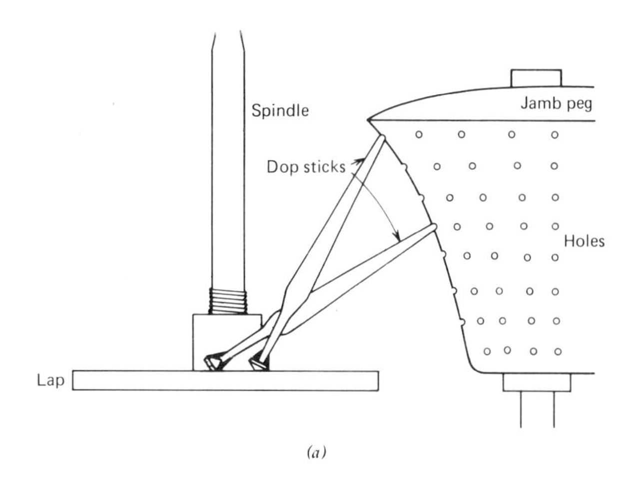

FIG. 12.20. (*a*) The jamb peg, a device for controlling the angle of cut when placing facets on a gem. Adapted from J. Sinkankas, 1962 *Gem Cutting*, D. Van Nostrand Co., Inc. (*b*) A lapidary using a jamb peg in cutting a faceted gem. Courtesy of Gerhard Becker, Idar-Oberstein, West Germany.

Cabochons

Cabochons are fashioned using the grinding, sanding, and polishing steps just described. The first step is usually to saw a slab of the desired thickness. The outline is then marked with a template and the stone is further roughed into shape by sawing. Sawing may be omitted in the case of small pieces of more valuable gem material such as opal and star sapphire. The stone is then brought into final dimensions by grinding, then lapped or sanded, and polished.

Faceted Gems

Faceted gems are cut by holding the preformed stone, mounted on a dop stick, against a lap until a flat spot, or facet, has been ground. After a facet is ground the stick is successively rotated and/or tilted until the shape of the stone is complete. The facets are polished by repeating the procedure with a polishing lap. All the facets on the top of the stone are ground and polished. The stone is then reversed on the dop stick and the bottom facets are ground and polished.

A simple device known as a jamb peg is used by most professional lapidaries to establish the angle at which the various sets of facets are cut (Fig. 12.20). The jamb peg has a large number of holes in it to hold the pointed end of the dop stick. This controls the angle of cut, but placing facets around the gem by rotating the dop stick is done entirely by hand. Obviously, this requires a high degree of skill and experience.

A mechanical "faceting head" (Fig. 12.21), designed primarily for amateur gem cutters but also

FIG. 12.21. A mechanical faceting head, also used for controlling the angle at which facets are cut. Courtesy of MDR Manufacturing Co., Inc., Culver, California.

used by professionals, holds the dop rigidly in precisely determined angular positions.

In Sri Lanka gems are still cut, as they were hundreds of years ago, by holding the stone against a hand-powered abrasive wheel (Fig. 12.22). This commonly results in a poorly cut stone that is usually recut before reaching the western market.

FIG. 12.22. Gemstone cutting in Sri Lanka (Ceylon) using a hand-powered abrasive wheel.

Carving and Engraving

Carving and engraving of gemstones is generally done by holding the piece in the hands against tools held in a horizontal arbor. Modern-day equipment is essentially the same as that used hundreds of years ago except for replacement of hand or foot power by electric motors. The tools used may be steel burrs or disks charged with diamond dust, small silicon carbide burrs and disks such as those used by dentists, or metal tools used with loose grit (Fig. 12.23). A flexible shaft grinder (Fig. 12.24) may also be used.

Tumbling

In recent years the polishing of irregularly shaped pieces of gem material by tumbling has become an important lapidary operation. Today millions of tumbled gems are produced each year by both professionals and amateurs. Tumbling is a method whereby baroque-shaped polished gems can be mass produced, and the gem rough need not be of the highest quality to produce attractive gems. The use of tumbled gems has greatly increased since strong epoxy-type cements have become available,

FIG. 12.23. Gem carving in Kofu, Japan.

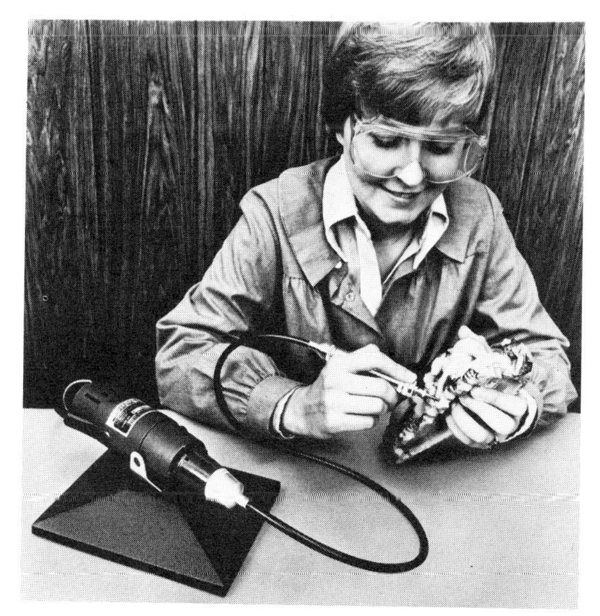

FIG. 12.24. A flexible shaft grinder being used to carve a gemstone. Courtesy of the Dremel Mfg. Division, Emerson Electric Co., Racine, Wisconsin.

because it is no longer necessary to groove or drill the tumbled gems to set them in jewelry.

The principle of gem tumbling is to place the irregular pieces of gem material, with abrasive and water, in a barrel or drum, which is rotated slowly. The resultant abrasive action slowly wears off rough corners, and if kept up long enough all the pieces of rough gemstone will be worn smooth. The process may take as long as a week. A coarse abrasive is used at first, then a finer abrasive, and finally polishing is done, which may take another several days to a week.

A wide variety of gem tumblers have been designed. Most have hexagonal barrels, although round barrels are also satisfactory. There are many commercially available tumblers, one of which is shown in Figure 12.25.

Other Operations

Other lapidary operations include making beads and spheres, drilling, and mosaic and inlay work. These operations are similar to, or modifications of, those just described.

FIG. 12.25. A tumbler for grinding and polishing irregular shaped pieces of gem material. Courtesy of the Covington Engineering Corp., Redlands, California.

Diamond Cutting

There are six basic steps in fashioning a diamond crystal (called "rough" in the trade) into a finished gem. They are marking, grooving, cleaving, sawing, girdling, and faceting. Not all these steps are necessarily used in fashioning every stone—it depends on the size, shape, and quality of the rough.

Marking is done by a planner, who examines each piece of rough to determine how it should be cut to yield the greatest value. He must determine whether it should be cut into a single large stone or several smaller ones, and the shape of each. To do this he must consider the size and shape of the rough, the number and location of imperfections, and the direction of the cleavage, "grain" (Fig. 12.26).

If it is decided to cleave the stone then the diamond crystal goes to the cleaver. Large diamonds are often preshaped by cleaving into pieces suitable for sawing. The cleaver draws a line showing the octahedral cleavage plane selected and cuts a small groove along that line using as a tool a sharp-edged piece of diamond set in a holder. He then mounts the diamond in a holder (dop) and places a blunt steel knife into the groove. The cleaver then strikes the back of the knife with a mallet, and—if all goes well—the diamond splits along the cleavage plane as planned. This is a risk-filled operation, and if the direction of cleavage has

FIG. 12.26. Diamond crystals marked for sawing. Courtesy of Diamond Information Center, New York.

not been accurately located the diamond can shatter instead of splitting cleanly (Fig. 12.27).

Most preliminary shaping of diamonds is done by sawing (Fig. 12.28). The sawing is done with a thin, rapidly revolving, diamond-charged phosphor–bronze disk. The disk is usually about 7.5 cm in diameter and 0.06 mm thick, and is run at a speed of about 4000 rpm. The edge of the saw is impregnated with diamond powder, and the diamond, with the proposed saw cut marked in ink, is placed in a clamp and lowered onto the edge of the saw. The crystal must be marked so that the sawing takes place across the cleavage. In the case of well-formed octahedral crystals the cut is usually made to produce two equal halves. The saw can cut through a 1 carat rough diamond in 4–8 hours.

The next process is girdling, also called bruting and rounding (Fig. 12.29). The diamond to be girdled is placed in the chuck of a lathe. While it turns in the lathe a second diamond, held in a dop on the end of a stick, is held against it, and the diamond is rounded into a cone shape. The diamond is then reversed in the lathe and a truncated double cone formed. Fancy-shaped stones are girdled by placing them off center in the lathe chuck. Diamond dust produced in the operation is carefully collected and used to charge diamond saws and laps.

The final step in completing the stone is faceting. The facets are ground and polished in a single operation, performed on a revolving horizontal lap, or "scaife" (Fig. 12.30). The lap is made of cast iron. Its surface is charged with diamond powder, and the speed of rotation is between 2400 and 3000 rpm. The stone is held either in soft solder in a dop or in a mechanical holder. The dop is clamped in a "tong" or "tang," which with its two feet on a bench and the diamond resting on the lap forms a tripod (Fig. 12.30).

If the finished gem is to be a standard round brilliant, the first polisher cuts the table facet, the eight main facets above the girdle, the eight main facets below the girdle, and the culet. The diamond then goes to the "brillianteer," who puts on the

FIG. 12.27. Diamond cleaving. Shown being cleaved is the Star of Sierra Leone, which weighed in the rough 968.9 carats. Courtesy of Harry Winston, Inc., New York.

FIG. 12.29. Girdling a diamond by placing it in a lathe and rounding it into a cone shape.

FIG. 12.28. (*a*) The Star of Sierra Leone being cut on a diamond saw. (*b*) A battery of diamond saws. Courtesy of Harry Winston, Inc., New York.

FIG. 12.30. Faceting a diamond. The facets are ground and polished on a lap, or 'scaife.'' The diamond may be mounted in a dop or held in a clamp. Courtesy of Harry Winston, Inc., New York.

remaining 41 facets. The operation of faceting is not always divided as above. Often a single craftsman finishes the entire stone, especially if its shape is other than a round brilliant.

In diamond there are significant variations in hardness in different directions. The optimum directions for polishing are those parallel to a crystallographic axis. With reference to Figure 12.31, the faces of the cube (*h*) are parallel to two crystallographic axes. On each cube face there are two optimum polishing directions, and facets parallel to the cube faces of a diamond crystal are easy to polish. Faces of the dodecahedron (*d*) are parallel to one crystallographic axis and each dodecahedron face has one optimum polishing direction. Octahedron faces (*o*) are equally inclined to three crystallographic axes and these faces are the most difficult to polish. Indeed, it is almost impossible to polish a facet parallel to an octahedron plane, and almost impossible to saw a diamond if the plane of the cut varies more than a few degrees from that of a cube face.

Diamonds are cut so that the table is approximately parallel to a face of the cube, octahedron, or dodecahedron. In the diamond cutting industry these are referred to respectively as the four-, three-, and two-point faces and the corresponding diamonds are referred to as four-point, three-point, and two-point stones. Figure 12.32 shows the optimum polishing directions for the crown and pavilion facets for each.

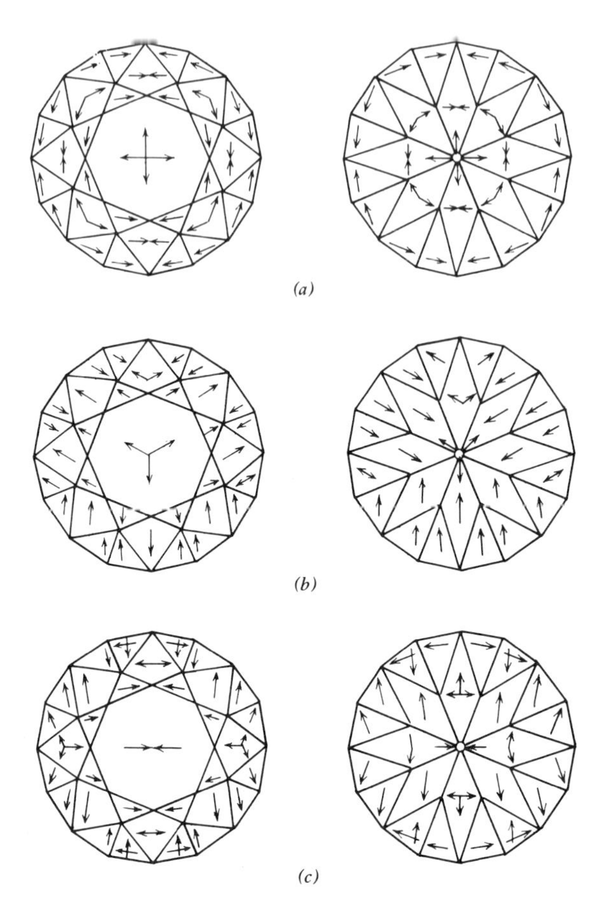

FIG. 12.32. The optimum polishing directions for the crown and pavilion facets of a brilliant cut diamond. (*a*) Four-point stones. (*b*) Three-point stones. (*c*) Two-point stones. Adapted with permission from E. H. Kraus and C. B. Slawson, 1947, *Gems and Gem Materials*, McGraw-Hill Book Co., Inc., New York.

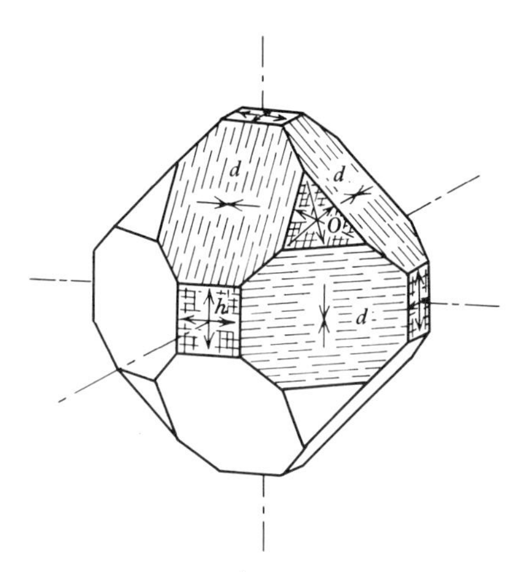

FIG. 12.31. The optimum polishing directions for diamond are those parallel to a crystallographic axis. Facets parallel to a cube face are easiest to polish. Adapted with permission from E. H. Kraus and C. B. Slawson, 1947, *Gems and Gem Materials*, McGraw-Hill Book Co., Inc., New York.

In actual practice the diamond cutter may deviate considerably from these orientations because of the desire to obtain as large a stone as possible and for other reasons. Facets are usually not parallel to a crystallographic axis, but the most favorable direction for polishing is the one that most nearly approximates parallelism to an axis. Hence in one direction across a facet, say from right to left, the rate of polishing may be more rapid than from left to right. In theory the direction should be toward a possible face of a dodecahedron. In practice the sense of direction is determined by trial, and the expertness of the diamond polisher depends to a great extent on his ability to recognize these properties of diamond and apply them quickly and accurately.

References and Suggested Reading

Kennedy, G. S. et al., 1967. *The Fundamentals of Gemstone Carving.* Edited by Parry D. Kraus. The Lapidary Journal, San Diego.

Kraus, E. H. and C. B. Slawson, 1947. *Gems and Gem Materials.* McGraw-Hill Book Co., Inc., New York. Chapter 8.

Parsons, C. J. and E. J. Soukup, 1957. *Gem Materials Data Book.* Gemac Corp., Mentone, Calif.

Quick, L. and H. Leiper, 1959. *How to Cut and Polish Gemstones.* Chilton Co., Philadelphia.

Sinkankas, J., 1962. *Gem Cutting.* D. Van Nostrand Co., Inc., Princeton, N. J.

Sperisan, F. J., 1961. *The Art of the Lapidary.* Bruce Publishing Co., Milwaukee.

13 DESCRIPTIVE GEMOLOGY

Introduction

This chapter describes the individual gem minerals as well as those organic substances and synthetic materials used in jewelry and for ornamental purposes. Compared to the nearly 2500 known mineral species, the number of minerals considered here is small. Nevertheless, included are some that are extremely rare, found at only one or two localities, and other more common minerals whose gem qualifications are marginal.

On the following pages the naturally occurring gems are divided into three groups: (1) important gemstones, (2) less important gemstones, and (3) other gems and ornamental stones. Following these are organic gems and synthetic gems.

For easy reference, the presentation of data in the descriptions of naturally occurring gems in groups 1 and 2 and of organic gems follows a common scheme. After a brief introductory statement, the headings and the data given under each are as follows:

Crystallography.

The crystal system, habit, common forms, and twinning.

Physical Properties.

Cleavage, **H** (hardness), **S.G.** (specific gravity), *luster* and transparency, and *color.*

Optical Properties.

The designation of isotropic, uniaxial, or biaxial and the optic sign for uniaxial and biaxial; refractive index (or indices), birefringence, pleochroism, and dispersion.

Chemical Properties.

The chemical formula and the elements that may substitute for those in the formula.

Luminescence and Absorption Spectra.

The reaction to ultraviolet light and the major absorption lines are given if characteristic.

Diagnostic Features.

The outstanding properties and tests that aid in identifying a gem and distinguishing it from others.

Occurrence.

A brief statement of the mode of occurrence, associations, and principal localities where the gem is found.

Name.

The origin of the name.

IMPORTANT GEMSTONES

The important gemstones are considered in order of decreasing importance but admittedly the order is arbitrary, for each gemologist has his own ordering which may vary from time to time. According to one major gem dealer the order of importance of individual gems in the trade today is as given below. It will be noted that these are not according to mineral species for in the list we find pearl and two varieties each of beryl, corundum, and chrysoberyl.

MAJOR GEMS IN DECREASING ORDER OF IMPORTANCE

Diamond	Jadeite	Tourmaline
Emerald	Cat's eye (chrysoberyl)	Quartz
Sapphire (blue)	Alexandrite	Nephrite
Ruby	Aquarmarine	Turquoise
Pearl	Topaz	Garnet
Opal		

Since we consider the group of important gemstones on the basis of mineral species, the arrangement is necessarily different from the above. With several gem minerals added, and pearl omitted since it is included with organic gems, the list is revised as follows:

IMPORTANT GEMSTONES

Diamond	Chrysoberyl	Garnet
Beryl	Topaz	Zircon
Corundum	Tourmaline	Peridot
Opal	Quartz	Spinel
Jade	Turquoise	Feldspar

Diamond

Diamond is generally acknowledged as the most precious of gems. One reason for this is that it is the hardest known natural material. It also has a very high refractive index and dispersion, properties that account for the extraordinary brilliance and fire of a well-cut diamond. Because of its unsurpassed hardness diamond also has many important industrial uses.

There are four varieties of diamond, gem diamond and three industrial types. Gem diamonds are well crystallized, transparent stones. *Bort,* widely used in industry, is poorly crystallized, gray to brown, and translucent to opaque. *Ballas* is the name given to spherical aggregates of many small diamond crystals. *Carbonado* is an opaque, black or gray, tough compact variety of industrial diamond.

Diamond is the only gem mineral that is mined by highly mechanized, large-scale methods. Annual production amounts to more than 50 million carats, of which only about 20% are of gem quality (Fig. 13.1).

De Beers Consolidated Mines Limited, through its Central Selling Organization (CSO), markets about 80% of the world's diamond production. De Beers operates mines in southern Africa, and through the Diamond Corporation, a wholly owned subsidiary, contracts with other producers for the purchase of their production. These, and other diamonds purchased on the open market in Africa and elsewhere, are sold through the Central Selling Organization.

Crystallography.

Isometric. Crystals are usually octahedral in appearance. Flattened and elongated crystals are common, and curved faces are frequently observed (Figs. 13.2*a* and *b*). Twins of the spinel type, usually flattened parallel to the twin plane (called macles), are common (Fig. 13.2*c*). All of these types are present in the diamond rough of Figure 13.3. Rarely massive.

Two types of diamond are known, called type I and type II. The electrical conductivity of type I is poor, but of type II is good. The two types also differ in their infrared and ultraviolet characteristics. Two

FIG. 13.1. Sorting diamond rough. Rough diamonds are sorted by shape, quality, color, and size. Each diamond can be sorted into one of over 2000 different categories. Courtesy of Diamond Information Center, New York.

varieties of type II have been noted, type IIa and type IIb. Most blue diamonds are type IIb.

Physical Properties.

Cleavage perfect octahedral (four directions, referred to as grain by diamond cutters). **H** 10. Actually the hardness difference between corundum (**H** 9) and diamond is at least three times greater than between topaz and corundum (see Fig. 5.5). The hardness of diamond also varies with direction (see Chapter 10). **S.G.** 3.52. *Luster* adamantine; uncut crystals have a characteristic greasy appearance. Transparent to opaque.

Color usually colorless or pale yellow; also shades of red, orange, green, blue, and brown to black. Deeper shades of color are rare and such stones are termed "fancy." Colorless or pale blue stones are considered best, but these are rare; most gem diamonds are tinged with yellow.

Optical Properties.

Isotropic. Anomalous double refraction, caused by internal strain, is often seen. **R.I.** 2.42; disp. 0.044. Both these properties are high and account for diamond's brilliance and fire.

Chemical Composition.

Pure carbon, C. Will burn in an atmosphere of oxygen, and in air if heated to a very high temperature. A trace of nitrogen is sometimes present and is the cause of the color of some yellow diamonds.

The arrangement of the carbon atoms in the diamond structure is shown in Figure 13.4. The atoms are in tetrahedral coordination and each atom is held to four others by strong covalent bonds. The structure of the carbon polymorph, graphite, is shown in Figure 13.5. In graphite the carbon atoms are in sheets, closely packed and strongly bonded; but the sheets are held to each other by the weak

(a)

(b)

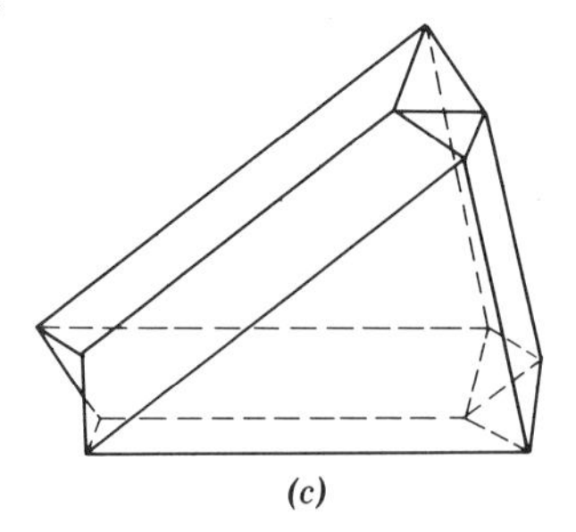

(c)

FIG. 13.2. Diamond crystals. (*a*) Etched octahedron with striated faces of hexoctahedron. (*b*) Curved faces of hexoctahedron. (*c*) Twinned octahedron flattened parallel to an octahedron face.

FIG. 13.3. Diamonds in the rough. Courtesy of Diamond Information Center, New York.

FIG. 13.5. Model showing the graphite structure. Sheets of strongly bonded carbon atoms are held to other sheets by the weak van der Waals' bond.

van der Waals' bond. Diamond is hard, lustrous, and transparent; graphite is soft, dull, and opaque.

Luminescence and Absorption Spectra.

Many, but not all, diamonds fluoresce under both long- and shortwave u.v., with the effect generally stronger under longwave. The colors seen may be blue, green, yellow, and occasionally reddish. According to Liddicoat (1975) about 15% of gem diamonds fluoresce strongly under longwave u.v. A type of diamond known as a "Premier" fluoresces a strong, light blue, in some cases with sufficient intensity that the fluorescence may be observed in daylight. Many diamonds show phosphorescence as well as fluorescence.

Yellow diamonds usually show a number of lines in their absorption spectrum, the strongest being in the deep violet at 4155 Å. Others are at 4530, 4660, and 4780 Å. The absorption spectrum of irradiated diamonds is discussed below under treated stones.

FIG. 13.4. Model Showing the structure of diamond. Each carbon atom is bound to four carbon neighbors by strong covalent bonds giving a close-knit, strongly bonded structure.

Inclusions.

Crystal inclusions in diamond are common, and may be diamond, peridot, pyrope garnet, chrome diopside, and others (See Plate III and Fig. 13.6). An included diamond crystal oriented differently from the host crystal is called a *knot* by diamond cutters. Also present may be black carbon inclusions, al-

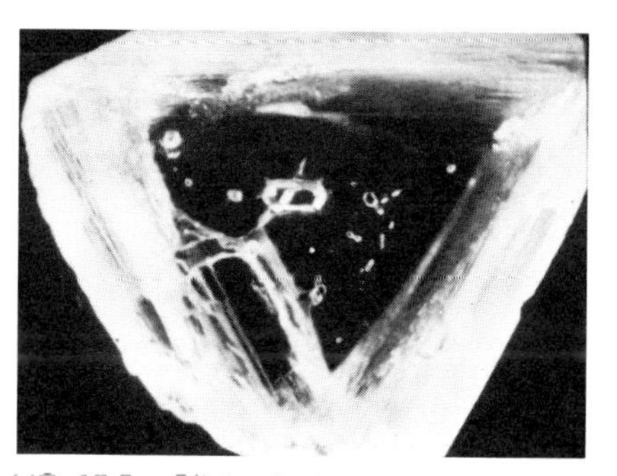

FIG. 13.6. Olivine inclusions in diamond.

though many inclusions that appear black are actually cleavages seen in reflected light. *Clouds* of submicroscopic gas-filled spaces having a wispy or cottony appearance are also common.

Diagnostic Features.

The high luster, single refraction, **S.G.** 3.52, and extreme hardness distinguish diamond from natural and synthetic substitutes. Examination under magnification often shows incipient cleavage cracks, especially around the girdle. *Naturals,* or portions of the original crystal surface often left on the girdle, are a characteristic feature (Fig. 13.7). The extreme sharpness of the facet junctions and total absence of scratching or wear marks are characteristic of diamond.

Synthetic and Treated Stones.

As described in Chapter 8, synthetic gem quality diamonds have been made in the laboratory, but at present there is no commercial production. The gem quality diamonds made by the General Electric Company in 1970 (Fig. 13.8) may be distinguished from natural stones by the following characteristics:

FIG. 13.7. Cut diamonds showing "naturals," that is, portions of the original crystal surfaces on the girdle. Courtesy of the Gemological Institute of America.

(1) the synthetics show white dustlike inclusions not seen in natural stones; (2) under shortwave u.v. all G.E. diamond crystals except yellow fluoresce strongly in yellow and green colors, and all but the yellow stones phosphoresce; (3) no absorption lines or bands are seen in the stones; and (4) all but the yellow stones are semiconductors, whereas among natural stones only blue ones are conductive.

The detection of treated diamonds presents a serious problem, especially those in which the color has been altered by radiation treatment. For many years green diamonds occasionally reached the market colored by exposure to radiation from radium compounds (a practice that has been discontinued). Such an irradiated diamond retains a dangerously high degree of radioactivity which can be detected by a Geiger counter or by placing the stone on a photographic film and observing the resulting autoradiograph (Fig. 13.9). Diamonds treated by cyclotron or nuclear reactor have no dangerous residual radioactivity.

Short exposure to high-energy particles in a cyclotron, or to neutrons in a nuclear reactor, induces a green color which, with continued exposure, will turn black. The color is caused by electrons being displaced from their normal positions in the structure, forming color centers (see *color* in Chapter 6).

Heat treatment of green cyclotron- or nuclear-reactor-treated diamonds changes the color to yellow or brown. Because of shallow penetration of particles in a cyclotron, the color in a cyclotron-treated stone is a thin skin and may show color zoning. Diamonds that have been irradiated by neutrons in a nuclear reactor have color throughout. Other colors are sometimes produced. Some diamonds have been colored blue by high-energy electron bombardment.

The distinction between treated and natural colored diamonds, especially the yellow or "canary" stones, is difficult. Following is a summary of tests developed by Robert Crowningshield of the Gemological Institute of America for detection of irradiated diamonds.

Neutron-bombarded green diamonds pose the most difficult problem. They have color throughout and cannot be detected by color zoning. Neither do

FIG. 13.8. Gem quality synthetic diamond crystals made at the General Electric Research and Development Center. The black material is graphite, the substance from which the diamonds were made. Courtesy of General Electric Co., Schnectady, New York.

they show lines in their absorption spectrum that are characteristic of yellow treated stones. The GIA has observed that the only known natural green stones they have seen have "naturals" on their girdles, and has concluded that any green diamond without a natural is suspect.

Irradiated diamonds of yellow color are distinguished by the presence of a narrow absorption band at 5920 Å, as well as bands at 4980 and 5040 Å. The line at 5920 Å, considered to be the key to the test, is often difficult to detect, and may temporarily disappear if the stone under examination is heated too strongly by the spectroscope lamp. Large stones may not show the 5920 Å line, but presence of the lines at 4980 and 5040 Å is considered to be proof that the stone has been treated.

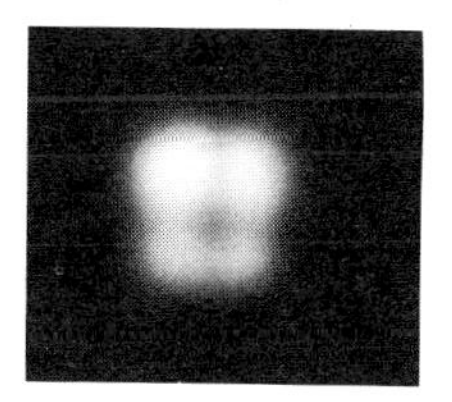

FIG. 13.9. Autoradiograph of a 10 carat diamond colored green by exposure to radiation from a radium compound at least 70 years earlier. The dark cross results from two strips of thin lead foil placed between the diamond and the photographic film. That the lead was partially penetrated by the radiation illustrates the high retention of radioactivity. Exposure time 18 hours.

Another method of treating diamonds to improve their color, especially to mask a pale yellow color, is to coat the stone with a fluoride coating of the type used for coating camera lenses. The coating is applied to the pavilion facets, or sometimes only at the girdle. It may be detected visually if thick enough to produce a characteristic iridescence. Such a coating is insoluble in ordinary solvents, and if it cannot be seen under the microscope the only proof is to remove it by boiling in concentrated sulfuric acid.

Simulation.

Diamond is simulated by a variety of materials, both natural and synthetic, including zircon, synthetic white sapphire and spinel, synthetic rutile, strontium titanate, YAG, GGG, cubic zirconia, doublets, and glass. Of these zircon, synthetic sapphire, and synthetic rutile are doubly refractive. The remainder are singly refractive, like diamond, but are readily distinguished by standard tests.

Factors Affecting Value.

The value of a cut diamond is determined by its color, clarity, cut, and carat weight, sometimes referred to as the four "C's."

The cut, or "make" of a diamond is extremely important. A finely proportioned and finished diamond can be worth as much as 50% more than one of similar weight and color that is poorly cut. In "make", the precision with which the facets are placed, the degree of polish, and proportions are all important. In 1919 Marcel Tolkowsky, a European cutter, worked out the proportions for a brilliant cut diamond to yield the best compromise between high degree of brilliancy and fire. These proportions, somewhat modified, are the now widely used "American" cut. They are given in Figure 13.10.

Color grading of diamonds is the process of placing them on a scale from colorless to yellow. Minor differences, not detectable to the untrained eye, are important. Several color grading systems are in use, and in Figure 13.11 three widely used

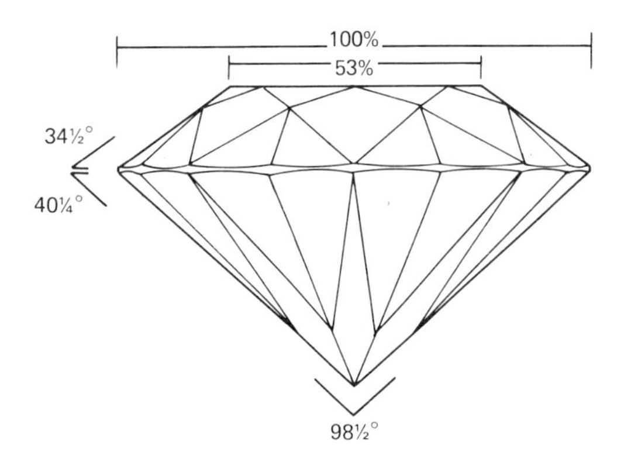

FIG. 13.10. Proportions of an American cut diamond.

systems are compared. Terms such as river, top Wesselton, and cape are derived from sources in South Africa; for example "river" came to mean exceptionally transparent and colorless stones because the river, or alluvial, diggings produced many stones of this quality. The other systems shown in Figure 13.11 are those adopted by the Gemological Institute of America and the American Gem Society.

Many companies have adopted their own nomenclature systems, using terms such as white, commercially white, AA, A, B, and C, or 0, 1, 2, 3, etc. Unfortunately, none of these systems may be directly compared with another. The term "blue-white" is still used, usually to misleadingly describe a diamond tinged with yellow.

Color grading diamonds may be visual or instrumental. The visual method is the oldest and most widely used. Unfortunately, it is subject to errors of several types, such as lighting conditions and training and experience. Some of these errors can be minimized using a set of master stones that have been instrument-graded for visual comparison (Fig. 13.12).

The instrument most widely used for color grading is the filter-type colorimeter, by means of which the degree of yellowness of a diamond is measured quantitatively (Fig. 13.13). Here, too, there are many problems and research is continuing to improve color grading methods, especially to develop a method using a spectrophotometer, a much more precise and sophisticated instrument than the colorimeter.

Clarity is determined by the presence or absence of blemishes and flaws. Some segments of the diamond trade consider only internal flaws when determining clarity. Others consider that external blemishes, such as scratches, nicks, and naturals are detrimental to the "clarity" of a stone. The Federal Trade Commission ruling on this point is as follows:

(a) It is unfair trade practice to use the word "perfect" or any other word, expression, or representation of similar import, as descriptive of any diamond that discloses flaws, cracks, carbon spots, clouds, or other blemishes or imperfections of any sort when examined in normal daylight, or its equivalent, by a trained eye under a ten-power, corrected diamond eye loupe or equal magnifier.

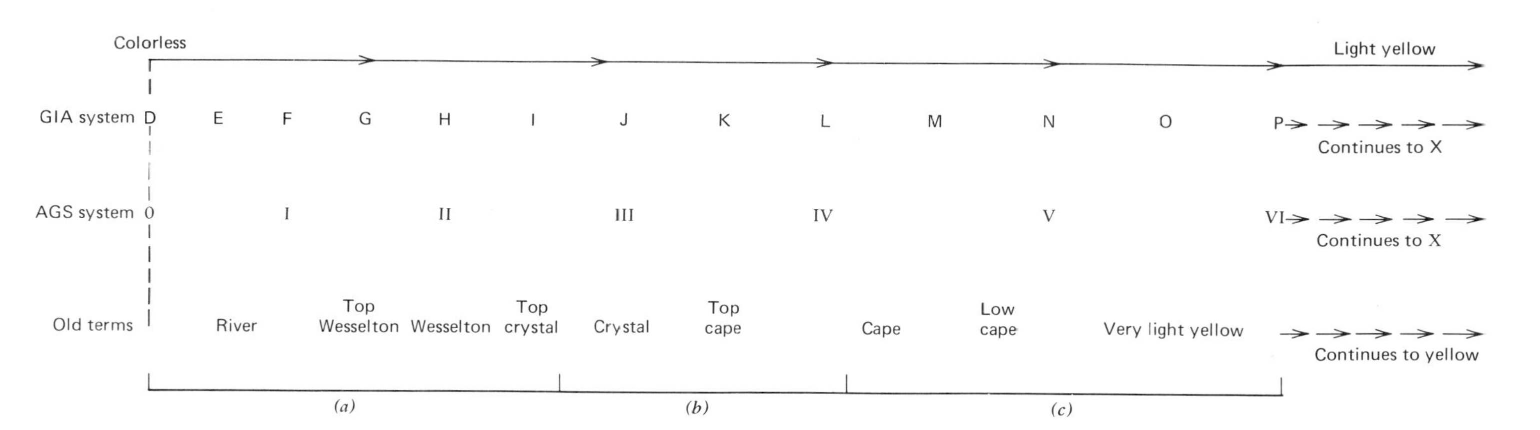

FIG. 13.11. Comparison of three systems currently used for color grading diamonds. From *Jewelers' Circular—Keystone,* 1966. (*a*) Stones in these grades will "Face up" colorless (i.e., slight traces of color will not be apparent in mounted stones except to the trained eye). (*b*) Small stones in this range will "face up" colorless when mounted but larger stones will be tinted. (*c*) Mounted stones in these grades will display a yellowish tint even to the untrained eye.

FIG. 13.12. The GIA Diamondlite for color grading diamonds by comparing them with master stones whose colors are known. Courtesy of the Gemological Institute of America.

There are many systems of nomenclature in use for grading a diamond for clarity. Most use descriptive terms such as flawless (equals perfect), imperfect, and terms for the intermediate grades such as VVS, meaning very very slightly imperfect. The scale used by the Gemological Institute of America is shown in Figure 13.14.

Determination of the clarity of a diamond is much less subjective than that of color. It is best done using a wide-field binocular microscope with dark field illumination.

Occurrence.

The primary source rock of diamond is kimberlite, a type of peridotite. The diamonds were formed at a depth of greater than 150 miles in the earth's crust and forced to or near the surface in pipelike bodies with a roughly circular or elliptical cross section (Fig. 13.15). Erosion of these kimberlite pipes has

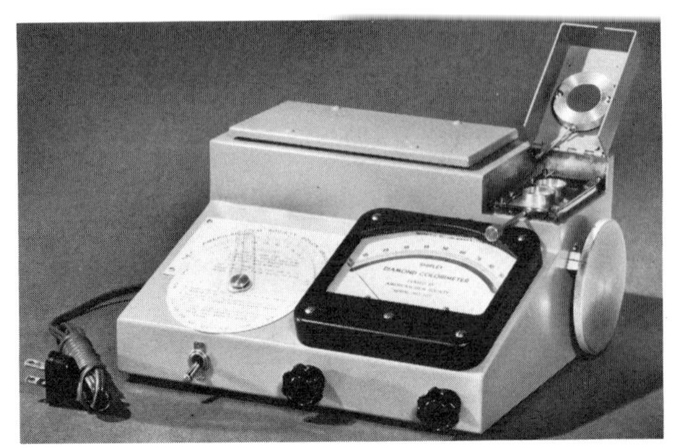

FIG. 13.13. The American Gem Society colorimeter for color grading diamonds. Courtesy of American Gem Society.

resulted in alluvial diamond deposits in river gravels and in some cases along the seacoast (Fig. 13.16).

All the diamonds known in ancient times came from alluvial deposits in India. Diamonds were discovered in Brazil in 1725, also in alluvial deposits, and that country was the chief source of the mineral for about 150 years. At present Brazilian production is only about 250,000 carats/year. The most important deposits are in the states of Minas Gerais and Bahia.

The first African diamonds were discovered in 1866 in the gravels of the Vaal River, South Africa. In 1871 diamonds were discovered *in situ* in kimberlite pipes near the present city of Kimberley. The diamonds in this district were first found in altered kimberlite called "yellow ground." As the pipes were worked to greater depth the diamonds were found in fresh kimberlite called "blue ground" (See Plate II for diamond in Matrix). The principal mines in the Kimberley district are the Kimberley, DuToitspan, De Beers, Wesselton, and Bultfontein. The mines were first worked as open pits and then, as they became deeper, by underground mining methods. The most famous, the Kimberley mine (Fig.

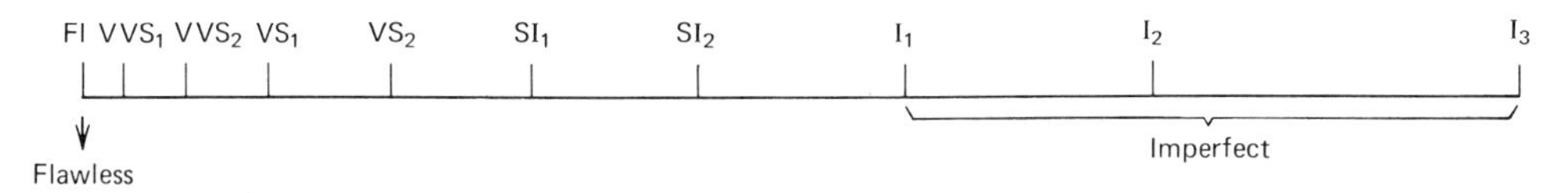

FIG. 13.14. The GIA system of nomenclature used for grading diamonds in terms of their inclusions. From left to right: flawless, two grades of "very very slightly included" (VVS_1 and VVS_2), two grades of "very slightly included" (SI_1 and SI_2), and three grades of imperfect (I_1, I_2, and I_3).

13.17) was worked as an open pit to a depth of 400–500 ft, and eventually to a depth of 3500 ft before it was closed.

In following years diamond deposits were discovered in other parts of Africa, principally in Zaire, Ghana, Sierra Leone, Angola, Tanzania, and South-West Africa. Most recently, important diamond mines have been opened in Botswana and Lesotho. For many years, and until recently, the continent of Africa produced 98% of the world's diamonds. This figure has now dropped to about 90% with the rapid development of deposits in the U.S.S.R. Elsewhere in the world, minor diamond production comes from Venezuela, Brazil, Guyana, and Indonesia.

Diamonds have been found sparingly in various parts of the United States. Small stones have occasionally been found along the eastern slope of the Appalachian mountains from Virginia to Georgia. The largest of these are the "Dewey" diamond, 23.75 carats, found in 1855 at Manchester, Virginia, and the "Punch Jones" diamond, 34.46 carats, found in 1928 in Peterstown, West Virginia. In 1969 or 1970 an 18.2 carat crystal was found in Princeton, Louisiana. Diamonds have also been reported from California and Idaho, and in the glacial drift south of the Great Lakes, in Wisconsin, Michigan, Illinois, Indiana, and Ohio. In 1906 a diamondiferous kimberlite pipe was discovered in the United States near Murfreesboro, Arkansas. This pipe has been mined sporadically but is at present unproductive. Now as the Crater of Diamonds State Park, it is open to the public, who for a fee are allowed to search for diamonds. The largest diamond ever found in the United States is the "Uncle Sam," 40.23 carats, found in 1924 at Murfreesboro, Arkansas.

FIG. 13.16. Diamond mining at the Consolidated Diamond Mines of South-West Africa located along the Atlantic coast north of the Orange River. (*a*) Removal of sand. (*b*) Final cleanup after reaching bedrock.

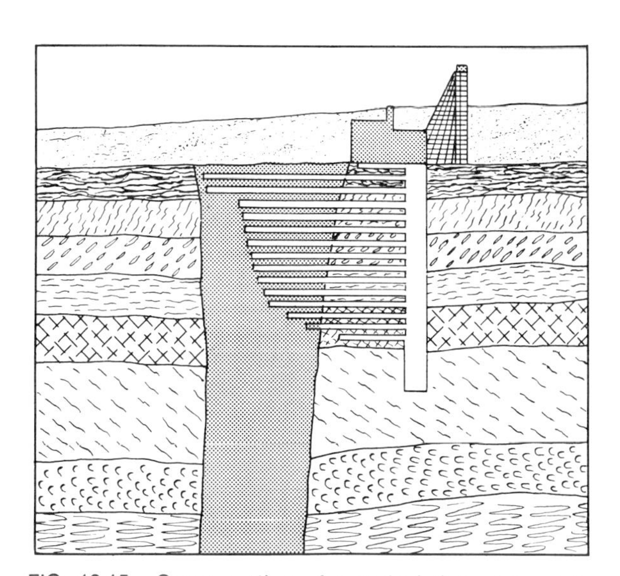

FIG. 13.15. Cross section of a typical diamond-bearing kimberlite pipe. After open pit mining becomes impractical deep mining is conducted from a vertical shaft in barren rock outside the pipe.

Name.

The name diamond is a corruption of the Greek word *adamas,* meaning invincible.

FIG. 13.17. The "Big Hole" at Kimberley. This crater shows the shape of the diamond-bearing kimberlite pipe that was one of the first diamond mines in South Africa. It is 1500 ft across, 3600 ft deep, and is filled with water that is now within 500 ft of ground level. Kimberley was mined out in 1914, but in 40 years of operation it produced 14,504,567 carats of diamonds. Courtesy of Diamond Information Center, New York.

Beryl

Beryl is a mineral with several gem varieties covering a wide range of color, including emerald, one of the most valued of all gems. Beryl is also the most abundant mineral containing beryllium, and is the chief source of this rare light element (about two-thirds the weight of aluminum) with many important industrial uses.

Crystallography.

Hexagonal. Characteristically in vertically striated prismatic crystals with the first-order prism and basal pinacoid the only forms (Fig. 13.18*a*). Some gem crystals that have grown into open spaces may show faces of several additional forms (Fig. 13.18*b*). Beryl containing cesium is often in tabular crystals flattened on the basal pinacoid (Fig. 13.19). Crystals of common beryl are frequently large; at Albany, Maine a tapering crystal 27 ft long weighed more than 25 tons.

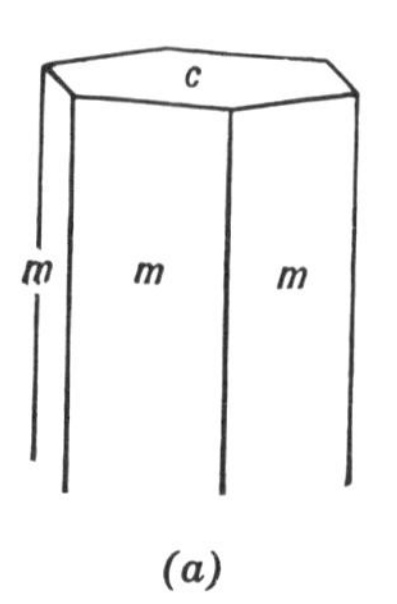

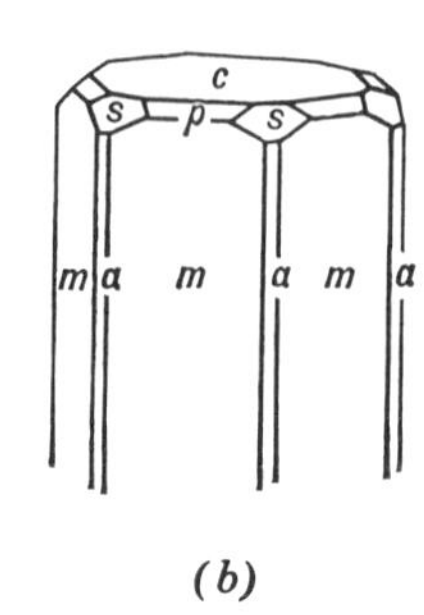

FIG. 13.18. Beryl crystals. The elongated habit is characteristic of most beryl crystals.

Physical Properties.

Cleavage imperfect parallel to the basal pinacoid; **H** $7\frac{1}{2}$–8. **S.G.** 2.67–2.82. *Luster* vitreous. Transparent to translucent.

Color commonly bluish-green or light yellow; also deep emerald-green, golden yellow, pink, white, and colorless. Color serves as the basis of several gem varieties of beryl. They are *emerald,* deep green; *aquamarine,* blue to sea green; *golden beryl,* golden yellow (called *heliodor* in South-West Africa); *morganite,* pale pink to deep rose and pink-orange, and *goshenite,* colorless. Some of these colors also occur as cat's eye gems, especially aquamarine and golden beryl (see Plate VII).

Optical Properties.

Uniaxial (-). The refractive index of beryl varies depending on minor variations in chemical composition. The average **R.I.** for beryls of various colors is given below.

	R.I.			
	ω	ϵ	Biref.	S.G.
Aquamarine	1.580	1.575	0.005	2.71
Emerald	1.583	1.577	0.006	2.72
Yellow	1.575	1.570	0.005	2.70
Pink to red	1.594	1.585	0.009	2.82

Dispersion for all beryls is low, about 0.014. Deeper-colored beryls have distinct pleochroism ($\epsilon > \omega$): *emerald,* ω yellow-green, ϵ blue-green; *aquamarine,* colorless and light to darker blue; *morganite,* light red and violet-red.

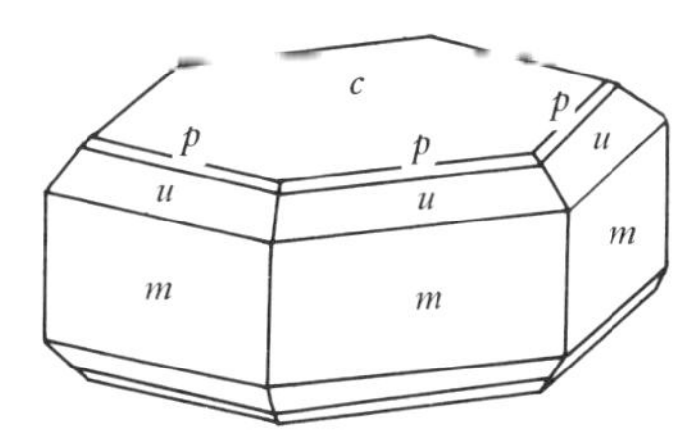

FIG. 13.19. Beryl crystal flattened on the basal pinacoid, typical of cesium beryl.

Chemical Composition.

Beryllium aluminum silicate, $Be_3Al_2Si_6O_{18}$. The structure of beryl projected on the basal pinacoid is shown in Figure 13.20. The large, nearly circular openings represent channelways that extend parallel to the c axis and provide space for large ions to enter. In pink, rose, and peach-colored beryl cesium is housed in these channelways. Its presence causes an increase in refractive index and specific gravity. The color of emerald is generally attributed to the presence of chromium but trace amounts of vanadium may be partly responsible.

Luminescence and Absorption Spectra.

Of the gem varieties only emerald shows fluorescence under ultraviolet light, from very weak to sometimes strong red. All beryl of other colors is essentially nonfluorescent. Most synthetic emerald fluoresces a strong dull red.

Similarly only emerald has a distinctive absorption spectrum, that characteristic of chromium-bearing minerals. There is a difference in the spectrum for the ordinary and extraordinary rays. For the ordinary ray there are two lines in the red at 6830 and 6800 Å, a third at 6370 Å, and a line in the blue at 4775 Å. For the extraordinary ray the doublet at

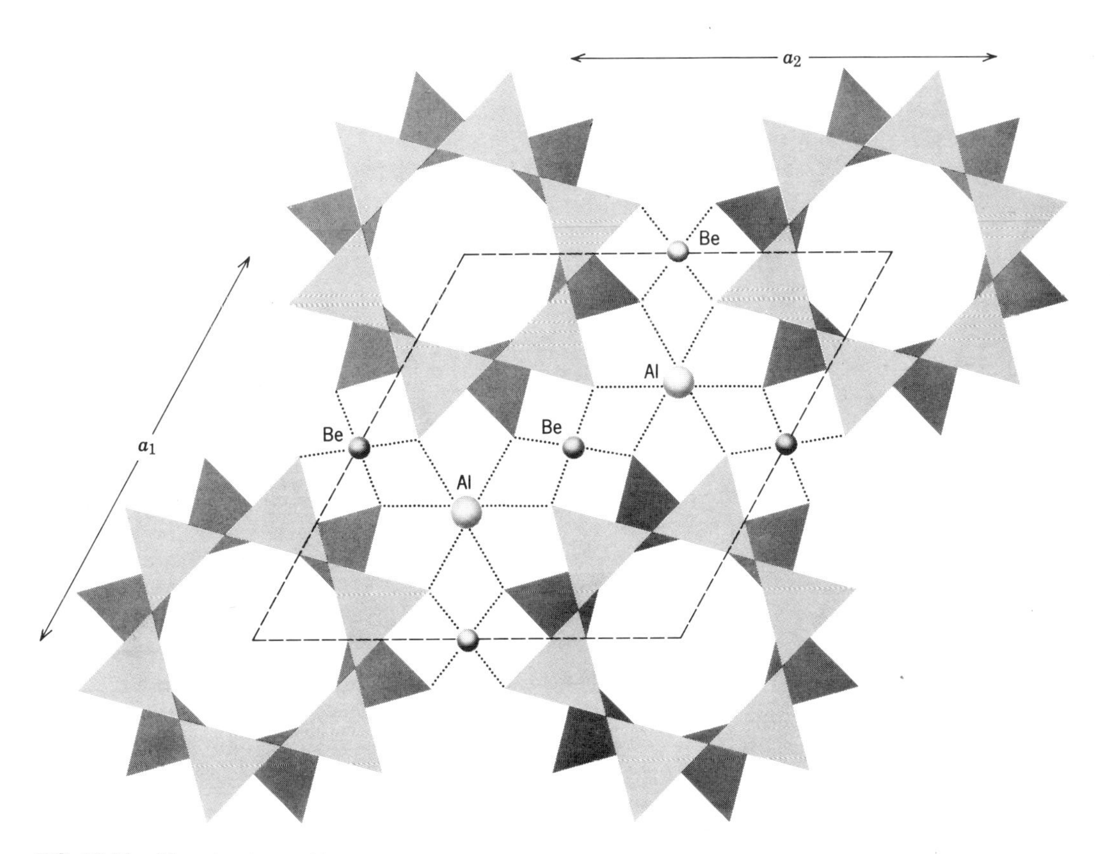

FIG. 13.20. The structure of beryl projected on the basal pinacoid. Dashed lines outline the unit cell.

6830 and 6800 Å is present but the 6370 Å line is missing, and there are no lines in the blue.

Inclusions.

Emerald contains both crystal inclusions and three-phase inclusions (irregular cavities filled with solid, liquid, and gas) (Fig. 9.1). Crystal inclusions of pyrite and calcite are present in emeralds from Colombia, and emeralds from some other localities have needle-shaped inclusions of tremolite. A badly fractured appearance is common in emerald. Inclusions in synthetic emerald are important aids in identification and are described under synthetic emerald (see *synthetic gems* below). Other gem varieties of beryl are relatively free of inclusions.

Diagnostic Features.

Beryl, because of its wide range of color, may be confused with more gems than any other species. However, standard tests, especially refractive index and specific gravity, permit easy identification.

Colombian and Siberian emerald when viewed through the Chelsea filter show a red color not present in similar appearing glass, sapphire, and most tourmaline. However, the test should be used with caution for, when observed through the filter, synthetic emerald appears deep red, and some natural emerald, on the other hand, shows no red.

Synthesis.

Synthetic emerald is being made at present by different processes, such as the flux growth method, by Chatham and Gilson, and the hydrothermal method by Linde. The Lechtleitner process, consisting of a hydrothermal overgrowth of emerald on colorless beryl, is also used.

The properties of the synthetics made by the processes differ somewhat from one another, and from natural emerald, as do the inclusions. To identify synthetic emeralds it is necessary to determine **R.I., S.G.,** and fluorescence, and to examine the inclusions under magnification (see synthetic emerald under *Synthetic Gems* below, and Plate XII).

Occurrence.

Beryl occurs mainly in pegmatites associated with granitic rocks. Most of this beryl, firmly embedded in the rock, is of nongem quality; the gem material is usually found as crystals growing into cavities within the pegmatite. As emerald it is also found in mica schists and gneisses, and in Colombia is associated with limestone.

High-quality aquamarine and morganite come from a number of localities in the vicinity of Governador Valadares and Teófilo Otoni, Minas Gerais, Brazil. Other noted localities for aquamarine, green beryl, golden beryl, and morganite are in Madagascar, South-West Africa, and in various places in the U.S.S.R.

In the United States fine aquamarine comes from Mt. Antero, Colorado, as well as various localities in Maine, New Hampshire, Massachusetts, Connecticut, and North Carolina. Golden beryl has been found in Maine, Connecticut, North Carolina, and Pennsylvania. The finest morganite comes from Pala, Hiriart Hill, and Mesa Grande, San Diego County, California. In recent years these deposits have been reopened, and the finest morganite yet found was taken from the White Queen claim on Hiriart Hill.

The most notable localities for emerald are in Colombia. There are several mines in two districts, Chivor and Muzo, and from these much of the world's emerald rough has come. Emerald has also been found in the U.S.S.R., Austria, Brazil, South Africa, and Rhodesia. In the United States some fine-quality material has come from North Carolina, at Hiddenite, Shelby, and Crabtree. The deposits near Hiddenite are the most important. The first emerald crystal was found here in 1875 and the greatest number of crystals were found in 1882, including one 21.5 cm (8½ in.) long. Little mining was done in the ensuing years, but in 1969 the area was intensively explored by American Gems, Inc. and many fine crystals found. Among these was the largest (by weight) ever found there, measuring 54 mm ($2\frac{1}{8}$ in.) in diameter by 73 mm ($2\frac{7}{8}$ in.) long. Also found was a 59 carat crystal from which was cut a fine-quality gem weighing 13.14 carats, the finest emerald ever found in a North American deposit.

Name.

From the Ancient Greek, *beryllos,* originally applied to all green stones but later used only for beryl.

Corundum

Corundum gives us two of our most important gemstones, ruby and sapphire. Even though cut stones of these two gems have no apparent resemblance to one another, they are color varieties of the same mineral. They are chemically and physically the same, and differ only in the kind and amount of impurity elements they contain. Indeed, it is these chemical impurities that give them their value, for pure corundum is colorless.

Common corundum is a relatively abundant mineral, and because it is surpassed only by diamond in hardness, tons are mined each year for use as an abrasive. This superior hardness also contributes greatly to the value of this mineral as a gem. Oddly enough, corundum is composed of the two light elements aluminum and oxygen, but the close packing and strong bonding of the atoms in the crystal structure results in high hardness and specific gravity.

Ruby and sapphire have been highly prized since ancient times, and much lore and legend are associated with them. Today, with diamond and emerald, they are the most valuable of all gemstones.

Crystallography.

Hexagonal-rhombohedral. The habit of crystals varies with variety and locality. Ruby crystals are generally, but not always, tabular, consisting of a short prism terminated by a basal pinacoid, with or without small pyramid and rhombohedron faces, as in figure 13.21c. Sapphire crystals usually are prismatic or pyramidal in habit, often with tapering hexagonal pyramids, as in Figure 13.21*a* and *b*. The crystals frequently show deep horizonal striations. Usually rudely crystallized or massive; also coarse to fine granular.

Physical Properties.

Fracture conchoidal. Although there is no cleavage there is frequently basal and rhombohedral parting, the latter giving nearly cubic angles. **H** 9. **S.G.** 4.02. *Luster* adamantine to vitreous. Transparent to translucent.

Color various; usually some shade of brown, yellow, pink, red, or blue, but may also be colorless, gray, green, and purple. Color zoning is often pronounced but only rarely in ruby.

Optical Properties.

Uniaxial (-). $\omega = 1.770$, $\epsilon = 1.762$. These are the usual values, but occasionally some dark red and green stones give values of ω as high as 1.778. Biref. 0.008; disp. 0.018. Except in the yellow varieties, strongly pleochroic with $\omega > \epsilon$; *ruby,* intense purplish-red and light orange-red, or in darker stones violet and orange; *blue sapphire,* light greenish-blue and dark violet-blue; *orange sapphire,* orange and yellow-brown; *green sapphire,* green to blue-green and yellow-green.

Chemical Composition.

Aluminum oxide, Al_2O_3. The color of ruby is due to a small amount (up to 4%) of chromium replacing aluminum in the crystal strucutre. A small amount of iron with the chromium imparts a brownish-red

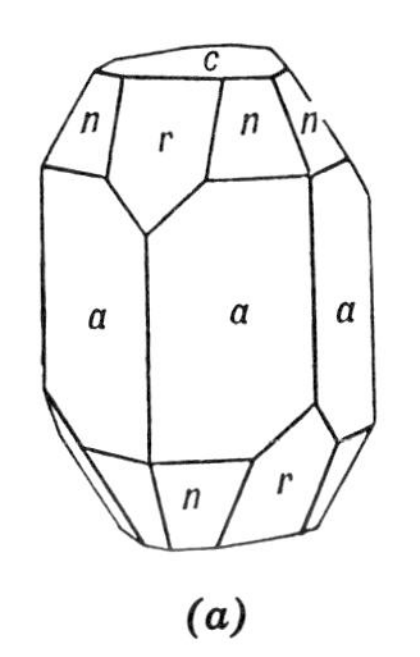

(*a*)

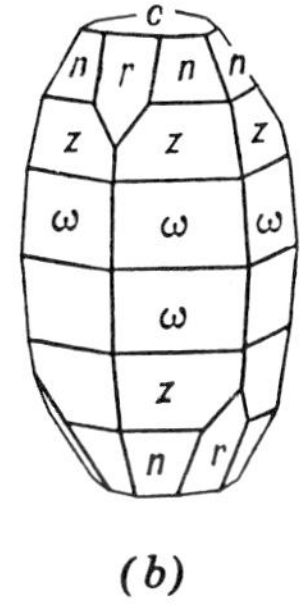

(*b*)

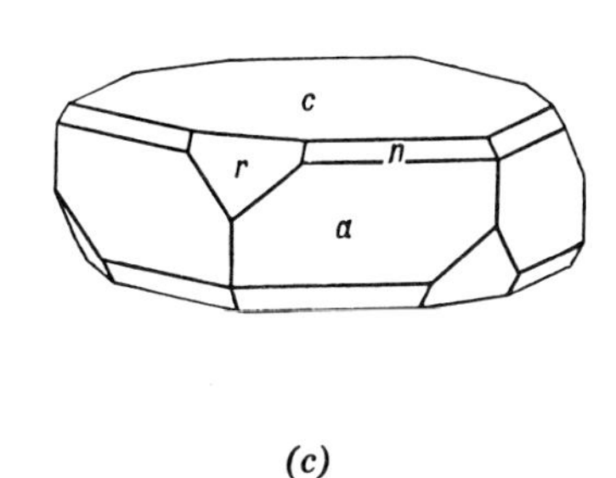

(*c*)

FIG. 13.21. Corundum crystals. (*a*) and (*b*) Sapphire. (*c*) Ruby.

color. The color of blue sapphire is due to the presence of small amounts of iron and titanium.

Figure 3.10 shows the structure of corundum as seen looking parallel to the *c* axis, that is, projected on the basal plane.

Luminescence and Absorption Spectra.

Corundum that is colored by chromium (ruby and pink and violet sapphire) fluoresces a strong red under both long- and shortwave u.v. Blue sapphire is essentially nonfluorescent, except some from Sri Lanka, Kashmir, and Montana that appear red or orange under longwave u.v. Most iron-rich green and yellow sapphires are nonfluorescent, but yellow sapphires from Sri Lanka may fluoresce orange-yellow under ultraviolet light and X-radiation.

The absorption spectrum of ruby is characterized by lines in the red: a doublet at 6962 and 6928 Å, and weaker lines at 6680 and 6592 Å, in the orange part of the spectrum. In the blue part of the spectrum there are three lines: a doublet at 4765 and 4750 Å, and a third at 4685 Å.

Blue sapphires show only an absorption line at 4500 Å. Iron-rich green and blue-green stones have three absorption bands in the blue region of the spectrum, at 4710, 4600, and 4500 Å.

Inclusions.

Inclusions are an interesting and important feature in ruby and sapphire. In some cases their nature alone allows positive identification. In others the locality may be determined by the nature of the inclusions.

The crystal inclusions seen in ruby and sapphire may be characterized as follows:

Silk. Needlelike crystals of rutile, arranged in three sets that intersect at 60° angles, and lie in planes perpendicular to the *c* axis of the crystal.

Zircon Crystals. Rounded crystals surrounded by a halo of black fractures.

Spinel Crystals. Small octahedral crystals, seen especially in rubies from Burma and sapphires from Sri Lanka.

Fingerprint inclusions are patterns of liquid- and gas-filled cavities that take their name from the fingerprint-like pattern (see Plate III.7).

Other included crystals commonly seen in ruby and sapphire are mica, hematite, garnet, and corundum.

Diagnostic Features.

Corundum occurs in a wide range of colors. Hence there are many gems that may resemble it, such as garnet, chrysoberyl, spinel, zircon, tourmaline, quartz, topaz, spodumene, and benitoite, as well as synthetic corundum, synthetic spinel, glass, and doublets.

Of these only benitoite, chrysoberyl, rhodolite and almandine garnet, and synthetic corundum have similar optical and physical properties. Benitoite has much higher birefringence and lower specific gravity; chrysoberyl has lower refractive index and specific gravity; garnet is isotropic; and synthetic corundum has characteristic bubble inclusions and curved growth lines.

Synthesis.

Corundum is synthesized on a large scale, mostly by the Verneuil technique (see Chapter 10). Therefore, synthetic ruby and sapphire in many colors, as well as synthetic star ruby and star sapphire, are often seen and great care must be taken to correctly identify them (see Plate XII).

Gem Varieties.

Color differences give rise to several gem varieties of corundum. *Ruby* is deep red. All other colors are *sapphire* (see Plate IV), although the term sapphire alone is generally understood to mean blue, and the other colors are described by a prefix such as *pink sapphire* and *yellow sapphire*. Because much gem corundum originated in the Orient, the term *oriental* has been prefixed to the names of gems having a characteristic color (other than red and blue) to describe gem corundum. Thus *oriental amethyst* is purple and *oriental topaz* is yellow. This term is misleading and should be discontinued.

The most desirable color for ruby is a dark, somewhat purplish-red, often called *pigeon's blood red*. The most desirable color of sapphire is a velvety cornflower blue, called *Kashmir blue,* from the locality. Other locality names have come to be used to describe color, whether or not the gem came from that locality; thus *Burma Ruby,* dark, somewhat purplish-red; *Siam ruby,* dark, slightly brownish- or orange-red; *Ceylon ruby,* lighter red and more transparent; *Siam sapphire,* dark blue; and *Burma sapphire,* fine royal blue. *Padparadscha* is a rare orange-yellow to orange variety of sapphire.

Star ruby and *star sapphire* exhibit the phenomenon of asterism better than any other gems. In them the six-rayed star is caused by light reflected from oriented, needlelike inclusions of rutile (see Fig. 6.7 and Plate II). To bring out this phenomenon the stones are cut as cabochons and must be oriented so that the *c* or optic axis of the crystal is perpendicular to the base of the stone. Illogically, the trade uses star ruby to describe not only gems with a true ruby color, but often also pink, purple, and violet stones, most of which should be called star sapphires.

Occurrence.

Corundum is common as an accessory mineral in metamorphic rocks such as crystalline limestone, mica schist, and gneiss. It is found also as a constituent of certain igneous rocks, as syenites and nepheline syenites, and in some pegmatites. It is also frequently found in detrital soil and stream gravels, where it has been preserved through its hardness and chemical inertness.

The most famous locality for fine rubies is Mogok, Burma. The stones are found in soil resulting from the weathering of a metamorphosed limestone. The rubies of Thailand are found near Bangkok, where they occur in a clay derived from the decomposition of a basalt. The rubies of Sri Lanka are found with other gemstones in gravels in the Ratnapura and Rakwana districts. Star rubies are also found at these localities, as well as in India in poor-quality stones. More recently good-quality ruby has been found associated with green zoisite in Tanzania. In the United States small rubies of fair quality are found in the vicinity of Franklin, North Carolina. The most important deposit is that of Cowee Creek, about 6 miles north of Franklin. In Georgia the Lauel Creek mine in Rabun County has produced semitransparent material of good color.

Sapphires are found associated with the rubies of Burma, Thailand, and Sri Lanka. The finest sapphires come from the Zanskar district of Kashmir. Good-quality sapphires are also found in Australia, in central Queensland, and near Inverell, New South Wales. In the United States the only important occurrences of gem sapphire are in Montana. Small waterworn crystals of several colors were early found during placer mining for gold in the sands of the Missouri river in the vicinity of Helena. During the end of the nineteenth century some systematic mining was carried out to recover sapphire from stream gravels, but in later years sapphire production was a by-product of gold mining. In 1967 a portion of the Eldorado Bar deposit was opened to collectors on a fee basis.

After the discovery of placer sapphires, violet-blue, tabular crystals with a steely luster were found embedded in a lamprophyre dike at Yogo Gulch. Early in the twentieth century mining was active at Yogo Gulch but ceased in 1929, although several attempts to reopen the mine were made in the following years. The mine is now being worked, apparently successfully, by the Sapphire International Corporation.

Name.

Corundum is probably derived from the Indian name of the mineral, *Kauruntaka.* Sapphire means blue and when first used probably denoted lapis lazuli. The origin of the word is unknown but reached us through the Latin form, *sapphirus.* Ruby means red, and is derived from the Latin *ruber.*

Opal

Opal is relatively abundant in the form of the milky white, *common opal,* a nongem material. We are concerned here only with the relatively rare precious opals that have long been valued for their subtle beauty resulting from an internal display of colors. It is difficult to improve on Pliny's description written nearly 2000 years ago in which he states, "There is in them a softer fire than in the carbuncle, there is the brilliant purple of the amethyst; there is the sea-green of the emerald—all shining together in incredible union." From the time of Pliny the great value of opals, frequently equal to that of diamonds, continued through the eighteenth century. But early in the nineteenth century a superstition arose that opals were unlucky stones and brought misfortune on the owner. As a result, there was a decrease in popularity with a corresponding lessening of value. Today the superstition is forgotten and opals once more occupy their preeminent position among gemstones.

Crystallography.

Amorphous. One of the few noncrystalline minerals and thus should be more properly classified as a mineraloid. Yet in spite of the lack of crystal structure, there is an internal order in opal.

It has recently been demonstrated that opal is composed of closely packed tiny spheres of amorphous silica. In common opal there is a random packing of spheres of irregular size resulting in the milky white "potch" opal. In precious opal spheres of the same size are tightly packed in an orderly three-dimensional array. Diffraction of light from these planes of ordered spheres results in the play of color in which rests the charm and beauty of opal (see under *Color* in Chapter 6, and Fig. 6.4).

Physical Properties.

H 5½–6½. **S.G.** 2.0–2.2; fire opal 2.0, black and white opal 2.11. *Fracture* conchoidal. *Luster* vitreous to greasy. Translucent, more rarely transparent.

Color: several types of precious opal are distinguished on the basis of their body color, but it is the play of color in precious opal that makes it "precious" (see Plate X). Unlike the color of most gems that results from the absorption of various wavelengths of white light, the flashing colors are caused by diffraction, giving them a high spectral purity and intensity. Held in highest esteem is *black opal* in which the internal color flashes are seen vividly against a dark body color which may be black, dark blue, or dark green. In *white opal* the brilliant play of color is less spectacular as seen against a white to gray body color. *Fire opal*, which may or may not have a play of color, is translucent to transparent and is so named because of its red to orange-red body color. *Hyalite* is a colorless, transparent opal that occurs in globules resembling drops of water. It does not usually show a play of color but when it does, it is highly prized as *water opal* with brilliant colors flashing from a colorless interior.

Optical Properties.

Isotropic. **R.I.** 1.44–1.46.

Chemical Composition.

Silicon dioxide with water, $SiO_2 \cdot nH_2O$. n in the formula indicates an indefinite amount of water which is usually between 4 and 9% by weight, but may be as high as 20%. The refractive index and specific gravity decrease with increasing amounts of water.

Luminescence and Absorption Spectra.

Under both long- and shortwave u.v. black opals are nonluminescent; white opals fluoresce a pale blue or pale green and may phosphoresce; fire opal may fluoresce a greenish color. Much of the common opal yields a green fluorescence and phosphorescence. Absorption spectra are of little importance.

Diagnostic Features.

The subtle play of color is diagnostic. In the simulated opal (see below) each flash of color comes from a specific, sharply defined platelet. To distinguish fire opal or hyalite lacking a play of color, it is necessary to determine the refractive index or specific gravity.

Synthesis.

Opalescence is sometimes confused with play of color, but it merely refers to the milky appearance of common opal. "Opal glass," which simulates this appearance, has been made for a long time. But recently Pierre Gilson in Switzerland has synthesized black and white opal that is nearly identical to the natural gem material in chemical and physical properties including the brilliant internal flashes of color.

Simulated Opal.

Present in the market today is an opal-appearing stone called *Opal-Essence* that is manufactured by John S. Slocum of Rochester, Michigan. Its internal color reflections give a striking resemblence to those of precious opal. Although it has the same hardness as opal, it can be distinguished by higher **R.I.** (1.49–1.51) and **S.G.** (2.41–2.50). The process by which it is manufactured is secret but microscopically it appears to be a glass or glass-like matrix enclosing platelike crystals from which the color flashes emanate. The matrix is isotropic but the included variously colored, platy crystals are uniaxial with strong pleochroism (see Plate XII).

Occurrence.

Opal is found in near-surface deposits formed by circulating ground water or from low-temperature hydrothermal solutions. It occurs lining and filling cavities in the rocks and may replace wood buried in volcanic ash. Although common opal is the petrifying material of most opalized wood, precious opal is found thus in some localities.

The finest gem opals of the ancients are believed to have come from India. But from Roman times until late in the eighteenth century, a region of Hungary (now part of Czechoslovakia) was the principal source. Today Australia is the leading producer. The first major finds of Australian opals were made in Queensland in 1872. These were followed by discoveries in New South Wales of the White Cliffs area (1899) and Lightning Ridge (1908). Later other discoveries of opal were made in South Australia and Western Australia. Precious opal is found at all localities and at Lightning Ridge it has replaced shells, wood, and crystals of the mineral glauberite having the appearance of small fossilized pineapples. Several important fire opal localities are in the state of Queretaro, Mexico.

In the United States common opal, especially wood opal, is abundant in many western localities, but only Idaho and Nevada have produced important quantities of gem quality material. The Idaho locality, near Moscow, was discovered in 1890 and was worked for only a few years, during which time a small quantity of fine gem material was produced. A second Idaho deposit, near Spencer, was found in the 1960s and since 1968 has been open to collectors on a fee basis. The most important United States locality for gem opal is Virgin Valley, Nevada. Important quantities of opal were produced from 1908 to 1920 and again starting in 1949. The Virgin Valley material is of outstanding beauty but almost invariably cracks and has not been widely used in jewelry.

Name.

The name opal originated in the Sanskrit *upala,* meaning stone or precious stone.

Jade

Jade is a gem name that does not signify a single mineral species, but two distinct and unrelated species, *jadeite* and *nephrite.* Both are silicates, but the former is a member of the pyroxene mineral group, whereas the latter is in the amphibole group.

About 1000 BC the Chinese used for making weapons and ritual ornaments a tough green mineral found in Khotan which they called *yu,* now known as nephrite. Jadeite was not known before the eighteenth century and was brought from Burma for carving by Chinese artisans. Nephrite was also early used by inhabitants of the Swiss lake dwellings and by the Maoris of New Zealand. Jadeite carvings excavated in Mexico and Central America show that this gem played an important part in the religious and social life of the Aztec and Mayan civilizations.

Today both nephrite and jadeite are widely used gem materials both for carving and in jewelry (see Plate I). The emerald-green, translucent form of jadeite, sometimes referred to as imperial or gem jade, is more valuable.

Jadeite

Jadeite is a sodium aluminum silicate and a member of the pyroxene mineral group.

Crystallography.

Monoclinic. Rarely in isolated crystals. Usually in compact massive aggregates of interlocking, more or less equidimensional crystals.

Physical Properties.

Cleavage prismatic, in two directions at angles of 87° and 93°, but not seen in the normal massive form which is extremely tough and difficult to break.

H 6½–7. **S.G.** 3.3–3.5. *Luster* vitreous, pearly on cleavage surfaces; translucent to opaque.

Color white, apple-green to emerald-green, pink, brown, red, orange, yellow, blue, violet, and black; often white with spots of green.

Optical Properties.

Biaxial (+). Because of its aggregate nature most specimens give an average **R.I.** of 1.66. For single crystals (Cloverdale) α = 1.640, β = 1.645, γ = 1.652; for diopside-jadeite (Guatemala) α = 1.654, γ = 1.669; biref. 0.012.

Chemical Composition.

Sodium aluminum silicate, $NaAlSi_2O_6$. May form a solid solution series with diopside ($CaMgSi_2O_6$), in which calcium and magnesium substitute in the jadeite structure for sodium and aluminum; this material is called *diopside-jadeite*. Jadeite and acmite ($NaFeSi_2O_6$) may also form a solid solution with ferric iron replacing aluminum. This dark green to black variety is known as *chloromelanite*. A small amount of chromium is the coloring agent in emerald-green, imperial jade.

Luminescence and Absorption Spectra.

Under longwave u.v. lighter-colored jadeites have a weak, whitish fluorescence. The darker varieties are nonfluorescent.

The absorption spectrum of jadeite shows a strong line at 4375 Å and weaker bands at 4500 and 4330 Å. In emerald-green jadeite the 4375 Å line is suppressed, but lines characteristic of chromium are present at 6915, 6550, and 6300 Å.

Occurrence.

Jadeite occurs in nodular and lens-shaped masses in serpentine, formed by the metamorphism of sodium-rich rocks. The finest gem quality jadeite comes from deposits in the vicinity of Tawmaw and Hpakon, Upper Burma. Material of lesser quality is found near Manzanal, Guatemala, and in Japan near Kotaki in Niigata Prefecture. In the United States jadeite of poor quality has been found in California at several localities in the Central Coast Ranges.

Nephrite

Nephrite is a calcium magnesium hydroxy silicate and a member of the tremolite–actinolite series of the amphibole mineral group.

Crystallography.

Monoclinic. Nephrite is never in isolated crystals, but occurs as interlocking masses of fibrous crystals.

Physical Properties.

Amphiboles have perfect prismatic *cleavage* at angles of 56° and 124°, but the cleavage is not evident in nephrite, which is extremely tough and difficult to break. **H** 6–6½. **S.G.** 2.90–3.02. *Luster* vitreous; translucent to opaque.

Color white to dark green, gray, blue-green, yellow, brown, reddish-brown, black. The most important type is a translucent dark green known as "spinach jade."

Optical Properties.

Biaxial (−). Because of its aggregate nature most specimens give a single **R.I.** = 1.62. Single crystals of tremolite–actinolite have α = 1.600–1.628, β = 1.613–1.644, γ = 1.625–1.655; biref. 0.022–0.027.

Chemical Composition.

Calcium magnesium hydroxy silicate, $CaMg_5Si_8O_{22}(OH)_2$; usually contains some ferrous iron substituting for magnesium. A member of the tremolite–actinolite series; when iron is present in amounts greater than 5% the mineral is called actinolite.

Luminescence and Absorption Spectra.

Nonfluorescent; absorption spectrum indistinct.

Occurrence.

Tremolite is most frequently found in impure, crystalline dolomitic limestones formed during metamorphism. Actinolite commonly occurs in crystalline schists and greenstones. *Nephrite* is found in many parts of the world and is much more common than jadeite. The earliest known source of nephrite was near Khotan in eastern Turkestan and this material was used by early Chinese carvers. Other important sources are near Lake Baikal, U.S.S.R.; New Zealand; Taiwan; and Frazier River, British Columbia. In the United States important deposits are found near Jade Mountain along the Kobuk River, Alaska, and near Lander, Wyoming.

Diagnostic Features.

In carved and polished pieces it is frequently difficult by inspection to tell nephrite from jadeite; a determination of **S.G.** offers the most certain means.

The two types of jade are not only confused with each other but with several other minerals, particularly with idocrase (californite), grossular (massive form), talc (soapstone), serpentine, prehnite, amazónite, and chalcedonic quartz. Some of these closely resemble true jade. They are widely used as jade substitutes, especially in carvings, and great care must be used in identification.

Most frequently used in *bowenite,* a translucent yellowish-green to green form of serpentine, with **H** = 5½, a hardness greater than is usual for serpentine. However, the **R.I.** of bowenite is near 1.52 and the **S.G.** is 2.58–2.62.

There are other materials that are carved that bear only a superficial resemblance to jade. *Pseudophite* ("Styrian jade"), a compact, massive form of chlorite, has **R.I.** near 1.57, **S.G.** 2.7, **H** 2.5. *Agalmatolite* ("pagoda stone") is a soft, waxy, compact material that may be pinite, pyrophyllite, or talc, and has **R.I.** 1.55 1.60, **S.G.** 2.75–2.80, and **H** 2½–3½. *Saussurite* is a tough, compact mineral aggregate consisting of a mixture of albite and zoisite, and has **R.I.** near 1.70 or near 1.50, depending on which mineral is predominant, **S.G.** about 3.0, and **H** near 6.5. *Verdite* is a green rock consisting of fuchsite (a chrome mica), and clayey matter, with **R.I.** near 1.58, **S.G.** about 2.9, and **H** 3.

Dyed jade, white jadeite that has been dyed to a fine "imperial" jade color, is widely distributed. Under magnification the color may be seen to be distributed along grain boundaries; identification may also be made with the spectroscope. Jadeite is also dyed a mauve color.

Name.

The name *jade* was derived from *piedra de hijada,* the Spanish name for jade, meaning "stone of the side" because it was supposed to cure kidney ailments if applied to the side of the body. The Spanish also referred to this material as *piedros de los rinones* (kidney stone), which was translated into Latin as *lapis nephriticus,* hence the word nephrite. Jadeite is so named because it was found to make up many jade specimens.

Chrysoberyl

Most commonly chrysoberyl is in transparent greenish-yellow crystals that are only rarely cut as gems. However, its rare, exotic varieties, cat's eye and alexandrite, rank high among precious gems.

Crystallography.

Orthorhombic. Crystals usually tabular and striated; also commonly in cyclic twins having a pseudohexagonal appearance (Fig. 13.22).

Physical Properties.

Cleavage prismatic distinct, also poor on side and basal pinacoids. Commonly breaks with a conchoidal fracture. **H** 8½. **S.G.** 3.73. *Luster* vitreous. Transparent to translucent.

Color yellow, greenish-yellow, brownish-yellow, brown, green, red (in incandescent light).

Alexandrite is a variety that appears emerald-green in daylight but red in artificial (incandescent, not fluorescent) light.

Cat's eye, also called *cymophane,* is a variety which when polished has an opalescent luster, and across whose surface plays a long narrow beam of light, changing its position with every movement of the stone (see Fig. 6.6). This chatoyant effect is best obtained when the stone is cut as a cabochon. A number of other gems of less value may exhibit chatoyancy, but in the jewelry trade the term cat's eye is reserved for chatoyant chrysoberyl alone.

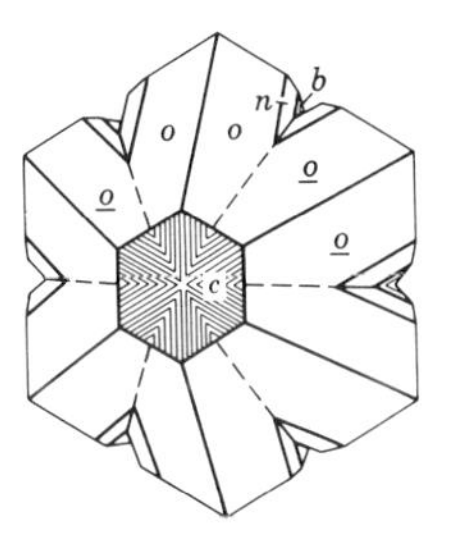

FIG. 13.22. Chrysoberyl, cyclic twin.

Cat's eyes are usually yellowish to brownish in color. Alexandrite cat's eyes are known but are very rare.

Optical Properties.

Biaxial (+); $\alpha = 1.745$, $\beta = 1.746$, $\gamma = 1.755$ (all $\pm$ 0.003); biref. 0.010; 2*V* variable, generally small; disp. 0.015. Pleochroism of yellow, yellow-brown, and brown stones is weak to distinct depending on color and depth of color. Alexandrite has strong pleochroism; α red, β orange, and γ green.

Chemical Composition.

Beryllium aluminum oxide, $BeAl_2O_4$. Iron is usually present and alexandrite contains a small amount of chromium which gives it its unusual color phenomena.

Luminescence and Absorption Spectra.

Chrysoberyls of yellow, green, and brown color show no fluorescence under ultraviolet light, whereas alexandrite fluoresces a weak red under both short- and longwave u.v.

Yellow and brown chrysoberyls have an absorption band due to ferrous iron at 4400 Å. The absorption spectrum of alexandrite is complex and, because of its strong pleochroism, varies with direction of light transmission through the stone. In its absorption spectrum lies the explanation of the color change as viewed in daylight and artificial light. Alexandrite contains trace amounts of chromium and the absorption spectrum shows broad bands in the yellow-green and violet portions of the spectrum, and narrow lines in the red and blue portions. The transmission of light in the red and blue-green parts of the spectrum are nearly equal, and any change in the nature of the incident light (as from daylight to incandescent) changes the color of the gem, because daylight is relatively rich in blue and green, and incandescent light contains more of the longer (red) wavelengths.

Inclusions.

The inclusions of greatest interest in chrysoberyl are those in cat's eye gems. Numerous minute tubelike cavities, all parallel to the vertical (c) axis of the crystal, cause the chatoyancy. To best show the single beam of light a cabochon is cut so the tubular inclusions are parallel to the flat surface of the stone and at right angles to the longest dimension.

Diagnostic Features.

Chrysoberyl may be confused with other yellow to yellow-green gems, and with synthetic alexandrite, synthetic corundum, and synthetic spinel.

Synthetic corundum and synthetic spinel that show a color change from daylight to incandescent artificial light are incorrectly called "synthetic alexandrite." The synthetic corundum has little resemblance to alexandrite since its color change is from grayish-blue in daylight to amethystine purple in artificial light. Some synthetic spinel has a color change more nearly like alexandrite but it is isotropic and not pleochroic.

Synthesis.

True synthetic alexandrite has been made using the flux growth technique and the crystal pulling method. The former material shows typical flux growth inclusions, and the latter may show curved growth lines.

Occurrence.

Chrysoberyl occurs in granitic rocks, pegmatites, and mica schists; it is also recovered with other gem minerals from alluvial deposits. Alexandrite was first discovered at Takovaya in the Ural Mountains in 1833 and the finest stones have come from that locality. The gem gravels of Sri Lanka have yielded the finest cat's eyes as well as good-quality alexandrite. Brazil is a source of yellow chrysoberyl as well as cat's eye and alexandrite. Burma, Rhodesia, and Zambia are also sources of alexandrite. Chrysoberyl, but not of gem quality, has been found in the United States in Maine, Connecticut, New York, and Colorado.

Name.

Chrysoberyl means golden beryl (a misnomer). Cymophane is derived from two Greek words meaning wave and to appear, in allusion to its chatoyant effect. Alexandrite was named after the Czarevitch, later Czar Alexander II of Russia.

Topaz

A great deal of confusion exists with both jeweler and layman regarding the name *topaz*. In ancient times all yellow stones were called topaz, but today the name designates only the mineral species described here, an aluminum fluosilicate. The widespread practice of using the name for yellow quartz (citrine) is incorrect and should be discontinued. The term *precious topaz* is sometimes used for true topaz, and citrine is often incorrectly referred to as quartz topaz, smoky topaz, Scotch topaz, false topaz, and Madeira topaz.

The name topaz may have been derived from the Greek word *topazos* or *topazion,* perhaps referring to the island in the Red Sea now known as Zebirget (formerly Saint John's Island), an important source of peridot. It is possible, therefore, that the gem we call peridot was formerly called topaz. The first use of the name topaz for the species now known by that name was by Henckel in 1737 when he described the deposits in Saxony.

Crystallography.

Orthorhombic. In prismatic, frequently multifaced, crystals terminated by dipyramids, first- and second-order prisms, and basal pinacoid (Fig. 13.23). Doubly terminated crystals are rare; usually only one end of the crystal is terminated, the other being a basal cleavage plane. Usually in crystals but also in granular masses. Crystals range in size from very small to some several hundred pounds in weight.

Physical Properties.

Cleavage perfect basal. **H** 8. **S.G.** 3.4–3.6, increasing with increasing fluorine content. *Luster* vitreous. Transparent to translucent.

Color Usually colorless to very pale blue, also medium blue, pale to deep yellow, brown, orange-brown (sherry-colored), greenish, pink, violet-pink, red (see Plate V). Many brown crystals fade on exposure to sunlight or turn pale blue. A brown color can be induced by X-radiation but is not permanent. Orange-brown crystals from Ouro Preto, Brazil turn colorless when heated to about 450°C but on cooling become pale pink to purple-red depending on the intensity of the original color. Most topaz of this color has been heat treated.

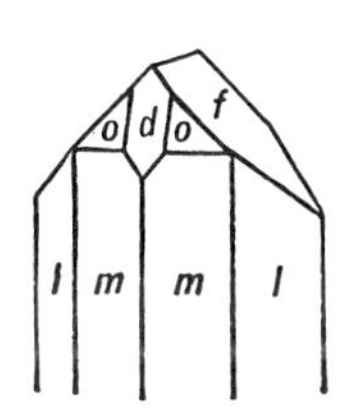

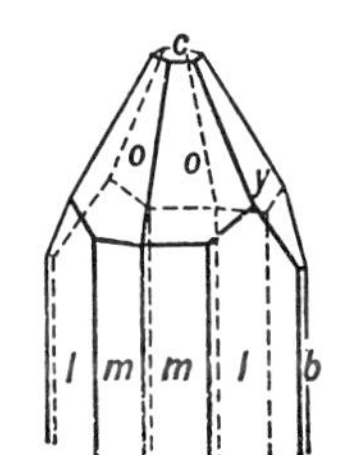

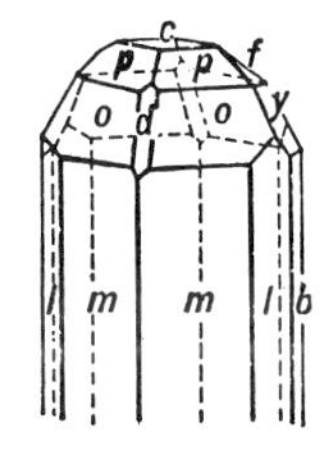

FIG. 13.23. Topaz crystals.

Optical Properties.

Biaxial (+). α = 1.607–1.630, β = 1.610–1.631, γ = 1.617–1.638. The minimum and maximum refractive indices for colorless, light blue, and light green stones are near 1.609 and 1.617; for yellow to brown and pink to red stones near 1.629 and 1.637. The birefringence is nearly constant at 0.008. Yellow topaz is pleochroic—brownish-yellow, yellow, and orange yellow. Blue topaz exhibits weak to distinct pleochroism depending on the depth of blue—colorless and light blue. In red topaz pleochroism is light red and yellow. The dispersion is 0.014.

Chemical Composition.

An aluminum fluosilicate, $Al_2SiO_4(F,OH)_2$. The ratio of F to OH is variable. Fluorine tends to raise the specific gravity and lower the refractive index, whereas hydroxyl acts in the reverse way.

Luminescence and Absorption Spectra.

Blue and colorless stones, the fluorine-rich type, have a weak yellowish or greenish fluorescence under longwave u.v., and under shortwave the fluorescence is still weaker. The orange-brown and pink stones, which are hydroxyl-rich, show strong fluorescence under longwave u.v., and a similar but weaker fluorescence under shortwave.

Only orange-brown topaz that has been heat treated to produce a pink color shows an observable absorption spectrum. In these stones there is a weak chromium doublet at about 6820 Å.

Inclusions.

The orange-yellow and orange-brown topazes from Ouro Preto, Brazil have as inclusions long tubelike cavities. The colorless, blue, and brown types usually have drop-shaped cavities containing two, and sometimes three, immiscible liquids, one of which may be carbon dioxide. Sometimes thin films of liquid occupy flattened cavities.

Diagnostic Features.

Topaz in the uncut form is recognized by its crystal form, its basal cleavage, its hardness (8), and its high specific gravity. The gemstones it resembles include quartz (citrine and rock crystal), chrysoberyl, grossular garnet, tourmaline, corundum, beryl, spodumene, synthetic corundum, synthetic spinel, doublets, and glass. Of the rare and unusual gems, topaz may be confused with danburite, apatite, scapolite, phenakite, euclase, orthoclase or labradorite, beryllonite, and brazilianite.

Tourmaline and glass are the stones that most closely resemble topaz in appearance and refractive index. Glass is isotropic whereas tourmaline has much lower specific gravity and higher birefringence. The rare and unusual gems of similar refractive index as topaz, andalusite, apatite, brazilianite, and danburite, are all much lower in specific gravity.

Synthesis.

Authenticated synthetic topaz has been made in laboratory experiments but not in commercial quantity. However, there are unconfirmed reports of the synthesis in the U.S.S.R. of colorless gem topaz that turns blue when irradiated. Synthetic corundum of appropriate color is sometimes incorrectly referred to as synthetic topaz.

Occurrence.

Topaz is formed through the agency of fluorine-bearing vapors given off during the last stages of the solidification of igneous rocks. It is found in cavities in rhyolite and granite, and is a characteristic mineral of pegmatite dikes, especially those carrying tin. It is found associated with tourmaline, cassiterite, apatite, fluorite, beryl, quartz, mica, and feldspar. It is also found in some localities as rolled pebbles in stream gravels.

The most important source of gem topaz of the highly prized orange-yellow and orange-brown colors is in deposits near Ouro Preto, Brazil, where small terminated crystals are found in clay with quartz crystals. Other notable occurrences are in the U.S.S.R. in the Nerchinsk district in Siberia as large pale yellow crystals, and in Mursinsk, Ural Mountains, as pale blue crystals; in Schenckenstein, Saxony with quartz; in Minas Gerais, Brazil; in Omi and Mino prorefectures, Japan; in San Luis Potosi, Mexico; in the gem gravels of Sri Lanka and Burma, and as large crystals in pegmatite near Mogok, Upper Burma; in South-West Africa as colorless to pale blue crystals; as fine blue crystals in Rhodesia; and in the tin mines of Nigeria in colorless waterworn crystals and pebbles. In the United States topaz is found at Pikes Peak, near Florissant and Nathrop, Colorado; Thomas Range, Utah; Streeter, Texas; San Diego County, California; New Hampshire and Maine; Virginia; and South Carolina.

Tourmaline

Tourmaline is unique among gem minerals in the great variety of colors in which it occurs. Moreover, in a single crystal there may be several different colors arranged in sharply contrasting zones (see variations in color under *Color* in Chapter 6 and Fig. 13.27). In addition to being a valued gem, tourmaline has several scientific and technological uses because of its strong piezoelectric effect.

Crystallography.

Hexagonal-rhombohedral. Tourmaline most commonly occurs in prismatic crystals striated parallel to the c axis. Three faces of the trigonal prism and six faces of the second-order hexagonal prism round into each other giving the crystal a trigonal cross section, sometimes resembling a spherical triangle (Fig. 13.24). Such a cross section is found in no other mineral and is therefore diagnostic. Tourmaline crystallizes in a low symmetry class lacking a symmetry center, which permits the presence of the trigonal prism. For the same reason the c axis is a polar axis and thus in doubly terminated crystals different forms are present at the two ends (Fig. 13.25). Although the mineral may occur in radiating

FIG. 13.24. Tourmaline crystals (elbaite) with quartz and albite feldspar, Pala, California.

groups and massive aggregates, most gem material is in well-formed single crystals.

Sometimes tourmaline grows in parallel aggregates of slender almost fibrous crystals. Cabochons cut from such material have a marked chatoyancy. The *cat's eye* effect may also result from the presence of inclusions or tubular cavities elongated parallel to the c axis.

Physical Properties.

Fracture conchoidal to uneven, no cleavage. **H** 7–7½. **S.G.** 3.0–3.25, gems 3.03–3.10. *Luster* vitreous. Transparent to translucent. Strongly piezoelectric and pyroelectric.

Color extremely variable (see Plate VIII). Color, specific gravity, and optical properties vary with chemical composition (q.v.).

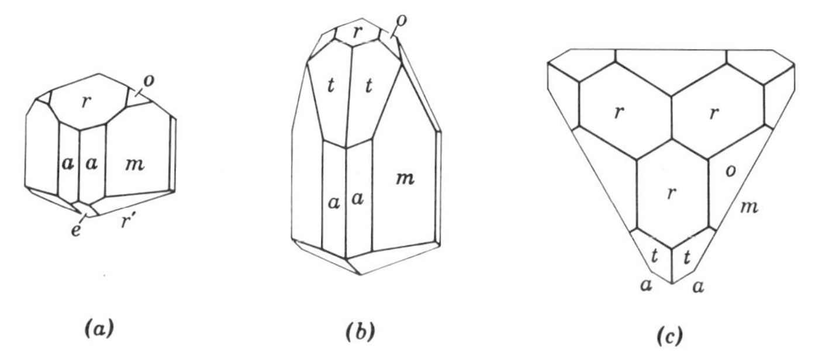

FIG. 13.25. Tourmaline crystals. (*a*) and (*b*) Doubly terminated crystals showing different forms at opposite ends of the *c* axis. (*c*) End view of crystal showing the trigonal arrangement of faces.

Optical Properties.

Uniaxial (−). The refractive indices vary and are given in the table of chemical varieties. Pleochroism is variable but commonly very strong, always with $\omega > \epsilon$. Because of the high absorption of the ordinary ray, deep-colored stones are usually cut with the table parallel to the optic axis. This permits more light to enter the stone and gives a pleasing effect from the dichroic colors as it is turned. Dispersion 0.017.

Chemical Composition.

Tourmaline is an extremely complex borosilicate. A portion of the ring-type structure is shown in Figure 13.26 and shows that ions of many elements may be present. A chemical formula for the mineral has long been uncertain. In 1890 John Ruskin's reply to the question, "And what is it made of?" was, "A little of everything ... the chemistry is more like a medieval doctor's prescription than the making of a respectable mineral." However, authorities today are essentially in agreement with the generalized formula, $XR_3Al_6B_3Si_6O_{27}(OH)_4$, in which X (Na,Ca) and R are the symbols for the major variable components that give rise to several mineralogical varieties shown in the table below.

In addition to the R components indicated, chromium and vanadium may be present coloring tourmaline emerald-green.

CHEMICAL VARIETIES OF GEM TOURMALINE*

	Elbaite	Liddicoatite	Dravite	Uvite
X	Na	Ca	Na	Ca
R	Al,Li	Al,Li	Mg	Mg
S.G.	3.03–3.10	3.02	3.03–3.15	2.96–3.06
ω	1.640–1.655	1.637	1.635–1.661	1.634–1.638
ϵ	1.615–1.630	1.621	1.610–1.632	1.619–1.620
Biref.	0.016–0.022	0.016	0.021–0.026	0.016–0.021
Color	Green, red, blue, yellow, colorless, brown		Brown-black	Brown

* The most common tourmaline is black *schorl* but it is rarely encountered as a gem. In it X is chiefly Na, and R is Fe,Mg.

It is the lithium-bearing elbaite and liddicoatite that occur in transparent crystals of many colors and are most commonly cut as gems. Names sometimes given to the color varieties of elbaite are red-pink *rubellite,* blue *indicolite,* green *verdelite,* and colorless *achroite*. But the preferred nomenclature for all is simply tourmaline with the color prefix.

Diagnostic Features.

The relatively high birefringence, 0.02±, and the strong dichroism serve to distinguish tourmaline from other similar appearing stones with the same range of refractive indices.

Occurrence.

Tourmaline is found in igneous and metamorphic rocks as an accessory mineral but rarely of gem quality. The exception is brown dravite that occurs in marbles, that is, in metamorphic limestones. The most common and characteristic occurrence of gem tourmaline is in granite pegmatites. Here it is found frequently associated with other gem minerals as spodumene, amblygonite, apatite, beryl, topaz, and quartz.

Localities where gem tourmaline has been recovered are many. The name *elbaite* is taken from the Island of Elba where essentially all the color varieties have been found. From Madagascar have come large crystals of liddicoatite, many of which show color banding parallel to pyramid faces (Fig.

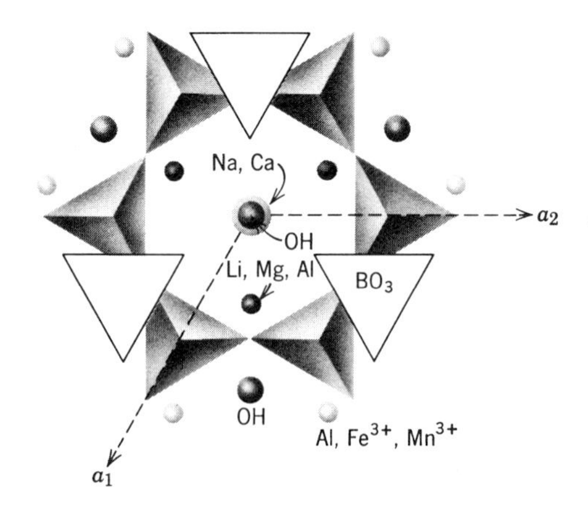

FIG. 13.26. Part of the structure of tourmaline, projected on the basal plane, shows the trigonal aspect of the structure.

FIG. 13.27. Liddicoatite from Madagascar cut parallel to the basal pinacoid showing symmetrical color zoning. Zoning parallel to a trigonal pyramid results in the triangular aspect. Courtesy of the Smithsonian Institution, Washington, D.C.

13.27). Many hundreds of pounds of this gem quality tourmaline have been manufactured into pressure gauges. Other famous localities are South-West Africa; Mozambique; the state of Minas Gerais, Brazil; Ural Mountains, U.S.S.R.; and alluvial deposits in Sri Lanka (Ceylon) and Burma. In the United States gem tourmaline is found at Newry, Paris, and Auburn, Maine; Haddam Neck, Connecticut; and Pala, Mesa Grande, Rincon, and Ramona, California.

Beginning in 1959 new mining activity at the classic localities of San Diego County, California and Oxford County, Maine produced important quantities of fine tourmaline. In San Diego County the Stewart Lithia and Tourmaline Queen mines yielded many specimens of gem material. A new discovery in 1972 of gem quality tourmaline at the Dunton mine, Newry, Maine (see Fig. 2.2) was one of the largest finds on record anywhere and made Maine the most important world source of high-quality green and red tourmaline.

Synthesis.

Tiny crystals of tourmaline, unsuitable for gems, have been formed hydrothermally by the recrystallization of glass formed from melted natural tourmaline. There are unconfirmed reports that green, gem tourmaline is being synthesized in the U.S.S.R.

Name.

From *turamali,* a Sinhalese word for zircon. The name has been applied to tourmaline since 1703 when a mislabeled parcel of stones arrived in Holland from Ceylon.

Quartz

Of all the minerals used as gems none compares with quartz in its diversity of occurrence and in abundance of varieties. It is found in igneous, sedimentary, and metamorphic rocks and in some is essentially the only mineral present. Because of its hardness (7), lack of cleavage, and chemical stability, it survives and maintains its identity as pebbles of stream beds and sand grains of sea beaches.

Quartz as the fine-grained variety, flint, was the first mineral used by man. From it he fashioned tools and weapons that started him on his long road to civilization. Later, it was undoubtedly colored varieties of quartz that were first used as "gems" for both personal adornment and for their presumed magical properties. The earliest work on minerals that has survived was written by Theophrastus about 300 BC. In this treatise, *On Stones,* quartz and its varieties are mentioned more often than any other mineral. And there is the observation that among the ancients there was no precious stone in more common use than *sardion,* a red chalcedony.

Quartz as a gem mineral is divided in the following discussion into two groups: coarsely crystalline varieties and fine-grained, microcrystalline varieties. However, all have the same chemical composition and crystal structure; their differences lie in methods of formation, grain size, and the impurities that give rise to varying colors. Although opal and quartz have a similar chemical composition, opal contains water and is amorphous. It is thus discussed separately.

Crystallography.

Hexagonal-rhombohedral. Quartz belongs to a low symmetry class lacking both symmetry planes and a symmetry center. As a result, the horizontal *a* axes are polar axes. Although most quartz is in formless grains in rocks, it is such an abundant mineral that well-formed crystals are common and vary in size from microscopic to individuals weighing several tons. They are usually prismatic with horizontally striated prism faces terminated by a combination of positive (*r*) and negative (*z*) rhombohedrons (Fig. 13.28). The two rhombohedrons are often equally developed giving the appearance of a hexagonal dipyramid (Fig. 13.29*a* and *b*). The tetrahedral units of the quartz structure are arranged in a helical fashion; in some crystals they spiral to the left (left-hand) and in others to the right (right-hand). External evidence of the "hand" is rarely seen but is evidenced in some crystals by the presence of face *x* (Fig. 13.29*c* and *d*), which is at the upper left of alternate prism faces in left-hand quartz and to the upper right of alternate prism faces in right-hand quartz. Quartz is almost invariably twinned according to two laws: (1) the Brazil twin in which right- and left-hand crystals interpenetrate and (2) the Dauphiné twin in which two right- or two left-hand crystals interpenetrate. Unlike most crystal twins, external evidence is rarely seen.

Equally important to the gemologist as the single crystals are the massive microcrystalline forms composed of tiny crystalline particles to which are given many varietal names.

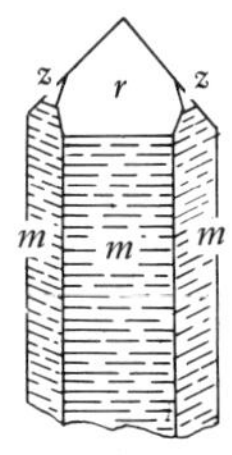

FIG. 13.28. Quartz crystal. Hexagonal prism (*m*) terminated by positive (*r*) and negative (*z*) rhombohedrons.

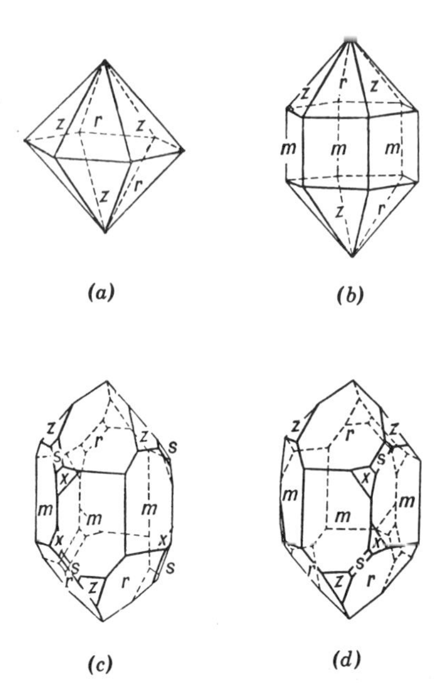

FIG. 13.29. Quartz crystals.

Physical Properties.

Fracture conchoidal, rhombohedral cleavage rare. **H** 7. **S.G.** 2.65 in crystals, 2.60 in fine-grained varieties. *Luster* vitreous in crystals; waxy to dull in fine-grained varieties. Transparent to translucent. Strongly piezoelectric.

Color usually colorless but may be any color due to traces of chemical impurities or to mechanical inclusions of colored minerals.

Optical Properties.

Uniaxial (+); $\epsilon = 1.544$, $\omega = 1.553$; in fine-grained varieties **R.I.** $\approx$ 1.53. Biref. 0.009; disp. 0.013. Pleochroism: absorption is normally $\epsilon < \omega$ but $\epsilon > \omega$ has been observed; amethyst weak to strong in shades of blue and violet; smoky quartz ϵ pale to dark brown, ω brown to nearly black; citrine ϵ colorless to pale yellow, ω yellow.

Quartz is optically active; that is, because of the lack of symmetry planes, the plane of polarized light moving along the optic axis is rotated to the left in left-hand crystals and to the right in right-hand crystals. For this reason a gemstone viewed parallel to the *c* axis between crossed polars will not go to extinction.

Chemical Composition.

Silicon dioxide, SiO_2. Of all minerals, colorless quartz is most nearly a pure chemical compound with constant physical properties. Trace amounts of other elements give rise to color varieties.

The structure of quartz, composed of SiO_4 tetrahedra arranged in a helical manner, is shown in Fig. 13.30.

Luminescence and Absorption Spectra.

Quartz is essentially nonluminescent and is without distinctive absorption spectra.

Diagnostic Features.

Transparent quartz has constant **R.I.** and **S.G.** and by these properties both colorless and colored stones can be distinguished from similar appearing gems. The double refraction distinguishes them from glass. Fine-grained varieties have both lower **R.I.** (1.535) and **S.G.** (2.60) than coarsely crystalline quartz. Also at high magnification between crossed polars, a thin flake or the thin edge of a microcrystalline quartz gemstone is seen to be composed of tiny birefringent particles.

Coarsely Crystalline Varieties.

Rock crystal. Colorless quartz, the commonest of the gem minerals. However, because of low refractive index and glassy appearance, it is rarely used as a major gemstone. Small colorless quartz crystals are common and are sometimes called "diamonds" modified by the locality name, as for example, "Herkimer diamonds" from Herkimer, New York. Large crystals are used for carving art objects and making crystal balls. A double image seen through a true crystal ball distinguishes it from the more common glass imitation.

Rock crystal is found abundantly throughout the world but Brazil has been the major producer of quartz for optical and technical purposes. Other important localities are in the Alps, Madagascar, and Japan. In the United States the best crystals have come from Hot Springs, Arkansas and Little Falls and Ellenville, New York.

Amethyst. Purple quartz varying from pale mauve to deep violet. (See Plate XI for illustrations of colored varieties of quartz.) Small amounts of iron as the coloring agent are distributed in layers usually parallel to rhombohedron faces and in most cut stones clear bands can be seen alternating with

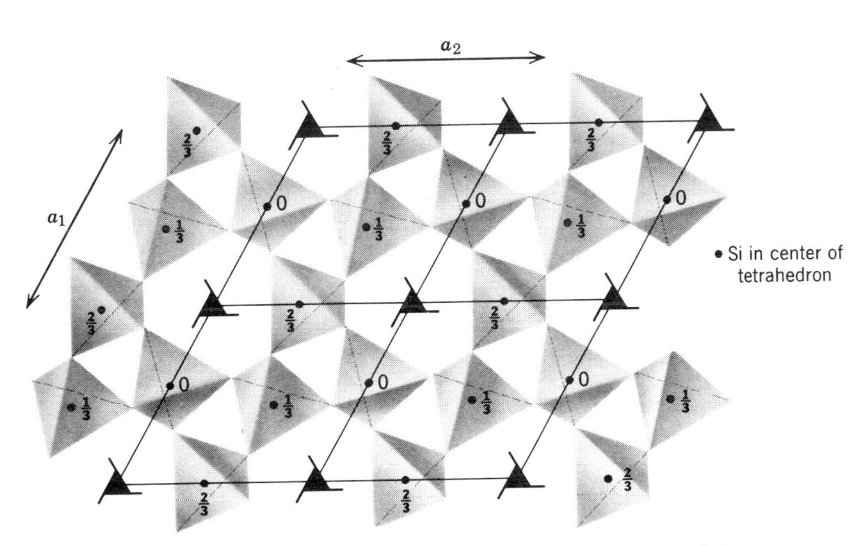

FIG. 13.30. Quartz structure model. Projection of the SiO_2 framework onto the basal plane. Four unit cells are shown. The numbers represent fractional heights of the centers (Si) of tetrahedra.

colored bands. Inclusions of iron oxide (goethite) are common. Some amethyst is strongly pleochroic: ϵ reddish-violet, ω pale blue.

Amethyst occurs in a variety of geologic environments but the most important is as crystals growing into cavities in lava flows. Such is its occurrence in Uruguay and Brazil, the principal commercial sources today. It is commonly a vein mineral and crystals of the finest quality were early found in veins in the Ural Mountains. It was called *Siberian amethyst* but today the term is applied to rich, deep-colored gems regardless of locality. Other important localities are in Zambia, Sri Lanka, Mexico, and Madagascar. Fine amethyst crystals, though not abundant, are found in many places in the United States; Maine, New Hampshire, North Carolina, Wyoming, and Colorado are a few of the most notable.

Citrine. Pale to dark yellow quartz. It is commonly mistaken for and sometimes sold as the more valuable topaz. In order to maintain the illusion, citrine is frequently sold under the names of "quartz topaz" or "topaz quartz." It is distinguished from topaz by **R.I.** and **S.G.** The impurity element in citrine, as in amethyst, is iron and much of the yellowish-brown quartz is produced by heat treating amethyst. Such heated stones retain the banding of the amethyst but lack a slight pleochroism present in natural citrine. Synthetic citrine is also nonpleochroic but lacks banding. Heat treating some amethyst from Brazil produces a clear green quartz. Natural citrine is uncommon and is found chiefly at the same localities as amethyst.

Smoky quartz or cairngorm stone. Quartz varying from almost black through brown to a smoky yellow and thus grades into citrine. Sometimes cut stones of smoky quartz are sold as "smoky topaz." Smoky quartz shows no dominant impurity element and the "smoke" is believed produced by exposure of rock crystal to radioactive material. On heating it becomes colorless but the color returns on exposure to radiation.

Smoky quartz is common. The best known and most productive area is in the Swiss Alps where veins have yielded many tons of beautiful crystals. Other notable localities are in the U.S.S.R.; Brazil; Madagascar; and Scotland from Cairngorm Mountain, from which it receives the name cairngorm stone. In the United States fine specimens have come from the Pikes Peak region of Colorado and from various localities in Maine and New Hampshire.

Rose quartz. Coarsely crystalline quartz with color varying from pale pink to deep rose-red which often fades in sunlight. Rose quartz is usually turbid, rarely transparent, and the color is presumably due to small amounts of titanium. In some crystals microscopic rutile needles are present oriented at right angles to the c axis and in three directions at 120° to each other giving a six-rayed star to a properly cut cabochon (see Fig. 6.8).

Rose quartz is a pegmatite mineral and is found in large amounts at many localities, but deep-colored flawless material is rare. Notable occurrences are in Brazil, Madagascar, and South-West Africa, and in the United States in Maine and South Dakota.

Milky quartz. Colorless quartz made milky white by the presence of myraid minute fluid inclusions. This is the typical vein quartz with which native gold is associated and, although not itself a gem material, may be cut in cabochons if it includes gold.

Quartz with inclusions. Inclusions of many minerals have been observed in coarsely crystalline quartz, but only a few are sufficiently abundant to color the mineral or large enough to be seen.

Rutilated quartz, sometimes called "Venus hair stone" is rock crystal containing reddish-brown to golden needles of rutile (Fig. 13.31). The needles are usually randomly oriented but may be oriented by the quartz structure in three directions at right angles to the c axis. Fibers of green actinolite and long slender crystals of black tourmaline are also found penetrating colorless quartz.

Quartz enclosing closely packed, parallel asbestos fibers has a chatoyancy when cut in cabochons. Such stones somewhat resemble the true chrysoberyl cat's eye and are called quartz cat's eye to distinguish them. The best examples of quartz cat's eye come from Sri Lanka and India.

Tiger's eye is a chatoyant quartz sometimes blue or gray but usually a golden-brown resulting from the presence of iron hydroxide, goethite. It is a pseudomorph after the blue asbestos, crocidolite,

FIG. 13.31. Rutilated quartz crystal, Brazil.

of South Africa in which the fibrous nature of the asbestos has been preserved in the replacing quartz. It is a common gem material used in beads, cabochon stones, and carvings.

Aventurine is the name given to quartz that contains small platy inclusions of mica which impart their color to the otherwise colorless mineral. A green color results from flakes of a chromium-bearing mica, and reddish-brown is due to spangles of mica of that color and resembles the aventurine varieties of feldspar. A brown glass, known as "goldstone," containing flakes of copper is manufactured to simulate the reddish-brown aventurine but it is softer and the reflections more flashy than in the natural material. Green aventurine quartz comes principally from India and is fashioned into beads, cabochons, and ornamental objects.

Microcrystalline Varieties.

Depending on the nature of their microcrystalline units, the fine-grained varieties of quartz can be divided into two types, fibrous and granular. Because of their fine textures, it is frequently difficult to distinguish between them without examination of small particles with a high-power microscope.

Fibrous varieties. *Chalcedony* is a general term given to all microcrystalline, fibrous varieties of quartz. However, because of color variations or banding, subvarieties of chalcedony are given special names. Specifically, *chalcedony* is a honey-yellow to gray, translucent material with a waxy luster. It commonly occurs in botryoidal masses with the fiber length at right angles to the hummocky surface. Chalcedony has a lower **S.G.** (2.60) than coarsely crystalline quartz because of a slight porosity. The porous nature of the mineral permits it to be colored by treating it with chemical solutions: black-brown, sugar and sulfuric acid; red, iron oxide; blue, potassium ferrocyanide and iron sulfate; yellow-green, hydrochloric acid; deep green, chromic acid (see Plate XI). Most chalcedony, usually cut cabochon, has been artificially colored. It is commonly found with agate as in Uruguay and Brazil. It is abundant in Oregon and California, and in Tampa Bay, Florida is found as pseudomorphs after corals and sponges.

Carnelian is a red chalcedony, colored by the iron oxide hematite, that grades into *sard,* colored by the brown iron hydroxide goethite. India is the most important source and much of the red mineral in the inlay work of the Taj Mahal is carnelian. Fine specimens have been found at many places in the United States particularly in Wyoming, Washington, Colorado, and Michigan.

Chrysoprase is an apple-green chalcedony colored by the presence of a nickel compound. It characteristically forms as a secondary mineral in veins in serpentine as at Frankenstein, Silesia; at Riddle, Oregon; and in Tulare County, California.

Agate is the most familiar of the subvarieties of chalcedony. It is commonly composed of alternating layers of different thickness, color, and porosity that tend to parallel the walls of the rock cavity in which it was deposited (see Fig. 13.32 and Plate I). The bands in many agates are concentric and completely fill the cavity. In others the banded portion only partially fills the cavity and the central portion may be a void or filled wholly or in part by coarsely crystalline quartz. The crystals in some such agates are amethystine. In some agates the bands show sharp reentrant angles which in section give mimetic patterns to which many names have been

FIG. 13.32. Agate geode with hollow center, cut and polished.

given. There is thus *fortification agate, landscape agate, eye agate,* and *brecciated agate* to mention but a few. Although white, milky, or gray are usually the natural colors of chalcedony in agate, yellowish-brown, brown-red, and rarely blue, lavender, and green are found.

Moss agate is a homogeneous, translucent chalcedony containing dendritic growths of black manganese oxide, or more rarely reddish-brown iron oxide, in a moss or fernlike pattern. In *iris agate* the fibers composing the bands are transparent and so slender and closely spaced that they act as a diffraction grating producing a delicate play of color in both transmitted and reflected light. Much *petrified wood* is agate, pseudomorphous after the wood, as at the Petrified Forest, Arizona.

Agate has a worldwide distribution but the most important commercial sources today are Uruguay and Brazil. Much of this material is sent for cutting to Idar-Oberstein, West Germany, itself a famous agate locality. The "thunder eggs" that are found abundantly in Oregon, California, and Nevada are chalcedony nodules frequently with colorful agate banding. The typical occurrence of agate is as fillings in gas cavities in lava.

Onyx is like agate but with alternating black and white layers in parallel planes. *Sardonyx* is onyx with red to orange layers alternating with white or black layers. Cameos are made from onyx with the carving in white standing in relief on the black background.

Granular varieties. The microcrystalline particles which make up the fine-grained granular varieties of quartz are roughly equidimensional rather than fibrous as in chalcedony.

Jasper is the principal fine-grained granular variety of quartz that is used for ornamental purposes. It is nearly opaque with a dull luster and usually admixed with other material, chiefly iron oxide. For this reason it is typically red to reddish-brown but because of other impurities may be green, dull blue, or black. *Riband jasper* is composed of bands of varying color that are usually broader and less well defined than in agate. The occurrence of jasper is in part similar to that of chalcedony, that is, as crusts or cavity and vein fillings. It also occurs as bedded deposits. With a decrease in impurities, jasper grades into *chert.* Chert and flint are the commonest of the granular microcrystalline varieties of quartz with no sharp distinction between them. Chert is usually light gray in bedded deposits, whereas flint is nearly black and occurs in nodules in chalk as at Dover, England. Both were used by early man for tools and weapons, but neither is considered a gem material.

Several fine-grained varieties of quartz may be either microgranular or microfibrous. *Plasma,* such a variety, is nearly opaque and colored green by included particles of green silicate minerals. It is used for carvings and mosaics. *Prase* is more translucent than plasma with a green color caused by included fibers of hornblende or by plates of chlorite. *Bloodstone* (heliotrope) is a plasma containing red spots of iron oxide or jasper resembling "blood" spots.

Because of their hardness, toughness, and chemical stability, all types of microcrystalline quartz are commonly found as waterworn pebbles in streams and on sea beaches.

Synthesis.

Quartz was synthesized as early as 1845 but it was 100 years later before large crystals were grown. Because of the scarcity of high-quality quartz for radio oscillators during World War II, a concerted effort was made to synthesize the mineral. Although little quartz was grown during the war, the experiments resulted in development of a hydrothermal method by which flawless crystals weighing 1 lb or more are today routinely manufactured (see Chapter 10).

Most of the synthesized quartz, amounting to many tons a year, is colorless and used for technical purposes. Recently, however, colored crystals have been made through the introduction of "impurity" elements into the hydrothermal solutions. As a result much of the citrine reaching the market today is synthetic and difficult to distinguish from the natural mineral. Other colors of synthetic quartz are smoky, clear green, and a bright blue having no natural counterpart (see Plate XI). Amethyst, but of poor quality, has also been synthesized.

Occurrence.

The major occurrences of each variety of quartz are given above under the description of the variety.

Name.

The name quartz is from a German word of ancient derivation at first applied to massive vein quartz but by the end of the eighteenth century used for all varieties.

Turquoise

Turquoise has been used as a gem since ancient times, certainly as early as 3000 BC in Egypt. It was also used by the early civilizations of Mesoamerica.

Crystallography.

Triclinic. Rarely in minute crystals, usually cryptocrystalline. Massive, compact. In thin seams, incrustations, and disseminated grains.

Physical Properties.

H 5–6. **S.G.** 2.6–2.8. Both properties variable depending on degree of compactness. *Luster* waxlike. Opaque to semitranslucent.

Color blue, bluish-green, green (see Plate XII).

Optical Properties.

Biaxial (+), $\alpha = 1.61$, $\beta = 1.62$, $\gamma = 1.65$, determined on single crystals from Virginia. Massive material gives only one refractometer reading near 1.62.

Chemical Composition.

A hydrous phosphate of copper and aluminum, $CuAl_6(PO_4)_4(OH)_8 \cdot 4H_2O$.

Luminescence and Absorption Spectra.

Under longwave u.v. fluorescence varies from faint greenish-yellow to bright blue. Nonfluorescent under shortwave u.v. The absorption spectrum, seen in reflected light, shows two lines in the blue-violet at 4300 and 4200 Å.

Diagnostic Features.

Of natural minerals variscite resembles poorer-quality turquoise of green rather than blue color, but variscite has lower **S.G.** and **R.I.** Chrysocolla may resemble turquoise but is much softer.

Simulation.

Turquoise is imitated by glass, plastic, enamel, and dyed chalcedony. Of these some glass imitations may be difficult to detect (see Plate XII).

Another widely used turquoise substitute is powdered turquoise, or mixtures of chemicals giving the same color, bonded in plastic. This material usually shows a molded appearance on the back, and is cut, rather than powdered, when scratched with a knife.

Natural turquoise of poor quality is also frequently treated to enhance its color and hardness. Methods used include impregnating with plastic, paraffin, and oil. Touching a red-hot needle to the material will aid in detection, either by revealing the oil or paraffin, or by the acrid odor coming from the plastic.

Most imitation turquoise contains copper compounds. A drop of hydrochloric acid placed on the back of such material quickly turns yellow-green, an effect not seen with natural turquoise.

Occurrence.

Turquoise is a mineral of secondary origin, usually found in the form of small veins traversing more or less decomposed volcanic rocks. The finest turquoise comes from near Nishapur in the province of Khoraman, Iran (Persia). The Sinai peninsula is a historically important source of Egyptian turquoise. Other occurrences are known in the Uzbeck Re-

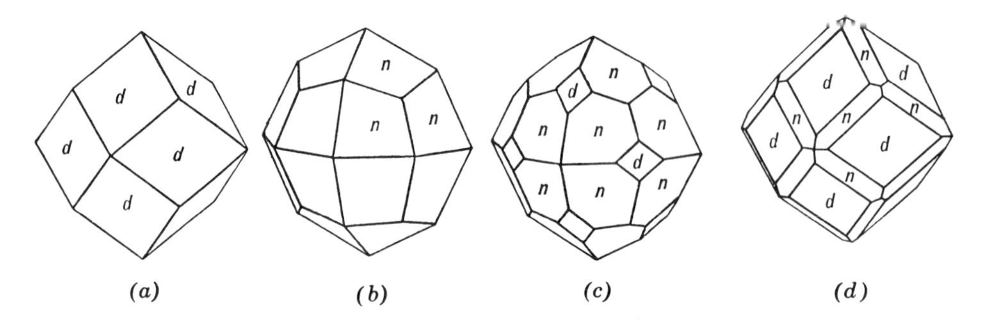

FIG. 13.33. Garnet crystals.

public, U.S.S.R., in northern Chile at Chuquicamata, and at Dayboro, near Brisbane, Australia.

In the United States excellent turquoise is or has been mined at several localities in Arizona, New Mexico, and Nevada.

Name.
Turquoise is French and means Turkish, the original stones having come into Europe from Persia through Turkey.

Garnet
The average person associates the term garnet with a single gem, reddish in color. In reality the term garnet refers to a group of closely related subspecies, many of which grade more or less into one another. As a result, garnet may be not red but brown, yellow, orange, white, green, and black. The orange and green varieties are particularly attractive and are important and valuable gemstones.

Crystallography.
Isometric. Common forms are the dodecahedron (Fig. 13.33*a*) and trapezohedron (Fig. 13.33*b*) often in combination (Fig. 13.33*c* and *d*). Figure 13.34 illustrates a well-formed dodecahedron embedded in the schist in which it grew. Although commonly in distinct crystals, garnet is also in rounded grains and coarse to fine granular.

Physical Properties.
H 6½–7½. **S.G.** 3.5–4.3, varying with composition. *Luster* vitreous to resinous. Transparent to opaque.

Color varying with composition, most commonly red, also pink, white, yellow, orange, green, brown, black.

Optical Properties.
Isotropic but often showing anomalous double refraction. **R.I.** 1.74–1.88, disp. 0.024–0.057, both varying with composition (see table below).

Chemical Composition.
The garnets are silicate minerals which conform to the general formula $A_3B_2(SiO_4)_3$. A may be calcium, magnesium, ferrous iron, or divalent manganese; B may be aluminum, ferric iron and, more rarely, chromium, titanium, zirconium, and vanadium (Fig. 13.35). The formulas of the chief subspecies are given below, with the specific gravity and refractive index of the more or less pure compound.

Subspecies	Composition	S.G.	R.I.
Pyrope	$Mg_3Al_2(SiO_4)_3$	3.60±	1.74±
Almandine	$Fe_3Al_2(SiO_4)_3$	4.05±	1.79±
Spessartine	$Mn_3Al_2(SiO_4)_3$	4.18±	1.80±
Grossular	$Ca_3Al_2(SiO_4)_3$	3.65±	1.74±
Andradite	$Ca_3Fe_2(SiO_4)_3$	3.83±	1.88±
Uvarovite	$Ca_3Cr_2(SiO_4)_3$	3.77±	1.87±

It is rare to find a garnet that corresponds to any one of these formulas. Most specimens show solid solution between the pure subspecies. For example, complete series extend between almandine and pyrope, between almandine and spessartine, and between grossular and andradite. However, only limited substitution may take place between the calcium garnets (grossular and andradite) and the iron, manganese garnets (pyrope, almandine, and spessartine).

FIG. 13.34. Garnet dodecahedron in schist, Salida, Colorado

Pyrope

Pyrope if pure would be colorless but it always contains iron and often some chromium, giving rise to red, purplish-red, and reddish-brown colors.

Most gem pyrope has **R.I.** between 1.73 and 1.75 and **S.G.** between 3.65 and 3.80. Since there is a complete solid solution series between pyrope and almandine there is no sharp dividing line between the two and these limits are arbitrary. Other properties are **H** 7¼, disp. 0.022.

Rhodolite is a name given to a pale rose-red to purple garnet, corresponding in composition to two parts pyrope and one part almandine. The properties of rhodolite are thus intermediate between pyrope and almandine: **R.I.** 1.76 (± 0.01), **S.G.** 3.84 (± 0.01). The absorption spectrum is similar to almandine (see below).

Pyrope is nonfluorescent under ultraviolet light. The absorption spectrum usually, like almandine, shows three bands at 5720, 5270, and 5050 Å. When chromium is present a weak doublet in the deep red may be seen at 6870 and 6850 Å.

Inclusions in pyrope are stubby crystals of low relief and needlelike crystals.

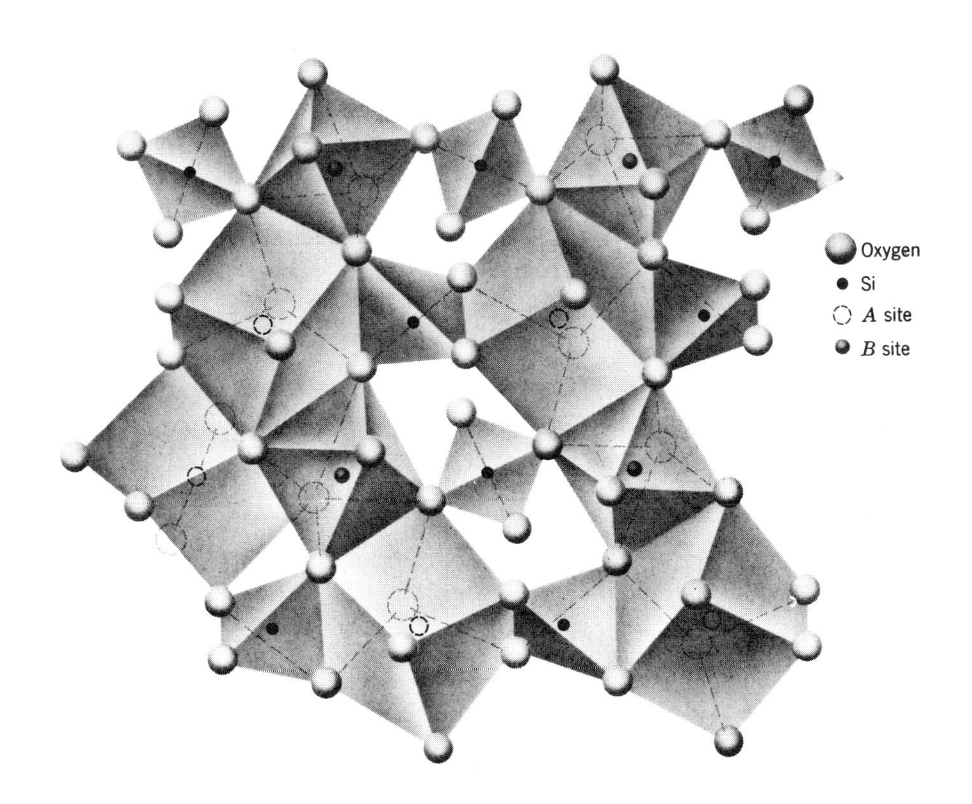

FIG. 13.35. Garnet structure, polyhedral model.

Pyrope may be confused with ruby, synthetic ruby, spinel, tourmaline, glass, and doublets. Ordinary tests will separate these with the exception of stones that have **R.I.** and **S.G.** overlapping with spinel, both being isotropic. However, most spinel fluoresces under ultraviolet light and pyrope does not. There are also characteristic differences in the absorption spectra and inclusions.

The best-known occurrence of pyrope is in Czechoslovakia, formerly Bohemia, hence the common name Bohemian garnet. Fine-quality pyrope is also found in kimberlite pipes, especially those in the Kimberley, South Africa area. Other localities are in Australia, Tanzania, Burma, Brazil, and the U.S.S.R. In the United States pyrope is found associated with kimberlite in Arizona and New Mexico. Rhodolite is found in North Carolina and Tanzania.

The name pyrope is from the Greek meaning firelike.

Almandine

Almandine never occurs pure in nature but for convenience, gems near the end of the almandine – pyrope solid solution series are called almandine. Most of the old gems known as *carbuncle* (a name once applied to all red gems) were almandine. The color range is from red-violet to dark brownish-red. The properties are **R.I.** 1.79 (± 0.03), **S.G.** 4.05 (± 0.12), disp. 0.027, **H** 7½.

Almandine is nonfluorescent in ultraviolet light. The absorption spectrum shows a strong band in the blue-green at 5050 Å and weak bands in the green at 5270 Å and the yellow at 5760 Å.

Crystal inclusions which may be zircon or hornblende are common in almandine. The needle-like crystal inclusions commonly present are possibly hornblende. This "silk" in almandine is in only two directions in any one plane (rather than in three directions as in ruby) and may produce a four-rayed star in cabochon-cut stones.

Almandine may be confused with ruby, synthetic ruby, spinel, glass, and doublets. Normal tests will differentiate these, and the absorption spectrum is characteristic.

Almandine is a very common mineral and occurrences are worldwide. Important gem localities are in India, Tanzania, Rhodesia, Madagascar, Sri Lanka, and Brazil. Star almandine comes from India and in the United States from Idaho.

The name almandine is from Alabanda, where in ancient times garnets were cut and polished.

Spessartine

Spessartine in gem quality is rare but attractive transparent yellow to yellow-brown and orange-brown material is known. The colors are much like hessonite, and the two varieties of garnet may be easily mistaken for one another.

The properties of spessartine are **R.I.** 1.80 (± 0.01), **S.G.** 4.18 (± 0.03), disp. 0.027, **H** 7.

Spessartine is nonfluorescent under ultraviolet light. The mineral, owing to its manganese content, shows weak absorption lines at 4950, 4890, and 4620 Å, and strong bands at 4320, 4240, and 4120 Å.

Inclusions in spessartine are characteristically featherlike clusters of liquid inclusions.

Spessartine may be confused with hessonite and other yellow to brown gems, such as zircon, topaz, quartz, beryl, sapphire, and chrysoberyl. All these with the exception of hessonite are anisotropic; spessartine is isotropic.

Spessartine is a widespread mineral. The principal sources of gem quality material are Burma, Sri Lanka, Madagascar, and Brazil. In the United States fine-quality material has been found at Amelia Court House, Virginia and San Diego County, California.

The name is after the Spessart district of Bavaria where the mineral was first found.

Grossular

The mineral name grossular is not well-known in the jewelry trade, but as an orange-yellow to orange-brown stone it has been a widely used gem for many years under the varietal names *hessonite* and *essonite*. More recently a transparent colorless to green grossular has become popular, and there is also a massive jade-like form marketed under the erroneous names "Transvaal jade" or "South African jade." Grossular occurs in a wide variety of colors including in addition to those already mentioned yellow, brown, violet-red, and orange-red.

The properties of grossular are **R.I.** 1.735–1.750, **S.G.** 3.61–3.75, disp. 0.027, **H** 7. Some translucent material has properties substantially lower than these: **R.I.** less than 1.72 and **S.G.** below 3.40. This material is *hydrogrossular,* a hydrous member of the garnet group with chemical composition expressed by the formula $Ca_3Al_2(SiO_4)_{3-x}(OH)_{4x}$.

Grossular is nonfluorescent under ultraviolet light, and the absorption spectra are unimportant.

The hessonite variety of grossular has very characteristic rounded crystal inclusions, often so abundant as to give the material a granular appearance. They are often accompanied by streaks that give the interior of the stones an oily look.

Hessonite may be confused with topaz, spessartine, quartz, beryl, sapphire, and chrysoberyl, but is readily identified by standard tests and its characteristic inclusions. The translucent forms of grossular may be confused with jadeite, nephrite, idocrase, and serpentine.

Grossular is a common mineral and is found in many localities worldwide, but few produce material of gem quality. The most important source of hessonite is the gem gravels of Sri Lanka. Fine gem quality material has recently been discovered at two widely separated localities. One (hessonite) is at the Jeffrey mine, Abestos, Quebec, Canada. The other is in the Tsavo National Park, Kenya and nearby areas of Tanzania. Crystals from the latter contain up to 1.6% vanadium oxide, coloring them a bright to dark green and have been called *tsavorite* from the locality. They make lovely gems but stones are small, rarely exceeding 2 carats. The massive jade-like material comes from 40 miles west of Pretoria, Republic of South Africa.

The name grossular comes from the botanical name for gooseberry, in allusion to the light green color of the original crystals.

Andradite

Andradite occurs in colors varying from yellow-green and green to black, but it is the green variety called *demantoid* that is most important as a gem. In fact demantoid is one of the rarest of gems and ranks in value with sapphire. Stones of over 4 carats are very rare. The color of demantoid is due to small amounts of chromium replacing iron in the crystal structure. A black variety of andradite, called melanite, has little importance as a gemstone. All the properties given here are for the yellow-green to green variety.

Demantoid has **R.I.** 1.875 (± 0.020), **S.G.** 3.84 (± 0.03), **H** 6½. The dispersion is 0.057, very high and greater than that of diamond. However, in demantoid the "fire" is partially masked by the strong body color of the gem.

Demantoid is nonfluorescent but appears reddish through the Chelsea filter. Usually it shows a strong absorption band in the violet part of the spectrum at 4430 Å, which appears as a cutoff because it is near the end of the visible spectrum. Deep green stones show a doublet in the red near 7000 Å, and two bands in the orange at 6400 and 6220 Å.

Demantoid has inclusions so characteristic as to assure positive identification with no further tests. These are actinolite (byssolite) fibers with a radiating pattern described as "horsetail" inclusions.

Demantoid may be confused with emerald, peridot, tourmaline, spinel, and sphene. However, standard tests, plus the characteristic "horsetail" inclusions, make identification relatively simple.

The main source of demantoid is in the Ural Mountains, U.S.S.R. A yellow to greenish-yellow andradite, sometimes called *topazolite* because of its resemblance to topaz, is found in the Ala Valley, Piemont, Italy, and at Zermatt, Switzerland.

The name andradite is for the Portuguese mineralogist, d'Andrada. The name demantoid is in allusion to its adamantine luster.

Uvarovite

Uvarovite occurs in brilliant, deep green crystals associated with chromium deposits, particularly in the Ural Mountains. The crystals are always small, too small for cutting, but were they larger, uvarovite would make a lovely gem. The mineral is named after Count Uvarov of Russia.

Zircon

Zircon is a common and widespread accessory mineral in igneous rocks of all types. It always occurs in well-formed crystals most of which are too small for use as a gem. But larger crystals, some as long as 12 in., are not uncommon. Because the mineral is mechanically strong and chemically stable, it weathers from the rocks in which it formed to be concentrated in stream gravels. It is from such accumulations that most gem zircon is recovered.

Both the luster and dispersion of zircon are high and, although not equalling, they approach diamond in these properties. Thus faceted gemstones of both colorless and colored zircon may be mistaken for diamond.

Zircon is an unusual mineral in that it occurs in two types. The "high type," also called normal for it is encountered in most gems, is crystalline; the "low type" is amorphous. The differences in the properties of the two types are pointed out below. But unless otherwise indicated, the following description is for the normal type.

Crystallography.

Tetragonal. Usually in crystals that most commonly show only the second-order prism and dipyramid (Fig. 13.36*a*). Other forms may be present as shown in Figures 13.36*b* and *c*. The evidence of the crystal forms is preserved in many partially rounded grains removed from stream gravels.

Physical Properties.

Cleavage poor prismatic; breaks usually with a conchoidal fracture. **H** 7–7½. **S.G.** 4.67–4.73. *Luster* adamantine. Transparent to translucent. For low type, **H** 6–6½. **S.G.** about 4.

Color reddish-brown, brown, orange, yellow, green, gray, colorless. Brownish-red crystals heated in air to about 900°C may turn colorless or a golden yellow; when heated in a reducing environment (without air) the crystals may become colorless or blue. Many of the colorless and golden yellow gems and all the blue stones have been produced by such heat treatment.

Optical Properties.

Uniaxial (+); $\omega = 1.92$, $\epsilon = 1.98$; biref. 0.06. Pleochroism weak; absorption $\omega > \epsilon$. The refractive indices and birefringence vary downward from these high values reaching a low (low type) with **R.I.** of 1.78 when the mineral is essentially isotropic. The low type may be slightly biaxial with a low birefringence (0.005). Between the high and low extremes are zircons that have been called "intermediate" with intermediate refractive indices; a **R.I.** of about 1.85 is most common. Dispersion of all zircon is about 0.040. Because of the high birefringence of normal zircon, one can observe a doubling of the back facets. To look for the doubling of the facets, the stone should be viewed from several directions for the most pronounced effect is seen when looking perpendicular to the optic axis.

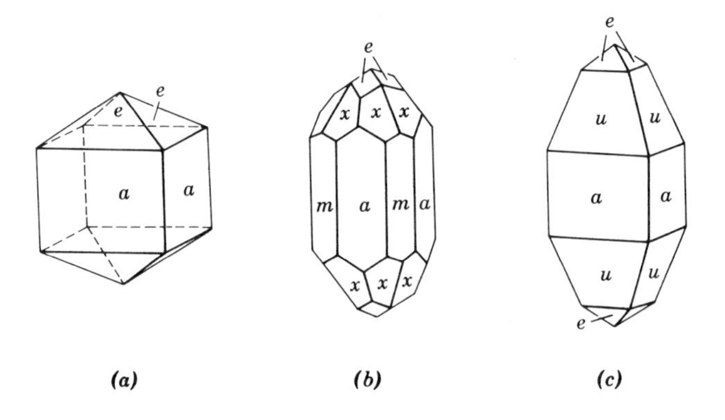

FIG. 13.36. Zircon crystals.

Chemical Composition.

Zirconium silicate, $ZrSiO_4$, but always containing some hafnium, usually 1–4%, and a little iron. In addition zircon may contain uranium and thorium. It is believed that all zircon was originally the normal, high type (Fig. 13.37) and that the low type is *metamict* having altered from it. That is, radiation from these radioactive elements has destroyed the crystal structure making the crystals amorphous. The breakdown of the crystal structure is accompanied by changes in physical properties. There is a lowering of hardness, **S.G.**, and **R.I.**, and sometimes a color change to green. Metamict zircons are frequently bounded by crystal faces and are thus amorphous pseudomorphs after the original crystalline mineral. On being heated to a high temperature (1450°C) the low type zircon is restored to the crystalline state with a return of the original physical properties.

Luminescence and Absorption Spectra.

Under long- and shortwave u.v. zircon may fluoresce yellow, but the intensity of the emitted light is variable and in some stones fluorescence is absent. The absorption spectrum varies with the source of the stone and the extent to which it is metamict. In the low type the bands are fewer and less distinct. The most constant band is in the red at 6535 Å. In many zircons other bands are uniformly distributed throughout the spectrum. The major bands are at 6910, 6225, 6590, 5895, 5625, 3575, 5150, 4840, and 4325 Å. Green zircon appears pinkish through the Chelsea filter.

FIG. 13.37. Structure of zircon consisting of SiO_4 tetrahedra and distorted ZrO_8 cubes.

Diagnostic Features.

Zircon may be confused in appearance with chrysoberyl, corundum and spinel both natural and synthetic, sphene, topaz, beryl, and glass. However, it can be distinguished from all these by its higher luster, higher **R.I.** (above the limit of the refractometer), higher birefringence as shown by the doubling of the back facets, and higher **S.G.** It is distinguished from diamond by its higher **S.G.** and double refraction.

Occurrence.

Zircon is a detrital mineral recovered principally from gravels. The most important gem producing localities are in southeast Asia in Laos, Cambodia, Thailand, and Upper Burma. Zircon is also found abundantly in the gem gravels of Sri Lanka. Colorless zircons from here have been called misleadingly *matura* (or *matara*) *diamonds*. Of less importance are several localities in New South Wales, Australia. Zircon is found in several places in the United States but rarely is it of gem quality.

Name.

The etymology of the word *zircon* is in doubt. It has been proposed that it comes from the Persian word *zargoon* meaning gold-colored.

Peridot (Olivine)

Peridot is the gem variety of the mineral olivine but it has not always gone under that name. It has been called chrysolite (a synonym for olivine) and "evening emerald," although its color is distinctly different from that of emerald. The ancient name was *topazion* derived from "Topazius," an island in the Red Sea, and until the eighteenth century was called topaz. At that time, for an unexplained reason, the name was transferred to another gem, our topaz of today and shortly thereafter *peridot* was used to indicate gem olivine.

The Red Sea island, today known as Zebirget (formerly Saint John's Island), is undoubtedly the source of the peridot known in Biblical times. For centuries the location of the island was lost but was rediscovered early in the twentieth century. Since then this ancient locality has produced many fine gems.

Crystallography.

Orthorhombic. Crystals usually show a combination of three prisms, *k, d, m,* the three pinacoids, *a, b, c,* and the dipyramid *e* (Fig. 13.38). Often flattened parallel to *a* or *b*. Most commonly found as grains either loose or embedded in rock.

Physical Properties.

Cleavage poor on front and side pinacoids. Commonly shows conchoidal fracture. **H** 6½–7. **S.G.** 3.34 (approx.) for peridot, but for the olivine series ranges from 3.27 to 4.37. *Luster* vitreous to oily. Transparent to translucent.

Color olive-green, characteristic of gems; may be brown, green, yellow (see plate V).

Optical Properties.

Biaxial (+). For most gem peridot $\alpha = 1.654$, $\beta = 1.672$, $\gamma = 1.690$ (all $\pm$ 0.01); biref. 0.036, sufficiently great to show a doubling of the back facets. $2V \simeq 90°$. Disp. 0.020. Pleochroism weak; β yellow, α and γ pale yellow.

Chemical Composition.

Magnesium, iron silicate, $(Mg,Fe)_2SiO_4$. The name olivine is given to the complete solid solution series that exists between the magnesium end member forsterite (Mg_2SiO_4) and the iron end member, fayalite (Fe_2SiO_4). Both **R.I.** and **S.G.** increase with increasing iron content. About 10% of the Fe + Mg in peridot is Fe and for it the formula can be written as $Mg_{1.8}Fe_{0.2}SiO_4$. Figure 13.39 is a polyhedral diagram showing the structure of olivine.

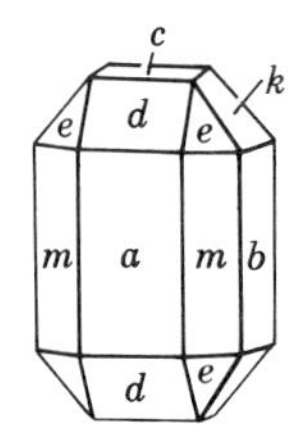

FIG. 13.38. Peridot crystal.

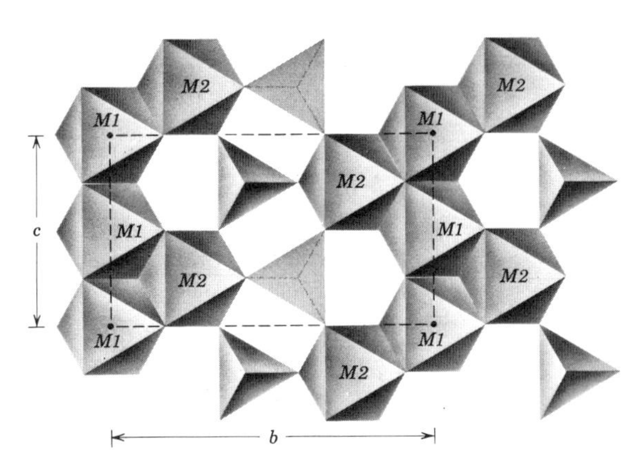

FIG. 13.39. Polyhedral model composed of octahedra and tetrahedra showing the structure of olivine. In peridot the *M*1 and *M*2 sites are occupied by Mg and Fe^{2+} randomly. The dashed lines outline the unit cell.

Luminescence and Absorption Spectra.

Peridot is nonfluorescent under ultraviolet light. The absorption spectrum shows bands characteristic of iron at 4930, 4730, and 4530 Å.

Diagnostic Features.

Although the color of peridot is characteristic, a cut stone may be confused with the following: chrysoberyl, zircon, and garnet, all of which have appreciably higher **R.I.** and **S.G.;** tourmaline with lower **R.I.** and **S.G.;** moldavite and glass, both of which are isotropic; sinhalite with similar **R.I.** but higher **S.G.** and more pronounced pleochroism.

Inclusions.

Stones from Zebirget may have reddish-brown inclusions of biotite mica; crystals from Burma may have clouds of microscopic hematite inclusions.

Occurrence.

The principal occurrence of gem peridot is on the Red Sea island, Zebirget, about 50 miles from the Egyptian port of Berenice. Peridot is found at the eastern end of the island where crystals appear to have grown into open fissures or cavities. High-quality stones have also come from Burma. Olivine is a common rock-forming mineral and it has been found in small rolled grains in many places, notably in Australia, Brazil, and Norway. In the United States peridot is recovered from the rock in which it was formed at Kilbourne Hole and Portillo Mar, New Mexico, and the San Carlos Indian Reservation, Arizona. It is also found in the lavas of Hawaii.

Name.

Olivine from the green color. Peridot from a thirteenth-century English word *peridote* or *peridota* and thus should be pronounced per' i dot'.

Spinel

Spinel in its red variety is a gem that has often been confused with ruby, as indicated by some of the names used for it, such as *ruby spinel, balas ruby,* and *rubicelle.* Many historic "rubies" were probably spinel. Two such instances are the "Black Prince's Ruby" and the "Timur Ruby" in the British Crown Jewels. Both these historic gems were thought to be ruby until comparatively recent times. One reason for the confusion is that the two minerals, spinel and corundum, are associated in gem gravels. Spinel, as does corundum, occurs in may colors, although red is the best known and most valuable.

Minerals of the spinel group make up a solid solution series with wide variations in composition. Unless otherwise noted the data given here are for the gem (magnesium) variety.

Crystallography.

Isometric. Usually in octahedral crystals (Fig. 13.40*a*); also in twinned octahedrons (Fig. 13.40*b*) known as the spinel twin.

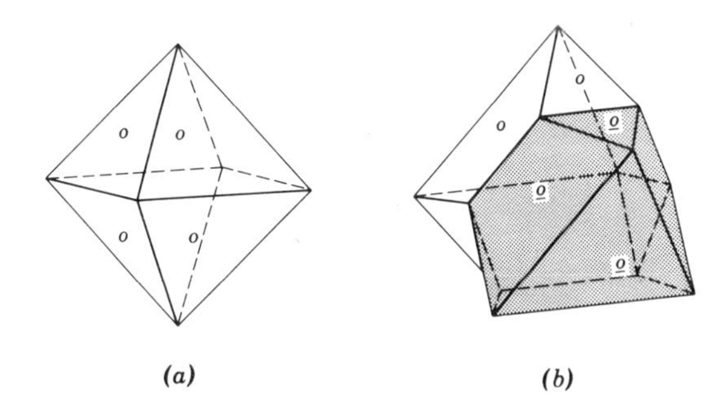

FIG. 13.40. Spinel crystals. (*a*) Octahedron. (*b*) Spinel twin.

Physical Properties

H 8. **S.G.** 3.57–3.72. *Luster* vitreous. Transparent to translucent

Color white, pink, red, lavender, blue, green, brown, black.

Optical Properties.

Isotropic. **R.I.** = 1.71–1.72 (for gem varieties). Disp. 0.020.

Chemical Composition.

Magnesium aluminum oxide, $MgAl_2O_4$. The general chemical formula for spinel may be written AB_2O_4, in which there may be nearly complete substitution of the divalent elements magnesium, ferrous iron, zinc, and manganese in the *A* positions, but only limited substitution of the trivalent elements aluminum, ferric iron, and chromium in the *B* positions. Figure 13.41 is a packing model illustrating the structure of spinel.

Given below are the end members of the spinel group and their chemical compositions.

THE SPINEL GROUP

Spinel	$MgAl_2O_4$	Magnetite	$FeFe_2O$
Hercynite	$FeAl_2O_4$	Franklinite	$ZnFe_2O$
Gahnite	$ZnAl_2O_4$	Jacobsite	$MnFe_2O$
Galaxite	$MnAl_2O_4$	Magnesiochromite	$MgCr_2O$
Magnesioferrite	$MgFe_2O_4$	Chromite	$FeCr_2O$

Transparent, gem quality spinels are all close to $MgAl_2O_4$ in composition.

Luminescence and Absorption Spectra.

Red and pink spinel fluoresce red in ultraviolet light, the effect being stronger under longwave than under shortwave. Pale blue and violet-blue spinels flu-

FIG. 13.41. Packing model illustrating the structure of spinel.

oresce green under longwave u.v., and purple spinels fluoresce a reddish to orangy color.

The color of red and pink spinels is due to trace amounts of chromium. The absorption spectrum shows a broad band in the yellow-green at 5400 Å and absorption in the violet. There are no lines in the blue portion of the spectrum and lines in the red are seen with difficulty. Blue spinel, colored by ferrous iron, has a strong band in the blue at 4580 Å, a narrower band at 4780 Å, and weaker lines in the orange, yellow, and green.

Inclusions.

Spinel often contains octahedral inclusions, either negative crystals or solid inclusions of spinel or magnetite. Needlelike inclusions ("silk") of the type commonly seen in corundum are rare in spinel, although they are sometimes present in sufficient abundance to cause asterism. These star spinels usually have four-rayed stars and are black.

Diagnostic Features.

Spinel may be confused with corundum but has lower **R.I.** and is singly refractive. Spinel has an **R.I.** and **S.G.** that may overlap with pyrope garnet, which is also singly refractive. Distinction may be made by fluorescence, a careful study of the inclusions, and by spectroscopy (see garnet).

Synthesis.

Synthetic spinel is made in quantity by the Verneuil method. Whereas in natural spinel the ratio of alumina (Al_2O_3) to magnesia (MgO) is 1:1, in synthetic spinel the ratio is usually about 2.5:1. This excess of alumina results in higher **R.I.** (1.73 or higher) and **S.G.** (3.64 compared to 3.60 in natural spinel).

Most synthetic spinel is produced as boules of colorless material, but it is made in many other colors, most of which are not found in nature. Deep blue synthetic spinel may be confused with sapphire, in a light blue color it imitates aquamarine, and light to dark green material may be confused with peridot and tourmaline.

Synthetic spinel may be identified by the presence of spherical gas bubbles, which are generally smaller and less numerous than in synthetic corundum. Another characteristic inclusion in synthetic spinel is a type described as "bread crumb." Curved growth lines are rarely seen. Strong anomalous double refraction, giving a "crosshatched" structure under crossed polars, is a very useful feature for identification.

A deep blue, coarsely crystalline sintered synthetic spinel is made to imitate lapis lazuli. The realism of this product is enhanced by the addition of particles of pure gold. Another form of synthetic spinel is made by reheating colorless material causing the excess alumina to separate, giving a cloudy appearance that imitates moonstone.

Occurrence.

Spinel is a common metamorphic mineral, and gem quality spinel is usually formed as a contact metamorphic mineral in limestone. It is found frequently as rolled pebbles in sands and gravels where it has been preserved because of its resistant physical and chemical properties.

Most gem spinel comes from the gem gravels of Burma, Sri Lanka, and Thailand, associated with corundum and other alluvial gem minerals.

Name.

Of uncertain origin, possibly derived from the Latin, *spina,* for thorn.

Feldspar

Feldspar is the most abundant mineral making up between 50 and 60% of the rocks of the earth's crust. However, it is not a single species with fixed chemical composition but a mineral group the members of which vary in chemistry and hence in many of their properties. There are high- and low-temperature varieties as well as intimate intergrowths of two varieties. Although most feldspar is in small grains as a rock-forming mineral, in places it occurs in larger crystals which because of opalescence, iridescence, or color are used as gem material.

The feldspars are all aluminum silicates which are commonly divided into two groups on the presence of other major elements: (1) the potash feldspars, orthoclase and microcline, containing potassium; and (2) the plagioclase feldspars, containing sodium and calcium.

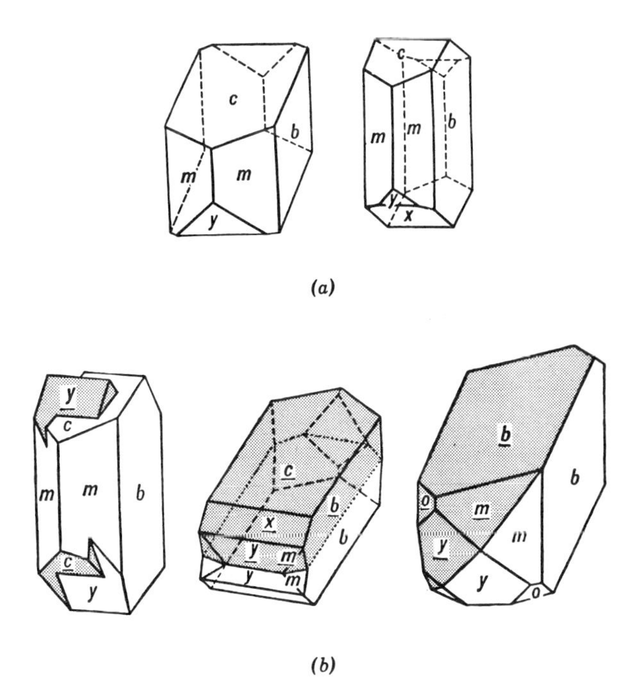

FIG. 13.42. Feldspar. (*a*) Single crystals. (*b*) Twinned crystals. From left to right: Carlsbad twin, Manebach twin, Baveno twin.

Orthoclase and Microcline

Orthoclase, the high-temperature form, and microcline, the low-temperature form of the same compound, have many properties in common and are here described together.

Crystallography.

Orthoclase monoclinic; microcline triclinic. Crystals of the two have similar appearance and several types of twinning in common (Fig. 13.42). Evidence of these twins is rarely seen in cut stones. However, because of its lower symmetry, microcline is also polysynthetically twinned according to the *albite law* with the side pinacoid the twin plane, and the *pericline law* with *b* the twin axis. This twinning may be seen between crossed polars.

Physical Properties.

Cleavage in two directions, good on the side pinacoid and slightly better on the basal pinacoid. In orthoclase the cleavages are at right angles; in microcline nearly at right angles. These cleavages are a characteristic common to all feldspars. **H** 6. **S.G.** 2.56. *Luster* vitreous. Transparent to translucent.

Color colorless, milky white, red, yellow, green. Microcline of rich green color is called *amazonite* (Fig. 13.43).

Optical Properties.

Biaxial (−), biref. 0.008; disp. 0.012. Orthoclase: $\alpha = 1.518$, $\beta = 1.524$, $\gamma = 1.526$. Microcline: $\alpha = 1.522$, $\beta = 1.526$, $\gamma = 1.530$.

Chemical Composition.

Potassium aluminum silicate, $KAlSi_3O_8$. Sodium may replace part of the potassium and in the variety *sanidine* as much as 50% of the potassium is replaced. At high temperature sodium can be accommodated in orthoclase crystals but as they cool the sodium ions are a poor "fit" in the structure and are exsolved as the sodium feldspar, albite. If the cooling time was long, the exsolved albite can be seen as parallel plates in the potash feldspar which has

FIG. 13.43. Microcline feldspar and smoky quartz, Crystal Peak, Colorado. The gray, blocky feldspar in the photograph is in reality a rich green amazonite.

inverted to microcline and the intergrowth is called *perthite*. It is called *microperthite* if the plates can be seen only microscopically and *cryptoperthite* if they can be detected only with X-rays. In the latter cases the host is orthoclase, there being insufficient time for inversion to microcline.

Gem Varieties.

Moonstone, the most important feldspar gem, is a translucent variety of high-temperature orthoclase, *adularia.* Its outstanding feature is a beautiful bluish sheen that moves across the surface of the stone. Moonstone is a microperthite or cryptoperthite and the optical effect, known as *schiller* or *adularescence,* is produced by the interference of light reflected from the submicroscopic intergrowth of orthoclase and albite. Moonstone is cut cabochon, and the best effect is seen when the flat base of the stone is parallel to the planes of the intergrowth. Some moonstones have a chatoyancy in which a band with pearly luster is present on a more transparent background. Albite and oligoclase feldspar also have moonstone varieties.

Sanidine, a colorless, glassy variety of orthoclase and another transparent orthoclase, *adularia,* are sometimes cut as gems. More important, but at present a rare material, is the golden yellow orthoclase colored by small amounts of iron. These yellow gems, which are usually faceted, are too soft (**H** = 6) for day-to-day wear as a ring stone.

Larvikite is a rock from Larvik, Norway widely used as a facing on public buildings. It contains crystals of a sodium-rich orthoclase in perthitic intergrowth with oligoclase feldspar which emit a blue to gray schiller from the planes of intergrowth. It is frequently confused with labradorite, for the color produced by light interference in the two minerals is similar.

Amazonite or *amazonstone* is a bright green, or sometimes a blue-green, nontransparent variety of microcline. It is frequently seen in carvings resembling jade and as cabochon gemstones is used in pendants and brooches. The green color is believed due to vacancies (color centers) in the crystal lattice resulting from substitution of lead (Pb^{2+}) for potassium (K^+), leaving a K^+ vacancy.

It has frequently been stated that green feldspar is always microcline. Recently, however, bright green orthoclase has been found at Broken Hill, New South Wales, Australia.

Luminescence and Absorption Spectra.

Under shortwave u.v. moonstone gives off a faint yellow glow and yellow orthoclase fluoresces a weak reddish-orange. Amazonite under longwave u.v. may glow a yellow-green, whereas green orthoclase from Australia is nonfluorescent. Colorless orthoclase, including moonstone, gives no characteristic absorption spectra but the yellow stones show broad bands at 4480 and 4200 Å that result from the presence of iron.

Diagnostic Features.

The refractive indices and specific gravity of both orthoclase and microcline are relatively constant and are characteristic. By these properties yellow orthoclase can be distinguished from other similar appearing stones and moonstone from an imitation made of heat treated white spinel. Moonstone imitations made of "opal" glass are isotropic. In cut stones the two cleavages at or nearly at 90° can sometimes be detected under the microscope; in carved amazonite the cleavages are usually visible.

Inclusions.

Moonstone may show straight elongated cracks crossed by irregular shorter cracks arranged in a ladderlike fashion.

Synthesis.

Feldspar has been synthesized in the laboratory but not of gem size or quality.

Occurrence.

Orthoclase and microcline as rock-forming minerals are found abundantly on every continent, but their gem varieties are far more limited. Adularia in fine crystals is found in veins in the Swiss Alps but is usually clear and lacking schiller. Beautiful moonstones have come from India, Burma, Tanzania, and Madagascar but the most important source is Sri Lanka, where it occurs in the gem gravels. In the United States moonstones have been found in several states, notably in Colorado, New York, New Mexico, Indiana, Virginia, and North Carolina.

Although yellow orthoclase has been reported from several localities, the only important source has been in Madagascar where it occurred in a pegmatite at Itrongay.

Amazonite also is a pegmatite mineral but unlike yellow orthoclase is found at many localities, particularly in India and in the State of Minas Gerais, Brazil (not near the Amazon, the river for which it is named). In the United States fine crystals have come from the Pikes Peak region of Colorado and Amelia Court House, Virginia.

Name.

The name orthoclase refers to the right-angle cleavage and microcline is derived from two Greek words meaning little and inclined, referring to the small deviation of the cleavages from 90°. Feldspar is derived from the German word *Feld* meaning field.

Plagioclase Feldspar

The plagioclase feldspars are sometimes called the *soda-lime feldspars* for they form a complete solid solution series from sodium-rich, albite, to calcium-rich, anorthite. Various subspecies names have been given arbitrarily to feldspars of intermediate composition, as follows:

PLAGIOCLASE FELDSPAR SERIES

	Albite (%)	Anorthite (%)
Albite, $NaAlSi_3O_8$	100–90	0–10
Oligoclase	90–70	10–30
Andesine	70–50	30–50
Labradorite	50–30	50–70
Bytownite	30–10	70–90
Anorthite, $CaAl_2Si_2O_8$	10–0	90–100

Crystallography.

Triclinic. Polysynthetic twinning according to the albite law, with the side pinacoid the twin plane, is almost universal. In rough material the twinning is evidenced by closely spaced striations on the basal cleavage (see Fig. 4.50*g*).

Physical Properties.

Cleavage two directions at nearly right angles, basal pinacoid excellent, side pinacoid good. **H** 6. **S.G.** 2.62 (albite) to 2.76 (anorthite) (see Figure 13.44). *Luster* vitreous to pearly. Transparent to translucent.

Color colorless, white, gray; less commonly greenish, yellowish, red. A beautiful play of color is present in certain varieties (see below).

Optical Properties.

Biaxial. Optic sign (+) in albite and labradorite; (−) in others. The refractive indices increase with increase in calcium. For example, in albite $\beta = 1.532$ but in anorthite $\beta = 1.584$ as shown in Figure 13.44. From the curves of the figure one can see the birefringence varies from 0.008 to 0.014. Disp. 0.012. In transparent plagioclase between crossed polars, the albite twinning can be seen as parallel bands that go to extinction in two different positions.

Chemical Composition.

A complete solid solution series from the sodium aluminum silicate, albite, $NaAlSi_3O_8$, to the calcium aluminum silicate, anorthite, $CaAl_2Si_2O_8$. Potassium

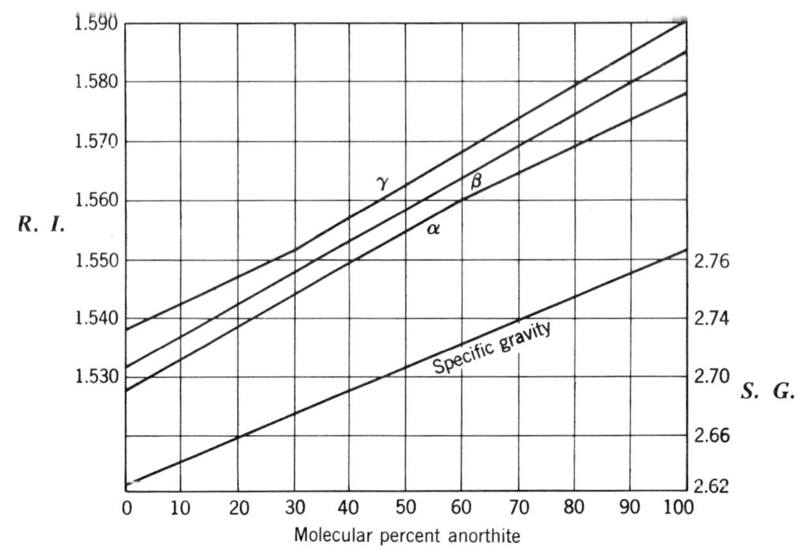

FIG. 13.44. Variation of refractive indices and specific gravity with composition in the plagioclase feldspar series.

may substitute for sodium toward the albite end of the series.

Diagnostic Features.

If seen, albite twinning and the two cleavages at nearly 90° are diagnostic. **R.I.** and **S.G.** are necessary to distinguish one member of the series from another.

Gem Varieties.

Albite in a white opalescent variety is sometimes cut as moonstones but such stones are uncommon. *Peristerite* is an albite with a white to creamy or gray background which, when light strikes it at certain angles, shows a lovely sky blue iridescence and is used as an ornamental stone (see *Play of Color,* in Chapter 6).

Oligoclase is known to exist in moonstone variety but it is rarely seen. Oligoclase is more commonly encountered as a gem in *aventurine feldspar* or *sunstone*. This variety contains platy inclusions of iron oxide which give the stones a golden or reddish color and emit red to yellow reflections.

Labradorite, usually a gray nondescript mineral, may in some specimens show a lovely play of color resulting from inclusions and exsolution of thin lamellae of different composition. As the stone is turned, colors of blue, green, yellow, and less commonly red play across the basal cleavage face or from polished planes nearly parallel to this face.

Luminescence and Absorption Spectra.

The plagioclase feldspars (albite, oligoclase, and labradorite) show no marked fluorescence and no characteristic absorption spectra.

Occurrence.

The plagioclase feldspars are even more abundant and widespread as rock-forming minerals than the potash feldspars, but as gem material are more restricted. Peristerite comes principally from Ontario and Quebec, Canada where it is found at several localities in both provinces. The most important source of sunstone is southern Norway, but it is found in several places in Canada and near Lake Baikal, U.S.S.R. The principal occurrence of labradorite exhibiting a play of color is on the coast of Labrador, where it is found in large cleavable masses. Facet grade labradorite comes from Warren Valley, Lake County, Oregon; near Alpine, Texas; Modoc County, California; and Millard County, Utah. Also seen as faceted stones but lacking a play of color is a straw-yellow labradorite from Chihauhau, Mexico.

Name.

Albite is named from the Latin, *albus,* meaning white. Oligoclase is derived from two Greek words meaning little and fracture, since it was believed to have poorer cleavage than albite. *Labradorite* receives its name from the locality, Labrador. Plagioclase is derived from the Greek word meaning oblique, in allusion to the oblique angle between the cleavages of all members of the group.

LESS IMPORTANT GEMSTONES

In this group are included 41 minerals that are used as gems but are less commonly encountered than the important gemstones of the preceding section. No effort has been made to rank them as to importance and they are considered alphabetically, followed by a brief discussion of the natural glasses.

LESS IMPORTANT GEMSTONES

Andalusite	Fluorite	Scapolite
Apatite	Gypsum	Scheelite
Aragonite	Hematite	Serpentine
Axinite	Idocrase	Sillimanite
Benitoite	Iolite	Sinhalite
Brazilianite	Kornerupine	Smithsonite
Calcite	Kyanite	Sodalite
Chrysocolla	Lapis Lazuli	Sphalerite
Danburite	Lazulite	Sphene
Diopside	Malachite	Spodumene
Dioptase	Phenakite	Staurolite
Enstatite	Pyrite	Variscite
Epidote	Rhodochrosite	Zoisite
Euclase	Rhodonite	Glasses (natural)

Andalusite

Andalusite has been an important industrial mineral used in the manufacture of spark plugs and other high-grade refractory materials but is seen only rarely as a gemstone.

Crystallography.

Orthorhombic. Usually in coarse prismatic crystals with a square cross section. Most gem material is recovered as waterworn pebbles.

Physical Properties.

H 7½. **S.G.** 3.16–3.20. *Luster* vitreous. Transparent to translucent.

Color red, green, yellow-green, brownish-green, reddish-brown.

Optical Properties.

Biaxial (−). $\alpha = 1.632$, $\beta = 1.638$, $\gamma = 1.643$ (all ±0.004); biref. 0.009–0.011; $2V = 85°$. Disp. 0.016. Pleochroism frequently strong with α red to reddish-brown, β and γ green to yellow.

Chemical Composition.

Aluminum silicate, Al_2SiO_5, polymorphous with kyanite and sillimanite (i.e., same composition). Although nearly a pure compound, the red varieties contain small amounts of iron and the green varieties small amounts of manganese.

Luminescence and Absorption Spectra.

Andalusite does not luminesce under shortwave u.v. but under longwave u.v. shows a weak green to yellow-green fluorescence. A band in the blue at 4550 Å may be present in brownish stones. Green stones show a sharp edge in the green at 5525 Å and less pronounced lines at 5495 Å and 5175 Å. In general there is strong absorption in the blue-violet region.

Inclusions.

An impure variety of andalusite called *chiastolite* contains black carbonaceous inclusions arranged in a regular manner which, in cross sections of elongated crystals, form a cruciform design that may vary from one end of the crystal to the other (Fig. 13.45). Sections cut from such crystals are polished and worn as amulets.

Diagnostic Features.

Brownish-red and green varieties of andalusite are difficult to distinguish from tourmaline, for the two minerals not only resemble each other in appear-

FIG. 13.45. Successive sections through a chiastolite crystal.

ance but also in **R.I., S.G.,** and pleochroism. The following differences exist between them: tourmaline—uniaxial, biref. 0.025, greatest absorption associated with the higher index; andalusite—biaxial, biref. 0.010, greatest absorption associated with the lowest index.

Occurrence.

Andalusite is most commonly found in schists as a product of regional metamorphism. Because it is hard and chemically stable, it weathers from the soft surrounding micaceous rock to be washed into neighboring streams. Most gem andalusite is recovered as waterworn pebbles from the gravels of such streams. Its two major occurrences are of this type, the gem gravels of Sri Lanka and the alluvial deposits of Brazil.

In the United States faceted gems of andalusite have been cut from crystals found in the vicinity of Custer, South Dakota. Chiastolite is found in Madera County, California, and in Worcester and Middlesex Counties, Massachusetts.

Name.

Andalusite from the province of Andalusia, Spain. Chiastolite from the Greek letter χ (*chi*), in reference to the black cross.

Apatite

Apatite is an abundant mineral and is one of the principal constituents of *phosphate rock,* mined in large quantities to make phosphate fertilizer. Although rarely existing in gem quality, apatite comes in a wide range of colors, and if it were not so soft would make a desirable gem.

Crystallography.

Hexagonal. Commonly in prismatic crystals, less often short prismatic or tabular (Fig. 13.46). Also in massive granular to compact masses.

Physical Properties.

Cleavage poor basal. **H** 5. **S.G.** 3.15–3.20. *Luster* vitreous. Transparent to translucent.

Color usually some shade of green or brown; also blue, violet, yellow, colorless.

Optical Properties.

Uniaxial (−). $\omega = 1.646$, $\epsilon = 1.642$; biref. 0.002–0.004; disp. 0.013. Pleochroism (dichroism) strong only in blue stones, ϵ blue and ω yellow.

Chemical Composition.

There is wide range in composition in the apatite mineral group, but the gem quality crystals are fluorapatite, a calcium fluo phosphate, $Ca_5(PO_4)_3F$.

Luminescence and Absorption Spectra.

Fluorescence varies with the color of the stone. Under longwave u.v. the fluorescence is as follows: yellow stones, lilac-pink; blue stones, violet-blue to sky-blue; green and violet stones, greenish-yellow. The absorption spectra also vary considerably. Yellow stones show two groups of closely spaced lines at 5800 and 5200 Å. Blue stones show broader bands, the strongest being at 5120, 4910, and 4640 Å.

Diagnostic Features.

Uncut crystals of apatite resemble beryl, but are readily distinguished by their lesser hardness. Cut stones can be distinguished from other gems with nearly the same refractive index by their very low birefringence.

Occurrence.

Apatite is found as a constituent of all classes of rocks, igneous, sedimentary, and metamorphic. It is

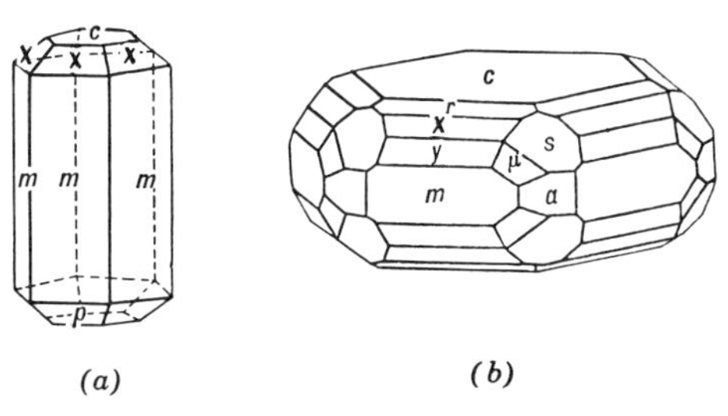

FIG. 13.46. Apatite crystals.

also found in pegmatites and hydrothermal veins, and with titaniferous magnetite (iron ore) deposits.

Gem quality blue apatite is found at Moguk, Burma and in the gem gravels of Sri Lanka, Yellowish-green crystals come from Spain, and a bluish-green variety from southern Norway. Well-formed crystals of yellow color are abundant at Cerro de Mercado, Durango, Mexico. Fine green gem quality material comes from Ontario and Quebec, Canada. In the United States violet crystals are found at Auburn, Maine; San Diego County, California; and near Keystone, South Dakota.

Name.
From the Greek word meaning to deceive, since the gem varieties were confused with other minerals.

Aragonite
Aragonite is the orthorhombic form of calcium carbonate, less common than calcite, the other polymorph. It is the principal constituent of pearl and of the mother-of-pearl of many shells. Some onyx marble is composed of aragonite.

Crystallography.
Orthorhombic. Crystals usually acicular, tabular in pseudohexagonal twins. Also found in reniform, columnar, and stalactitic aggregates.

Physical Properties.
Imperfect cleavage parallel to the length of the crystals. **H** 3½–4. **S.G.** 2.95 (harder and higher specific gravity than calcite). *Luster* vitreous. Transparent to translucent.

Color colorless, white, pale yellow and other lightly tinted colors.

Optical Properties.
Biaxial (−). $\alpha = 1.530$, $\beta = 1.682$, $\gamma = 1.686$; $2V = 18°$ (small); biref. 0.156. Disp. 0.015.

Chemical Composition.
Calcium carbonate, $CaCO_3$, polymorphous with calcite.

Diagnostic Features.
Like calcite it will effervesce in hydrochloric acid but is distinguished from calcite by its higher specific gravity and lack of rhombohedral cleavage. Cleavage fragments of columnar calcite are terminated by a cross cleavage which is lacking in aragonite.

Occurrence.
Aragonite is a widespread mineral little used as a gem material except as it occurs as pearl and as the pearly layer of shells.

Name.
From Aragon, Spain.

Axinite
Axinite is a rare mineral prized by mineral collectors for its groups of well-formed wedged-shaped crystals. Cut gemstones are uncommon.

Crystallography.
Triclinic. The mineral is interesting to crystallographers as the only representative of the crystal class of lowest symmetry, that is, no symmetry. Crystals are usually thin with sharp edges (Fig. 13.47).

Physical Properties.
Cleavage distinct in one direction. **H** 7. **S.G.** 3.27–3.35. *Luster* vitreous. Transparent to translucent.

Color commonly clove-brown, but most gems are blue or yellow and more rarely violet or green.

Optical Properties.
Biaxial (−). With variation of chemical composition there is a variation in **R.I.** but most gems have α =

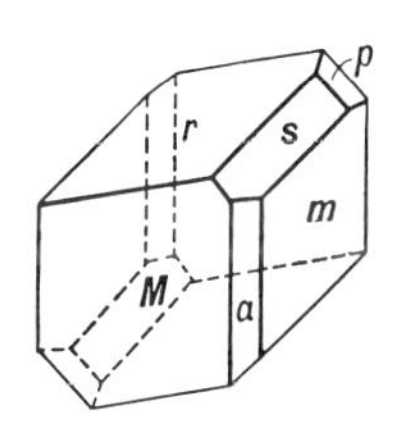

FIG. 13.47. Axinite crystal.

1.675, β = 1.682, γ = 1.685 all ±0.003; biref. 0.01; $2V$ = 70°. Pleochroism may be strong: α brown, β blue-violet, γ yellow-green. Disp. 0.014.

Chemical Composition.

A complex aluminum borosilicate with calcium, iron, and manganese, $(Ca,Fe,Mn)_3Al_2(BO_3)(SiO_4O_{12})$-(OH). There is a considerable range in composition with varying amounts of calcium, iron, and manganese.

Luminescence and Absorption Spectra.

Axinite is nonfluorescent in ultraviolet light. There is an absorption band in the green at 5120 Å and two in the blue at 4920 and 4660 Å.

Diagnostic Features.

Axinite may have the appearance of topaz, chrysoberyl, and corundum, all of which have appreciably higher **S.G.** Their refractive indices compared to axinite's are lower for topaz, and higher for chrysoberyl and corundum.

Occurrence.

Axinite occurs principally as a contact metamorphic mineral in cavities at the contacts of igneous and calcareous rocks. Notable localities are Bourg d'Oisans, Isere, France; St. Just, Cornwall, England; Obira, Japan; and Baja California, Mexico.

Name.

From the Greek word meaning axe, in allusion to the wedge-shaped crystals.

Magnesioaxinite, discovered as a gem, was described as a new mineral in 1976.

Benitoite

Benitoite, a sapphire blue gem, was discovered in 1906 in California. It was found to be not only a new mineral species, but a mineral crystallizing in the ditrigonal–dipyramidal crystal class, a class known to be possible, but until that time without a known example. Stones of over 2 carats are rare; the largest known, 7.5 carats, is in the Smithsonian Institution, Washington, D.C.

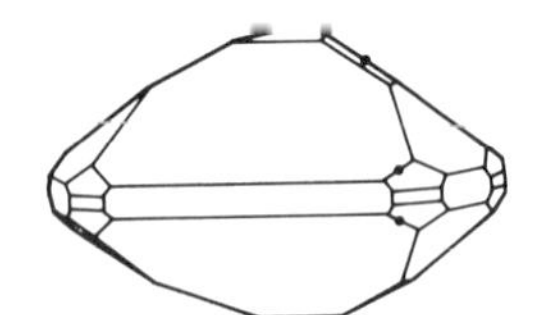

FIG. 13.48. Benitoite crystal.

Crystallography.

Hexagonal. Well-formed, tabular crystals with triangular cross section and terminated by trigonal pyramids (Fig. 13.48).

Physical Properties.

Fracture conchoidal. **H** 6½. **S.G.** 3.64. *Luster* vitreous; transparent to translucent.

Color colorless and light to dark blue.

Optical Properties.

Uniaxial (+). ω = 1.757, ϵ = 1.804. Biref. 0.047, disp. 0.044 are both unusually high. Pleochroism ϵ deep blue, ω nearly colorless.

Chemical Composition.

Barium titanium silicate, $BaTiSi_3O_9$.

Luminescence and Absorption Spectra.

Fluoresces light blue under shortwave u.v. The absorption spectra are unimportant.

Diagnostic Features.

Resembles sapphire but may be distinguished by its high birefringence, lower **S.G.,** and blue fluorescence.

Occurrence.

Occurs in fissures in altered serpentine with white natrolite and black neptunite. Only one locality is known, the Dallas Gem mine, San Benito County, California.

Name

For the locality.

Brazilianite

Several new minerals are described and named each year and occasionally a well-known mineral is found for the first time as gem material. It is far more rare to have a new mineral make its first appearance as gem crystals, but such was the case of brazilianite. In 1944 F. H. Pough, an American mineralogist, while in Brazil was shown a large yellow-green, transparent crystal assumed to be chrysoberyl. Dr. Pough noted that neither the hardness nor the crystal form agreed with chrysoberyl. Later through chemical, X-ray, and optical study it was demonstrated that the presumed chrysoberyl was a new mineral and it was named brazilianite.

Crystallography.

Monoclinic. Most commonly in crystals elongated parallel to the *a* axis but also in short prismatic striated parallel to *c*.

Physical Properties.

Cleavage perfect on the side pinacoid, elsewhere breaks with conchoidal fracture. **H** 5½. **S.G.** 2.98. *Luster* vitreous. Transparent to translucent.

Color chartreuse yellow to pale yellow in material used as gems; may be white or colorless (see Plate IX).

Optical Properties.

Biaxial (+); $\alpha = 1.602$, $\beta = 1.609$, $\gamma = 1.621$; biref. 0.019; $2V = 75°$. Disp. 0.014.

Chemical Composition.

Sodium aluminum phosphate, $NaAl_3(PO_4)_2(OH)_4$. A small amount of potassium may substitute for sodium.

Occurrence.

Brazilianite is found in pegmatites associated with other phosphate minerals. It has been found in two states in Brazil. The original find was in Minas Gerais and the second in Paraiba. Brazilianite not of gem quality also has been described from two localities in New Hampshire.

Name.

After Brazil, the country in which it was first found.

Calcite

Calcite is a very common mineral that occurs in a wide variety of colors, textures, and forms. It is the predominant or only mineral of limestone, marble, and onyx marble, all used as ornamental stones.

Crystallography.

Hexagonal-rhombohedral. Crystals common and extremely variable in habit. The calcite used as an ornamental material is usually massive. Although rare, some calcite crystallizes in a fine fibrous aggregate with a silky luster and is known as *satin spar,* the same name applied to fibrous gypsum (see gypsum, below).

Physical Properties.

Cleavage perfect rhombohedral (three directions, cleavage angle = 74°55′). **H** 3. **S.G.** 2.72. *Luster* vitreous to earthy. Transparent to translucent.

Color usually colorless or white, but also gray, pink, red, green, blue, yellow, brown.

Optical Properties.

Uniaxial (−). $\omega = 1.658$, $\epsilon = 1.486$; biref. 0.172; disp. ω 0.028, ϵ 0.014. Because of the very high birefringence has been called *doubly refracting spar* (Fig. 13.49).

Chemical Composition.

Calcium carbonate, $CaCO_3$, a composition it shares with the less common polymorph, aragonite.

Luminescence and Absorption Spectra.

Some calcite, but not all, fluoresces under ultraviolet light. The absorption spectra are unimportant.

Diagnostic Features.

Calcite may be distinguished by its hardness and perfect rhombohedral cleavage. It may be distinguished from dolomite, a calcium magnesium carbonate, by the fact that calcite effervesces freely in cold hydrochloric acid whereas dolomite does not. Distinguished from aragonite by having lower specific gravity and rhombohedral cleavage.

Varieties.

Iceland spar is the name given to colorless transparent crystals of calcite used for optical purposes.

FIG. 13.49. Double refraction in calcite.

It is cleavage fragments of Iceland spar that are used in the construction of the dichroscope (see Chapter 6). An occasional transparent crystal, either colorless or variously tinted, is cut as a faceted gemstone, but only for collectors. Because of the perfect cleavage and low hardness, faceted stones are not used in jewelry. The high birefringence results in a pronounced doubling of the back facets in cut stones.

Onyx marble, also called *Mexican onyx,* is a banded, translucent material formed as a compact spring or cave deposit. It takes a good polish and is extensively used for small carvings and ornamental objects. Onyx marble is often dyed green and sold incorrectly as "Mexican jade." It is easily distinguished from jade by its strong banding, lower hardness, and effervescence in hydrochloric acid.

Limestone is a sedimentary rock composed essentially of calcite. Although it has important industrial uses, its use as an ornamental material is limited. *Marble* is a metamorphosed limestone that is white when pure, but it may be variously colored by impurities. Since early Greek and Roman times, it has been used extensively for statuary and for many other ornamental purposes. Both limestone and marble are dyed many colors to imitate coral and other gem and ornamental materials.

Occurrence.

Calcite is one of the most common minerals. It occurs as crystals in veins and cavities and as a rock-forming mineral makes up enormous masses of limestone and marble. It occurs as cave deposits in the form of stalactites and stalagmites. Precipitated from both hot and cold spring waters, it forms porous deposits known as *travertine* or *tufa.*

Calcite is such a widespread and abundant mineral that it is impossible to specify the important localities.

Name.

From the Latin word *calx,* meaning burnt lime.

Chrysocolla

Chrysocolla is a green to bluish-green material that in itself is too soft to be a satisfactory gemstone. Most of the chrysocolla used in jewelry is mixed with quartz (chalcedony) and gains its durability from this mineral. The data given here are for pure chrysocolla.

Crystallography.

Essentially amorphous, possibly cryptocrystalline. Massive compact; in some cases earthy.

Physical Properties.

Fracture conchoidal. **H** 2–4. **S.G.** 2.0–2.4. *Luster* vitreous to greasy, earthy when impure.

Color green to bluish-green; brown to black when impure.

Optical Properties.

R.I. 1.50, but variable. Most chrysocolla used in Jewelry will give the **R.I.** of quartz.

Chemical Composition.

A hydrous copper silicate, $CuSiO_3 \cdot nH_2O$, but with a wide range in the amount of copper and silica as well as water.

Diagnostic Features.

Distinguished from similar appearing minerals by the low hardness, density, and refractive index.

Occurrence.

Chrysocolla is a secondary mineral found in the oxidized zones of copper deposits associated with azurite, malachite, cuprite, and native copper. It is present in numerous localities in the copper districts of Arizona and New Mexico. Other localities are in the U.S.S.R., Zaire, and Chile.

Name

Derived from two Greek words meaning gold and glue, the name of similar appearing material used to solder gold.

Danburite

Danburite is a rare gem mineral that has been cut into colorless, yellow, and light pink gems. It is a durable and attractive gem but little known because of its rarity.

Crystallography.

Orthorhombic. Prismatic crystals closely related to topaz in habit. Commonly in crystals.

Physical Properties.

Conchoidal fracture. No cleavage. **H** 7. **S.G.** 3.00. *Luster* vitreous. Transparent to translucent.

Color varies from colorless to pale to medium yellow and pale pink.

Optical Properties.

Biaxial, with the intermediate refractive index midway between the extremes ($2V = 90°\pm$); $\alpha = 1.630$, $\beta = 1.633$, $\gamma = 1.636$; biref. 0.006; disp. 0.016. Yellow stones are weakly pleochroic very light yellow and light yellow.

Chemical Composition.

Calcium borosilicate, $CaB_2Si_2O_8$.

Luminescence and Absorption Spectra.

Fluoresces light blue under ultraviolet light. Absorption spectra are unimportant.

Diagnostic Features.

Crystals of danburite are similar in appearance to topaz but lack cleavage. The **S.G.** of 3.00 readily distinguishes it from yellow topaz (3.52) and yellow quartz (2.65).

Occurrence.

Found in crystals in metamorphosed limestones and in low-temperature hydrothermal veins. The original locality is now buried under the city of Danbury, Connecticut. Gem quality material comes from Moguk, Burma; Madagascar; Japan; and Charcas, San Luis Potosi and Baja California, Mexico.

Name.

From the locality, Danbury, Connecticut.

Diopside

Diopside, like enstatite and jadeite, is a member of the pyroxene group of rock-forming minerals. It is a common mineral but rare in gem quality.

Crystallography.

Monoclinic. Prismatic habit, well-formed single crystals rare. Also granular massive, columnar, and lamellar. Frequently twinned polysynthetically.

Physical Properties.

Imperfect prismatic cleavage at 87° and 93°. Frequently shows parting on the basal pinacoid. (see Fig. 13.51 and 13.52 under enstatite). **H** 5–6, **S.G.** 3.25–3.35, increasing with iron content. *Luster* vitreous. Transparent to translucent.

Color colorless to light green, deepening to dark green with increasing iron content; also yellowish-green, blue. The violet-blue variety is called *violan* (or violane). *Chrome diopside* is a rare chrome-green variety.

Optical Properties.

Biaxial (+). α 1.66–1.75, $\beta = 1.67–1.73$, $\gamma = 1.69–1.75$, variable, increasing with iron content; for most gem material α is about 1.67, γ about 1.70; biref. 0.024–0.028. Darker colors show pleochroism: α pale green, β yellow-green, γ dark green.

Chemical Composition.

Calcium magnesium silicate, $CaMgSi_2O_6$. There is a solid solution series between diopside and *hedenbergite*, $CaFeSi_2O_6$, in which magnesium and iron can substitute for each other in all proportions. Chromium is present in chrome diopside.

Luminescence and Absorption Spectra.

Nonfluorescent. The absorption spectra are unimportant except for that of chrome diopside, which shows characteristic chromium absorption lines at 5080 and 5050 Å.

Inclusions.

Minute tabular inclusions in some iron-rich varieties cause some stones to be either four-rayed stars or cat's eyes.

Diagnostic Features.

Diopside may be confused with peridot, demantoid garnet, enstatite, tourmaline, chrysoberyl, emerald, and epidote. The **R.I.** and **S.G.** distinguish it from most of these. The birefringence distinguishes it from peridot and enstatite.

Occurrence.

Gem quality material comes from the gem gravels of Moguk, Burma and Sri Lanka. Chrome diopside comes from the Kimberley, South Africa diamond mines and Outokumpu, Finland. Violan is found at Saint Marcel, Piemont, Italy. Small gem quality crystals have come from DeKalb Junction and Gouverneur, New York.

Name.

Diopside comes from two Greek words meaning double and appearance, because the vertical prism zone can apparently be oriented in two ways.

Dioptase

Dioptase is a copper mineral occurring in clusters of small multifaced, emerald-green crystals that are more highly prized as mineral specimens than stones cut for them. A small group of uncut crystals mounted on a flat plate makes a lovely brooch.

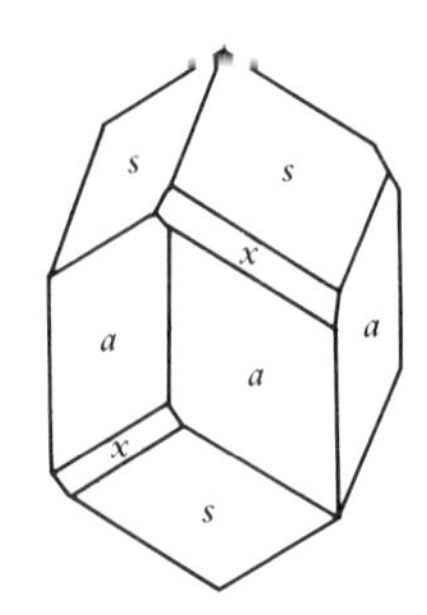

FIG. 13.50. Dioptase crystal.

Crystallography.

Hexagonal-rhombohedral. Commonly in small well-formed crystals showing faces of a hexagonal prism and rhombohedron (Fig. 13.50).

Physical Properties.

Cleavage perfect rhombohedral. **H** 5. **S.G.** 3.3. *Luster* vitreous. Translucent.

Color emerald-green.

Optical Properties.

Uniaxial (+). ω = 1.644–1.658, ϵ = 1.697–1.709. Biref. 0.053. Absorption $\omega > \epsilon$ in shades of green. Disp. 0.036.

Chemical Composition.

Hydrous copper silicate, $CuSiO_3 \cdot H_2O$.

Absorption Spectra.

There are two broad absorption bands, one in the yellow-green, the other in the blue-violet.

Diagnostic Features.

Characterized by its green color and high dispersion, 0.036; the dispersion of emerald is 0.014.

Occurrence.

Dioptase is a relatively rare mineral found in copper deposits. Notable localities are in the U.S.S.R., Zaire, Congo Republic (formerly French Congo), but in recent years the most spectacular specimens have come from Tsumeb, South-West Africa.

Name.

From the Greek words meaning through and to see, because the cleavages were distinguished on looking through the crystal.

Enstatite

Enstatite, like jadeite and diopside, is a member of the pyroxene group of rock-forming silicate minerals. It is a widespread mineral as a constituent of rocks such as peridotite and basalt, but is rarely of gem quality.

Crystallography.

Orthorhombic. Prismatic habit, well-formed single crystals rare. Usually massive, fibrous, or lamellar.

Physical Properties.

Good prismatic cleavage at angles of 87° and 93° (Fig. 13.51); also frequently has good parting on the front pinacoid, less common on the basal pinacoid (Fig. 13.52). **H** 5½. **S.G.** 3.2–3.5. *Luster* vitreous to pearly on cleavage surfaces. A variety with a submetallic, bronzelike luster is known as *bronzite*. Transparent to translucent.

Color grayish, yellowish, greenish white to olive-green and brown.

Optical Properties.

Biaxial (+). $\alpha = 1.650–1.665$, $\beta = 1.653–1.670$, $\gamma = 1.658–1.674$; biref. 0.010; $2V = 30–80°$. Pleochroism weak in the green type but strong in the brown with α pink to red, β yellow, γ green. The greater the iron content, the higher the indices and $2V$.

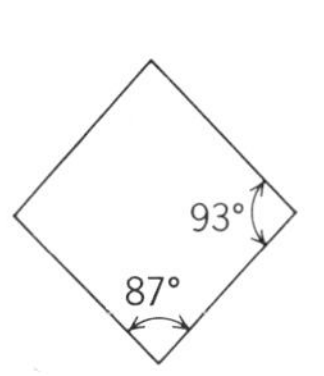

FIG. 13.51. Prismatic cleavage in enstatite, or in any pyroxene.

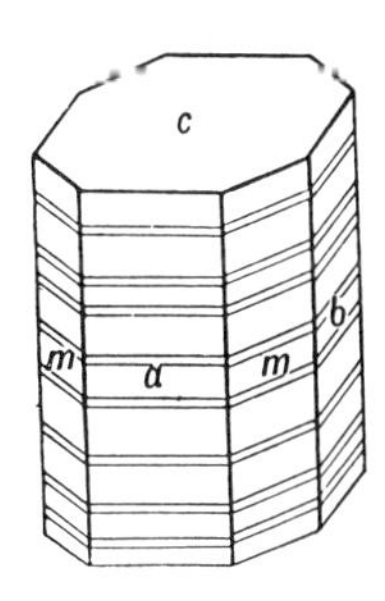

FIG. 13.52. Parting in enstatite on the basal pinacoid.

Chemical Composition.

Magnesium silicate, $MgSiO_3$. Rarely pure and usually contains iron. Ferrous iron may substitute for magnesium in all proportions up to the ratio of Mg:Fe = 1:1. If the amount of iron lies between 0 and 5% the mineral is called *enstatite;* if between 5 and 13%, the variety is called *bronzite;* if over 13%, it is called *hypersthene*.

Luminescence and Absorption Spectra.

Nonfluorescent. The absorption spectrum is characterized by a single sharp line at 5060 Å.

Inclusions.

Minute tabular inclusions in the iron-rich varieties may cause chatoyancy.

Diagnostic Features.

Enstatite may be confused with chrysoberyl, tourmaline, peridot, zircon, and diopside. The **R.I.** and **S.G.** distinguish it from most of these. The low birefringence of enstatite distinguishes it from peridot and diopside.

Occurrence.

Gem quality material, found mostly as waterworn pebbles, comes from Moguk, Burma and Sri Lanka.

Name.

The name enstatite comes from the Greek word meaning opponent because of its refractory nature. Bronzite is named for its bronze-like luster, and hypersthene is named from two Greek words meaning very and strong because its hardness is greater than that of hornblende, a mineral it resembles.

Epidote

Epidote is a common and widespread mineral but gem quality crystals have been found at only a few localities, and many of these are too dark to make attractive stones.

Crystallography.

Monoclinic. Crystals are elongated on the *b* crystallographic axis and frequently striated parallel to the length (Fig. 13.53). The mineral is also found in fine granular and fibrous aggregates.

Physical Properties.

Cleavage good on both front and basal pinacoids causing the mineral to break into splintery fragments. **H** 6–7. **S.G.** 3.30–3.45. *Luster* vitreous. Transparent to nearly opaque.

Color. Characteristically a pistachio-green to yellow-green to which the name *pistacite* has been given. May be black.

Optical Properties.

Biaxial (−); α = 1.715–1.751, β = 1.725–1.784, γ = 1.734–1.797; biref. 0.015–0.049. 2*V* = 70–90°. **R.I.,** birefringence, and 2*V* increase with increasing iron content. Pleochroism strong with absorption $\beta > \gamma > \alpha$. α pale yellow to pale green, β yellow to greenish-yellow, γ green. Disp. 0.030.

FIG. 13.53. Epidote crystals with quartz, Salzburg, Austria.

Chemical Composition.

Essentially calcium aluminum silicate but containing some iron, $Ca(Al,Fe)Al_2Si_3O_{12}(OH)$. A solid solution series extends from ironfree clinozoisite to epidote in which iron substitutes for some of the aluminum. The depth of color, specific gravity, and refractive indices all increase with increasing iron. Clinozoisite is the monoclinic polymorph of zoisite (q.v.).

Luminescence and Absorption Spectra.

Epidote is nonluminescent in ultraviolet light. A strong absorption band, present in the violet at about 4500 Å, varies in intensity depending on crystal orientation.

Diagnostic Features.

The yellow-green color is characteristic of most stones. The strong pleochroism is similar to that of deep-colored tourmaline but tourmaline has lower **R.I.** and **S.G.** as well as less dispersion.

Occurrence.

Epidote occurs commonly in metamorphic rocks as an alteration of dark minerals as pyroxene and amphibole. But most crystals, and hence most gem material, are found in contact metamorphic deposits in limestone. The finest crystals have come from the Austrian and Swiss Alps and Bourg d'Oisans, Isere, France. In the United States epidote is found in Virginia but the most notable locality is on Prince of Wales Island, Alaska.

A type of granite known as *unakite,* composed of quartz, red to pink feldspar, and green epidote, is used as an ornamental material and is sometimes cut as cabochons. It is found principally in North Carolina.

Name.

From the Greek word meaning increase, since the elongated basal face has one edge longer than the other.

Euclase

Euclase is a rare gem mineral that is seldom seen as cut stones because its easy cleavage makes it difficult to work, and fine crystals are more valuable as specimens than the stones that might be cut from them.

Crystallography.

Monoclinic. Prismatic crystals are characteristic (Fig. 13.54).

Physical Properties.

Cleavage perfect parallel to the side pinacoid. **H** 7½. **S.G.** 3.05–3.1. *Luster* vitreous. Transparent.

Color colorless to pale yellow, pale green, greenish-blue, blue. Most euclase cut as gems is pale blue.

Optical Properties.

Biaxial (+); $\alpha = 1.652$, $\beta = 1.655$, $\gamma = 1.671$; biref. 0.019; $2V = 50°$. Disp. 0.016.

Chemical Composition.

Hydrous beryllium aluminum silicate, $BeAlSiO_4(OH)$.

Luminescence and Absorption Spectra.

Euclase is nonfluorescent in ultraviolet light and in colorless stones absorption bands are lacking. But in colored stones a doublet is present in the red at 7050 Å and there are two bands at 4680 and 4550 Å in the blue.

Diagnostic Features.

Euclase resembles aquamarine in appearance but has higher **R.I.** and **S.G.** Microscopic examination of euclase may show traces of the highly perfect cleavage that will be lacking in beryl. Green spodumene may also be confused with euclase but it has slightly higher **S.G.** (3.18).

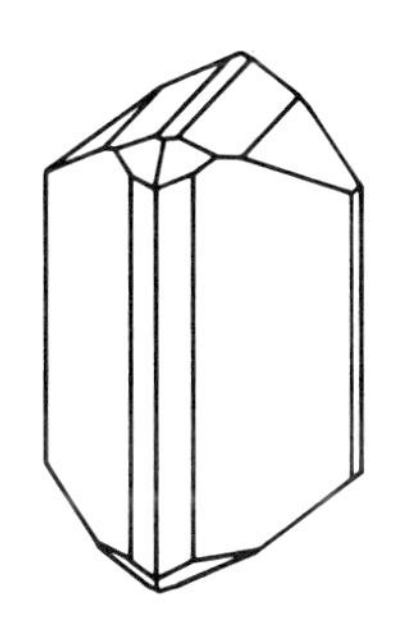

FIG. 13.54. Euclase crystal.

FIG. 13.55. Fluorite crystals, Cumberland, England.

Occurrence.

Euclase occurs in pegmatites and metamorphic rocks associated with other beryllium-bearing minerals. The principal occurrences are in Minas Gerais, Brazil and in the Ural Mountains.

Name.

From two Greek words meaning easy and fracture, in allusion to the perfect cleavage.

Fluorite

Fluorite or *fluorspar* is a most attractive mineral occurring in well-formed crystals of many different colors. Although it is sometimes cut as gems, the stones are not durable because of low hardness (4), but as an ornamental material it has been used for a long time in carvings.

Crystallography.

Isometric. It characteristically crystallizes in cubes both as single crystals and as twins composed of two interpenetrating cubes (Fig. 13.55); it rarely forms in octahedrons. Although it usually occurs in crystals or cleavable masses, it may be massive, granular, or in columnar aggregates.

Physical Properties.

Cleavage perfect octahedral (four equivalent directions). **H** 4. **S.G.** 3.18. *Luster* vitreous. Transparent to translucent.

Color varies greatly; commonly light green, yellow, blue-green, or purple; also colorless, white, rose, emerald green, blue, and light to very dark brown. In many crystals the color is not uniform but shows banding, sometimes in several colors, parallel to the cube faces. In a columnar variety known as *Blue John,* which occurs in nodules in Derbyshire, England, it has blue to dark purple bands alternating with colorless bands. The bands are irregular but roughly at right angles to the length of the columns. The material was used by the Romans for carving cups, bowls, and vases, and since then has been used for ornamental objects.

The color of fluorite has been attributed to crystal lattice defects, radioactive irradiation, and to the presence of hydrocarbons.

Optical Properties.

Isotropic. **R.I.** = 1.433 is essentially constant regardless of color. Some fluorite shows anomalous birefringence probably resulting from strain in the crystal lattice. Disp. 0.006.

Chemical Composition.

Calcium fluoride, CaF_2. Yttrium and cerium may be present substituting for calcium. The structure of fluorite is illustrated in Fig. 3.9.

Luminescence and Absorption Spectra.

Although fluorite gives its name to the phenomenon of fluorescence, not all fluorite fluoresces. Under longwave u.v. and to a lesser extent under shortwave u.v., some fluorite shows blue or green fluorescence. Absorption spectra are unimportant but in the green variety there is a strong band at 4270 Å.

Diagnostic Features.

The low **R.I.** and comparatively high **S.G.** are sufficient to distinguish fluorite from other stones of similar color. In carvings evidence of the cleavage can usually be seen. In some crystals one can detect under the microscope three phase inclusions, gas cavities with salt crystals and liquid.

Synthesis.

Clear colorless fluorite has many uses in optical equipment but is rare in nature. However, it is synthesized in large (5–10 cm in diameter) single crystals of optical quality. Also crystals in pale shades of red, green, yellow, and blue have been synthesized.

Occurrence.

Fluorite is a common mineral widely distributed. It is found in hydrothermal veins both alone and associated with ore minerals of lead and silver, in cavities in sedimentary rocks, and in pegmatites. The list of fluorite localities is long and the mineral is mined in many places for industrial purposes. Many of the finest specimens have come from England. An emerald-green fluorite comes from South-West Africa. In the United States there are many occurrences, but the most important deposits are in southern Illinois at Rosiclare and Cave-in-Rock. Fluorite carved by the American Indians was probably from this locality.

Name.

From the Latin *fluere* meaning to flow, in allusion to its use as a flux in smelting and because it melts easily.

Gypsum

Because of its low hardness (2), gypsum has limited use as a gem material. However, also because of its low hardness, it is easily worked and, as the fine-grained variety *alabaster,* has for centuries been carved into vases, urns, and ornamental oojects. It was popular with the ancient Egyptians and carvings of alabaster are found in the tombs of the pharaohs.

Crystallography.

Monoclinic. Crystals are common but *alabaster* and *satin spar* are the varieties used for ornamental purposes. Satin spar, composed of parallel fibers of gypsum, can be cut into cabochons and beads showing a beautiful chatoyancy. But if worn, cut stones soon become dull because of the low hardness.

Physical Properties.

Cleavage perfect on side pinacoid; the name *selenite* is given to clear transparent cleavage folia. There is also good cleavage on the front pinacoid and first-order prism. Cleavage is rarely seen in gypsum used for ornamental purposes. **H** 2. **S.G.** 2.32. *Luster* vitreous to dull. Translucent to nearly opaque.

Color. Crystals colorless; satin spar white; alabaster white or tinged with rose, brown, or red; may be colored artificially.

Optical Properties.

Biaxial (+). $\alpha = 1.520$, $\beta = 1.523$, $\gamma = 1.530$; biref. 0.01; $2V = 58°$; **R.I.** of alabaster 1.52–1.53. Disp. 0.005.

Chemical Composition.

Hydrous calcium sulfate, $CaSO_4 \cdot 2H_2O$.

Diagnostic Features.

Distinguished from most similar appearing material by its low hardness; that is, it can be scratched by the fingernail. In this way one can identify as gypsum a carving dyed green to simulate jade.

Occurrence.

Gypsum is a widespread mineral most commonly formed as a sedimentary rock deposited by evaporation from sea water or saline lakes. Occurrences of alabaster are more limited. The principal locality is the Volterra district of Italy, where it has been quarried since ancient Etruscan days. Ornamental alabaster also comes from several localities in England.

Name.

When gently heated, gypsum loses part of its water content, but if water is added to this calcined material it recrystallizes or "sets," forming plaster of Paris. The word gypsum is from the Greek name for the calcined mineral or plaster.

Anhydrite occurs with gypsum and, like gypsum, is calcium sulfate but lacks water. It has been used as a minor gem material. Its properties are as follows: orthorhombic; **H** 3, **S.G.** 2.9; white, blue, purple; **R.I.** 1.57–1.61. Three pinacoidal cleavages at right angles to each other are rarely seen.

Hematite

Most hematite is a pulverulent or compacted red mineral that serves as the world's major ore of iron. Only rarely is it found in crystals, black and hard, which may be faceted, fashioned into beads, or cut as cabochon stones. As intaglios it is a popular material for signet rings.

Crystallography.

Hexagonal-rhombohedral. Crystals are usually rhombohedral or tabular with rhombohedral truncations (Fig. 13.56). Desirable as mineral specimens are platy crystals grouped into rosettes known as iron roses. May also be in botryoidal shapes but is usually massive and earthy.

Physical Properties.

Conchoidal fracture. **H** 5½–6½. **S.G.** 5.26. *Luster* metallic. Opaque. These data are for crystals, the only hematite used for gem purposes.

Color black in crystals, red in other forms. Streak of black crystals red. The streak (color of powder) is obtained by drawing the edge of the stone across a piece of unglazed porcelain.

Chemical Composition.

Iron oxide, Fe_2O_3.

Diagnostic Features.

The red streak given by the black, highly reflecting, opaque stone is characteristic.

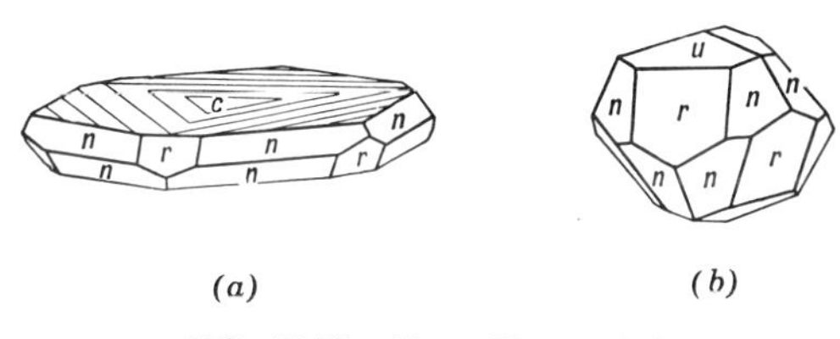

FIG. 13.56. Hematite crystals.

Occurrence.

Although hematite as an iron ore is found in millions of tons on every continent, in only a few places are crystals found that are used for gem purposes. The most important of these is Cumberland, England. Fine crystals have also come from the island of Elba, Norway, Sweden, and Brazil.

Name.

From the Greek meaning blood, in allusion to the red color of the powdered mineral.

Idocrase (Vesuvianite)

The mineral species known both as idocrase and vesuvianite (the latter name is now accepted mineralogical nomenclature) is known best for its massive green variety, called *californite,* although transparent material is known.

Crystallography.

Tetragonal. Crystals prismatic in habit, often terminated by a first-order dipyramid and base (Fig. 13.57). Also granular, massive.

Physical Properties.

H 6½. **S.G.** 3.35–3.45; 3.25–3.32 for massive californite. *Luster* vitreous. Transparent to translucent.

Color usually green or brown; also yellow, blue, red.

Optical Properties.

Uniaxial (−). $\omega = 1.705–1.752$, $\epsilon = 1.701–1.746$; biref. very weak, 0.004–0.006; disp. 0.019.

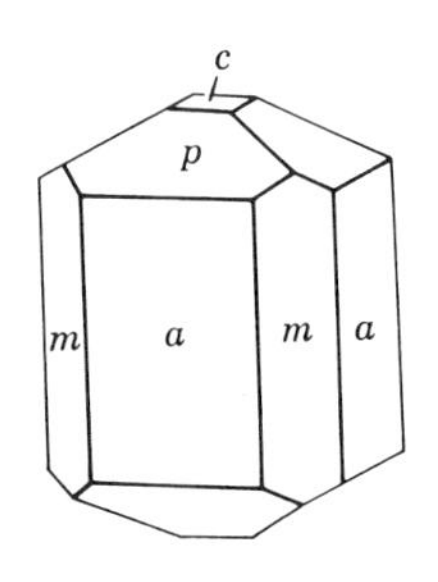

FIG. 13.57. Idocrase crystal.

Chemical Composition.

Hydrous calcium magnesium aluminum silicate, $Ca_{10}Mg_2Al_4(SiO_4)_5(Si_2O_7)_2(OH)_4$.

Luminescence and Absorption Spectra.

Nonfluorescent. Absorption spectra generally show a strong band in the blue at 4610 Å and a weak band at 5285 Å. The absorption band at 4610 Å is not always seen but is present in the californite variety, allowing distinction from jadeite.

Diagnostic Features.

The only commonly encountered variety of idocrase, *californite,* resembles jade and its substitutes. The **R.I.** of californite is higher than jadeite's and its **S.G.** slightly lower. Both properties are much higher than those of nephrite.

Occurrence.

Idocrase is usually found in crystalline limestones, formed as a result of contact metamorphism. Good crystals, but not usually of gem quality, have been found in numerous localities. *Californite* is found at several localities in California, the principal one being near Happy Camp, Siskiyou County.

Name

The name idocrase is from two Greek words meaning form and mixture, because the crystal forms present appear to be a combination of those found on different minerals. The names vesuvianite and californite are from the localities.

Iolite (Cordierite)

Iolite is the sapphire-blue gem variety of the mineral known by mineralogists as *cordierite.* It is also called *dichroite* because of its pronounced pleochroism. "Water sapphire," or "saphir d'eau" are misleading names early applied to the gem.

It is believed that the Vikings used transparent slices of iolite as a navigational aid. On cloudy days they could tell the sun's position by observing the partially polarized light from the sky through the strongly pleochroic plate.

Crystallography

Orthorhombic. Crystals are usually short prismatic (Fig. 13.58). The crystal structure is very similar to hexagonal beryl and thus both single crystals and twins are pseudohexagonal. Iolite usually occurs in irregular grains and masses that are commonly partially altered to soft minerals such as mica, chlorite, and talc.

Physical Properties

Cleavage poor on the side pinacoid, the mineral usually breaks with a conchoidal fracture. **H** 7–7½. **S.G.** 2.58–2.66; most gem iolite is toward the lower end of the range. *Luster* vitreous. Transparent to translucent.

Color may be yellow or green but only crystals of varying shades of blue or violet-blue are cut as gems.

Optical Properties

Biaxial usually (−) but may be (+). α = 1.522–1.568, β = 1.532–1.568, γ = 1.527–1.573; biref. 0.008–0.012; $2V$ = 40–90°; disp. 0.017. *Pleochroism* strong: α yellow, β blue to violet-blue, γ blue. α is parallel to the c axis so the deepest blue color is seen when the stone is viewed along this direction.

Chemical Composition

Essentially a magnesium aluminum silicate but some iron is present substituting for magnesium, resulting in the formula, $(Mg,Fe)_2Al_4Si_5O_{18}$. Both specific gravity and refractive indices increase with increasing iron. Only a small amount of iron is present in gem iolite.

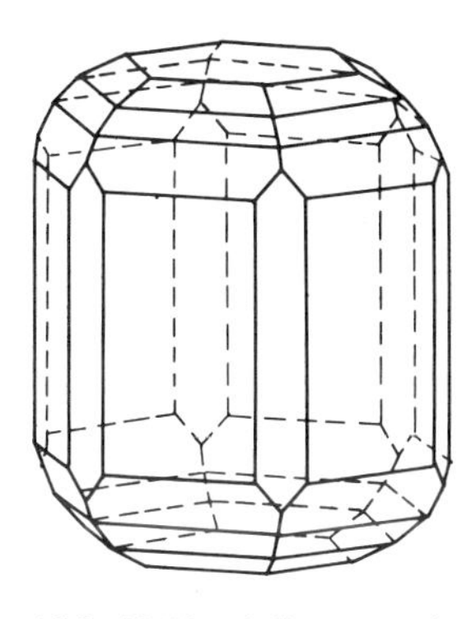

FIG. 13.58. Iolite crystal.

Luminescence and Absorption Spectra

Iolite does not fluoresce under ultraviolet light. Absorption bands are weak and, because of the strong pleochroism, vary with crystallographic direction. Ill-defined bands are present in the blue-violet region at 4920, 4560, and 4360 Å.

Diagnostic Features

The pleochroism is characteristic. Iolite and quartz resemble each other in chonchoidal fracture, hardness, **S.G.,** and **R.I.** and thus the two minerals can be confused easily. However, in general both **S.G.** and **R.I.** are slightly lower in gem iolite than in quartz. These two properties easily distinguish iolite from blue spinel and sapphire.

Inclusions

Common iolite is characteristically dusted through with fine tabular inclusions of an iron oxide, probably hematite. When such inclusions, oriented in parallel position, are present in a gem, they give a red color to the stone when viewed perpendicular to the platelets.

Occurrence

Iolite is a common mineral in gneisses and schists as well as contact metamorphic rocks. A characteristic metamorphic association is with sillimanite and spinel. It is also found in some granites and pegmatites. Gem quality material has come from India, Burma, Madagascar, and Tanzania, but the most important source is the gem gravels of Sri Lanka.

Name

Iolite is derived from two Greek words meaning violet and stone. *Cordierite* is named in honor of the French geologist, P. L. A. Cordier (1777–1861).

Kornerupine

Kornerupine is a rare mineral that has been faceted as gems, but is little known and of interest mainly to collectors.

Crystallography
Orthorhombic. In prismatic crystals; also in fibrous and columnar masses.

Physical Properties
Cleavage prismatic. **H** 6½. **S.G.** 3.27–3.45. *Luster* vitreous. Transparent to translucent.

Color colorless, yellow, pale green, brownish green, brown.

Optical Properties
Biaxial (−). α = 1.665–1.668, β near γ = 1.678–1.680; biref. 0.013; $2V$ = 20°; disp. 0.019. Pleochroism strong: α yellow, β brownish-yellow, γ green. To obtain the best color, stones must be cut with the table facet parallel to the length of the crystal.

Chemical Composition
A hydrous magnesium aluminum borosilicate, $Mg_3Al_6(Si,Al,B)_5O_{21}(OH)$.

Luminescence and Absorption Spectra
Nonfluorescent. Absorption spectrum shows several weak bands but only one of these, at 5030 Å, can usually be detected.

Diagnostic Features
In appearance kornerupine may be confused with dark greenish-brown tourmaline, peridot, beryl, topaz, sinhalite, enstatite, and quartz. Of these sinhalite is closest in refractive index but has a higher specific gravity.

Occurrence
Kornerupine was first found in a cordierite (iolite) gneiss at Fiskernaes, Greenland, but not of gem quality. Later gem material was found at Betroka, Madagascar, and in the gem gravels of Sri Lanka and Burma.

Name
For the Danish scientist A. N. Kornerup.

Kyanite
Kyanite is a common rock-forming mineral but is relatively rare in gem quality crystals. It is a polymorph of Al_2SiO_5, a composition it shares with sillimanite and andalusite.

Crystallography
Triclinic. Usually in tabular crystals elongated on the *c* axis (see Fig. 2.4).

Physical Properties
Cleavage perfect on the front pinacoid. **H** 5 parallel to the length of the crystals, 7 at right angles to the length. **S.G.** 3.55–3.66. *Luster* vitreous to pearly. Transparent to translucent.

Color usually blue, often with darker shades toward the center of the crystal; also colorless, green, gray.

Optical Properties
Biaxial (−). α = 1.712, β = 1.720, γ = 1.728; biref. 0.016; $2V$ = 82°; disp. 0.020. Pleochroism: α colorless, β violet-blue, γ cobalt-blue.

Chemical Composition
Aluminum silicate, Al_2SiO_5.

Fluorescence
Most kyanite fluoresces a dull red under longwave u.v.

Diagnostic Features
Kyanite is distinguished from similar appearing spinel by its double refraction and from sapphire by its lower **R.I.** Evidence of the excellent cleavage of kyanite may also aid in its identification.

Occurrence
As a regional metamorphic mineral, kyanite is found in schists and gneisses associated with garnet and staurolite. It is found in gem quality crystals at St. Gotthard, Switzerland; also in India, Burma, Kenya, Brazil, and in the United States in North Carolina and Georgia.

Name
Derived from the Greek word meaning blue.

Lapis Lazuli
Lapis lazuli is a gem material consisting of a complex mixture of several minerals belonging to the

feldspathoid mineral group, plus calcite, pyrite, and often small amounts of other rock-forming minerals such as augite, hornblende, and mica. Lapis lazuli is, therefore, a rock and not a mineral.

Lapis lazuli has been prized for centuries for its beautiful blue color. Prior to 1828 it was used as the paint pigment ultramarine, but since that date the pigment has been produced synthetically.

The chief constituent of lapis lazuli is *lazurite,* which gives it its deep blue color. The other feldspathoid minerals present are hauyne, nosean, and sodalite. All these minerals have closely similar properties; the following data are for lazurite.

Crystallography

Isometric. Crystals rare, usually dodecahedral, commonly massive, compact.

Physical Properties

Imperfect dodecahedral cleavage, not seen in massive material. **H** 5–5½. **S.G.** 2.4–2.45 for the pure mineral. Lapis lazuli has **S.G.** variable because of the presence of pyrite and other minerals, but usually near 2.75. *Luster* vitreous. Translucent.

Color deep azure-blue, greenish-blue.

Optical Properties

Isotropic. **R.I.** near 1.50.

Chemical Composition

A complex sodium calcium aluminosilicate with variable amounts of sulfate, sulfur, and chlorine; $(Na,Ca)_8(Al,Si)_{12}O_{24}(SO_4,S,Cl)$.

Luminescence and Absorption Spectra

Spots or streaks of orange fluorescence under longwave u.v., more pinkish under shortwave. Absorption spectra unimportant.

Inclusions

Lapis lazuli is characterized by small metallic yellow inclusions of pyrite. If not too prominent they are considered desirable as proof that the material is genuine.

Diagnostic Features

Lapis lazuli may be confused with sodalite, lazulite, azurite, dyed quartz, sintered synthetic spinel, and glass. It may be distinguished from these by standard tests and the presence of pyrite inclusions.

Simulation

Lapis lazuli is imitated by a dyed jasper (quartz) called "Swiss lapis." This material lacks the pyrite seen in lapis lazuli, and has an **R.I.** of 1.54.

A sintered synthetic spinel has been produced that has a good lapis color, but has **R.I.** of 1.725 and **S.G.** near 3.5. Some of this material has brassy specks introduced which are metallic gold or copper rather than pyrite.

Some pale-colored lapis lazuli is dyed to enhance its color, but the dye may be removed with a solvent such as acetone.

Occurrence

Lazurite is a rare mineral occurring usually in crystalline limestones as a product of contact metamorphism. The best-quality lapis lazuli comes from near Badakshan, Afghanistan. It is also found at Lake Baikal, Siberia, U.S.S.R.; in the Andes of Ovalle, Chile; and on Baffin Island, Northwest Territories, Canada. In the United States it is found in San Bernardino County, California.

Name

Lazurite is an obsolete synonym for azurite; hence the mineral was named because of its color resemblance to azurite. Lapis lazuli comes from the Persian word *lazhward,* meaning blue.

Lazulite

Lazulite is a minor gem mineral of an azure-blue color resembling lazurite.

Crystallography

Monoclinic. Usually massive or granular, rarely in crystals.

Physical Properties

Cleavage poor prismatic. **H** 5½–6. **S.G.** 3.0–3.1. Translucent.

Color azure-blue, may be colorless to bluish-green.

Optical Properties

Biaxial (−). $\alpha = 1.612$, $\beta = 1.634$, $\gamma = 1.643$; biref. 0.031; $2V = 60°$. Pleochroism α colorless, β light blue, γ deep blue.

Chemical Composition

A basic phosphate of aluminum, magnesium, and ferrous iron, $(Mg,Fe)Al_2(PO_4)_2(OH)_2$.

Diagnostic Features

Can be distinguished from other blue minerals by the optical properties.

Occurrence

Lazulite is found in quartz-rich metamorphic rocks and pegmatites, usually associated with kyanite, andalusite, corundum, and garnet. It is found in Austria, Sweden, Brazil, and in the United States in North Carolina, Georgia, and California.

Name

Lazulite is named, as are azurite and lazurite, in allusion to the azure-blue color.

Malachite

The bright green color of malachite makes the mineral a most attractive ornamental material. The color is rarely uniform and in cut slices different shades of green are arranged in wavy bands, sometimes concentric, much as the banding in agate (see Fig. 2.7). Large deposits of malachite have long been known in the Ural Mountains. And the Russians early used the mineral as they used rhodonite, that is, for carving art objects and in thin slices for inlays and as a veneer on table tops and urns. Many of the choice Russian malachite pieces are now on exhibit in the Hermitage in Leningrad.

Malachite is principally an ornamental stone but, in spite of its low hardness, is used in cabochons, beads, pendants, and brooches.

Crystallography

Monoclinic. Individual crystals are very small and rarely seen. Malachite may be in relatively large crystals pseudomorphous after azurite, a mineral with which it is often associated.

Physical Properties

H 3½–4. **S.G.** 3.90–4.03. *Luster* vitreous when polished, fibrous rough may be silky. Translucent.

Color bright green to pale green.

Optical Properties

Biaxial (−); $\alpha = 1.655$, $\beta = 1.875$, $\gamma = 1.909$; biref. 0.254; $2V = 43°$. Pleochroism: α colorless, β yellow-green, γ deep green. Data for crystals. The mean **R.I.** of fine-grained ornamental material is about 1.85 and cannot be determined on the refractometer.

Chemical Composition

Hydrous copper carbonate, $Cu_2CO_3(OH)_2$.

Diagnostic Features

Malachite is characterized by its bright green color and banded appearance. It is distinguished from other green ornamental stones by its low hardness and by its effervescence in hydrochloric acid.

Occurrence

Malachite is formed by the near-surface alteration of primary copper minerals and is thus associated with other secondary copper minerals, particularly with blue azurite. Sometimes the two minerals are so intimately intergrown that they can be cut and polished together, yielding the attractive green and blue stone called *azurmalachite*. The abundant Russian malachite is now nearly exhausted and the major source today is Katanga, Zaire. Other important localities are Tsumeb, South-West Africa and Broken Hill, Australia. In the United States malachite was formerly abundant in the Southwest, particularly at Bisbee and Morenci, Arizona.

Name

Derived from the Greek word for mallows in allusion to the green color.

Azurite is a blue copper carbonate, $Cu_3(CO_3)_2(OH)_2$, so-called because of its intense azure-blue color. It occurs at the same localities as malachite and, although less abundant, the two minerals are almost invariably associated. Azurite in places occurs in well-formed monoclinic crystals. Complete or partial pseudomorphs of malachite after azurite are common. Properties: one good cleavage, **H** 3½–4, **S.G.** 3.77, luster vitreous, transparent to translucent. Optics: biaxial (+), α = 1.730, β = 1.758, γ = 1.838; pleochroism in blue with $\gamma > \beta > \alpha$.

Phenakite

Phenakite is a relatively rare mineral and its crystals themselves are prized as specimens. Faceted gemstones are uncommon.

Crystallography

Hexagonal-rhombohedral. Commonly in lenticular crystals with only rhombohedrons present (Fig. 13.59); but also in short prismatic crystals terminated by rhombohedrons.

Physical Properties

Cleavage imperfect parallel to the second-order prism and rhombohedron (in all, six directions), but cleavage is seldom seen and the mineral usually breaks with a conchoidal fracture. **H** 7½–8. **S.G.** 2.79–3.00. *Luster* brilliant and vitreous. Transparent.

Color most commonly colorless but also yellow, brown, and rose-red.

Optical Properties

Uniaxial (+). ω = 1.654, ϵ = 1.670; biref. 0.016, disp. 0.015.

Chemical Composition

Beryllium silicate, Be_2SiO_4, essentially a pure compound.

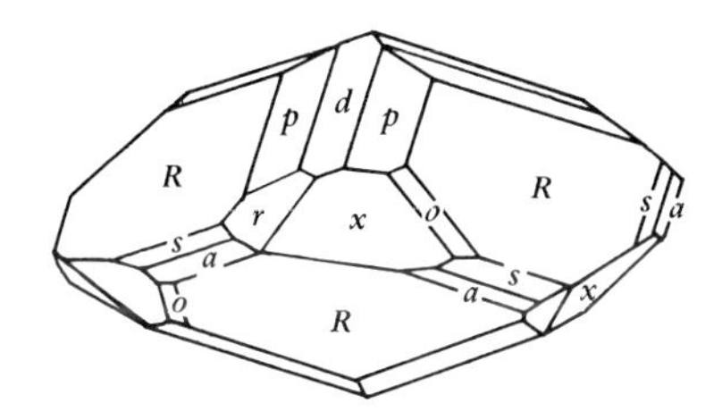

FIG. 13.59. Phenakite crystal.

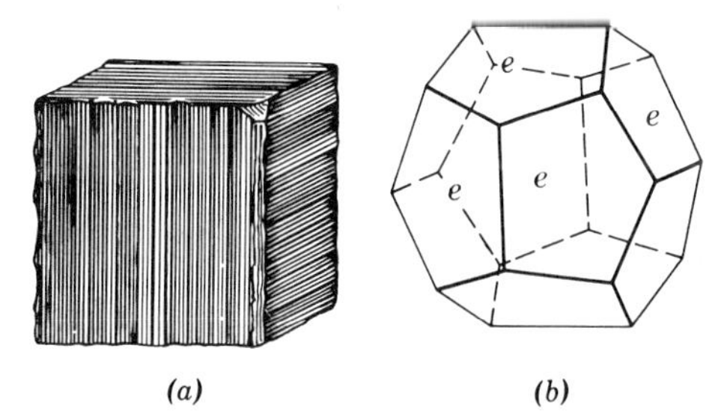

FIG. 13.60. Pyrite crystals. (*a*) Striated cube. (*b*) Pyritohedron.

Diagnostic Features

Phenakite resembles quartz and topaz. However, both its **R.I.** and **S.G.** are greater than those of quartz, whereas topaz has a higher **S.G.** but lower **R.I.**

Occurrence

Phenakite is associated with other beryllium minerals in pegmatites, cavities in granite, and in mica schists. It was early found in the Ural Mountains associated with emerald and chrysoberyl, and later it was found in Switzerland and Brazil. Gem material has also come from Tanzania and South-West Africa. In the United States phenakite has been found in Maine, Colorado, and New Hampshire.

Name

From the Greek word meaning a deceiver, because its crystals are frequently mistaken for quartz.

Pyrite

Pyrite is one of two metallic minerals used as gemstones; the other is hematite. It is the most common and abundant of the several minerals collectively known as "fool's gold." Another "fool's gold," marcasite, is polymorphous with pyrite and has similar properties. The two minerals were thus easily confused and the name *marcasite* was early given to and is still used today for cut stones which are in fact pyrite.

Crystallography

Isometric. Commonly in crystals as striated cubes and as pyritohedrons (Fig. 13.60). Much pyrite is granular and massive.

Physical Properties

Brittle with conchoidal fracture. **H** 6–6½. **S.G.** 5.02. *Luster* metallic, brilliant on crystal faces and polished surfaces. Opaque.

Color brass-yellow, streak (color of powder) greenish.

Chemical Composition

Iron sulfide, FeS_2. Although orthorhombic marcasite has this same composition, it is somewhat paler in color and has a slightly lower **S.G.** (4.89).

Diagnostic Features

Distinguished by its brass-yellow color, opacity, and brilliant mirrorlike surfaces.

Occurrence

Pyrite is the most common sulfide mineral with worldwide distribution formed under a great variety of geologic conditions. It may occur in isolated grains or crystals, or in massive aggregates of millions of tons. The "marcasite" used in jewelry is cut from single crystals of pyrite.

Name

Pyrite from the Greek word meaning fire, in allusion to the sparks emitted when struck with a hammer. Marcasite from an Arabic word, at one time generally applied to pyrite.

Rhodochrosite

Because of its lovely rose-red color, rhodochrosite is a most desirable ornamental stone. Its attractiveness is enhanced by its usual variegated nature in which red bands alternate with pale pink to almost white bands. Faceted stones of transparent material of uniform color are rare for uncut crystals themselves are prized by mineral collectors and may sell for a higher price than gemstones cut from them. Moreover, rhodochrosite is too soft to be worn as a faceted ring stone. In spite of its low hardness, banded rhodochrosite is cut as cabochons and extensively used in pendants and brooches.

Crystallography

Hexagonal-rhombohedral. Only rarely in rhombohedral crystals. Usually in banded aggregates of fine-grained crystals or in cleavable masses.

Physical Properties

Cleavage perfect rhombohedral (three directions). **H** 3½–4. **S.G.** 3.5–3.7. *Luster* vitreous. Translucent, only rarely transparent.

Color usually rose-red of varying shades; may be pale pink to brown.

Optical Properties

Uniaxial (−); $\omega = 1.816$, $\epsilon = 1.579$; biref. 0.237 (very high). Pleochroism: ω red, ϵ colorless to pale pink. These optical data for transparent crystals; **R.I.** of massive ornamental material not obtainable on the refractometer.

Chemical Composition

Manganese carbonate, $MnCO_3$. Calcium, iron, zinc, and magnesium may substitute for manganese resulting in change of color, **R.I.** and **S.G.**

Luminescence

The deep rose-red rhodochrosite does not fluoresce in ultraviolet light, but in the material from Argentina the light-colored bands fluoresce yellow in longwave u.v.

Diagnostic Features

Characterized by its rose-red color and ornamental material by its banded appearance. Rhodochrosite resembles rhodonite in color but it can be scratched by a needle and will effervesce in hydrochloric acid. Rhodonite can not be scratched and does not effervesce.

Occurrence

Rhodochrosite is a hydrothermal mineral associated with ores of the metals, particularly with those of silver. The most perfect crystals have come from Colorado silver mines. Fine large crystals have also come from Peru. Pale pink, cleavable rhodochrosite was found abundantly at Butte, Montana to be used as an ore of manganese, but little was used as ornamental material. Today the most important occurrence of rhodochrosite is at Capillitas in the province of Catamarca, Argentina. Here, in a former silver mine, staclactites and banded festoons are recovered to supply the world market with ornamental stone. Rhodochrosite also is found in the Kalahari manganese field of South Africa.

Name
Derived from two Greek words meaning rose and color in allusion to the rose-red color.

Rhodonite
Rhodonite is a mineral with a pleasing rose-red color and relatively high hardness, making it a desirable material for small carvings. It has been used for large carvings as well, for the sarcophagus of Alexander II of Russia was hewn from a single block. The Russians have been particularly skillful in using plates of rhodonite inlayed with other colored minerals for table tops and interior decorations. Translucent material is used in beads and cabochons but only rarely is the mineral transparent enough to be cut into faceted gems.

Crystallography
Triclinic. Crystals are rare and usually rough with rounded edges. The exception is at Franklin, New Jersey where large well-formed crystals have been found. Most commonly in massive aggregates or small grains.

Physical Properties
Cleavage good in two directions. **H** 6. **S.G.** 3.4–3.7, but gem material usually 3.6–3.7. *Luster* vitreous. Translucent to nearly opaque, only rarely transparent.

Color. Rose-red to reddish-brown. Veinlets or irregular patches of black manganese oxide are common.

Optical Properties
Biaxial (+); α = 1.716–1.733, β = 1.720–1.737, γ = 1.728–1.747; biref. 0.012–0.014; $2V$ = 60–75°. Pleochroism: α yellowish-red, β pale red, γ pale yellowish-red (seen only in transparent crystals). On the refractometer the common massive material gives a faint reading at about 1.73.

Chemical Composition
Manganese silicate, $MnSiO_3$. Some calcium is always present substituting for manganese. As calcium increases, the **R.I.** and **S.G.** decrease. In the variety *fowlerite,* found at Franklin, New Jersey, zinc substitutes for manganese.

Luminescence and Absorption Spectra
Rhodonite is nonfluorescent under ultraviolet light. In the absorption spectrum there is a broad band at about 5480 Å, a narrow line at 5030 Å, and a poorly defined band at 4550 Å.

Diagnostic Features
The rose-red color is characteristic. The color of rhodochrosite, manganese carbonate, and pink coral is similar to that of rhodonite; but both these are soft and easily scratched with slight pressure of a needle point. Rhodonite cannot be so scratched.

Occurrence
The principal source of rhodonite is near Sverdlovsk, U.S.S.R., on the eastern slope of the Ural Mountains. It is also found at Långban, Sweden; and in several localities in Australia, particularly at the Broken Hill mining district. In the United States massive material is found at Plainfield, Massachusetts and near Lancaster, California. Fine crystals of fowlerite have come from Franklin, New Jersey.

Name
From the Greek word for rose, in allusion to its color.

Scapolite
Although scapolite was recognized as a mineral species at the beginning of the nineteenth century, it was not until 1913 that gem material made its appearance. This was from Upper Burma and since then it has been found in several other localities.

Crystallography

Tetragonal. Usually in coarse prismatic crystals elongated on the *c* axis; frequently with a silky or fibrous appearance.

Physical Properties

Cleavage imperfect parallel to both first- and second-order prisms. There are thus four cleavage directions at 45° to each other. **H** 5–6. **S.G.** 2.55–2.74. For most gem material **S.G.** 2.60–2.66. *Luster* vitreous. Transparent to translucent.

Color colorless, yellow, pink, violet, greenish to bluish-gray.

Optical Properties

Uniaxial (−). $\omega = 1.55–1.60$, $\epsilon = 1.54–1.56$. For most gem material $\omega = 1.550–1.568$, $\epsilon = 1.540–1.548$. Biref. 0.01–0.02; disp. 0.017. Pleochroism: absorption $\omega > \epsilon$ with colors depending on the color of the stone. In violet and pink stones ω dark blue, ϵ pale purple; in yellow stones ω yellow, ϵ colorless.

Chemical Composition

Scapolite does not have a fixed chemical composition but is a complete solid solution varying from *marialite*, $Na_4Al_3Si_9O_{24}Cl$, to *meionite*, $Ca_4Al_6Si_6O_{24}CO_3$. With increase in calcium (meionite) there is a uniform increase in refractive index, birefringence, and specific gravity.

Luminescence and Absorption Spectra

Fluorescence in both long- and shortwave u.v. varies depending on the origin of the mineral. Some stones may fluoresce yellow to orange, others purple. The absorption spectrum is unimportant.

Inclusions

In some scapolite there are tubular cavities oriented parallel to the *c* axis of the crystal. When cut as cabochons, such material has a pronounced chatoyancy. Chatoyant pink scapolite has been called erroneously "pink moonstone."

Diagnostic Features

In appearance scapolite may be confused with topaz, tourmaline, and beryl but can be distinguished from them by its lower **R.I.** and **S.G.** However, the difference in these properties between beryl and scapolite is not great. To distinguish scapolite from quartz is more difficult for the ranges of both **R.I.** and **S.G.** of scapolite overlap these properties of quartz. However, the two minerals have different optic sign; quartz is uniaxial (+), scapolite is uniaxial (−). Also microscopic examination of scapolite may reveal cleavage that is absent in quartz.

Occurrence

Except for minor occurrences in pegmatites, scapolite is a metamorphic mineral. It is found in regionally metamorphosed rocks, but the best crystals are of contact metamorphic origin and occur in places where igneous rocks have intruded limestones. Gem material has come from Burma, Madagascar, Brazil, Tanzania, and Mozambique.

Name

From the Greek word meaning shaft, in allusion to the prismatic habit of the crystals.

Scheelite

Scheelite is far more important as an ore of tungsten than as a gem mineral but occasionally transparent crystals are cut as gemstones. Its low hardness limits its use in jewelry and faceted stones are mostly collectors' items.

Crystallography

Tetragonal. Crystals are usually simple second-order dipyramids. Commonly massive granular.

Physical Properties

Cleavage pyramidal (four directions) distinct. **H** 4½–5. **S.G.** 5.9–6.1. *Luster* vitreous to adamantine. Transparent to translucent.

Color white, orange-yellow, yellow, green, brown.

Optical Properties

Uniaxial (+); $\omega = 1.920$, $\epsilon = 1.934$; biref. 0.014. Disp. 0.025.

Chemical Composition
Calcium tungstate, $CaWO_4$. Most scheelite contains some molybdenum. Traces of the rare earths praseodymium and neodymium, jointly called "didymium," are usually present.

Luminescence and Absorption Spectrum
In shortwave u.v. scheelite fluoresces pale to deep blue, a constant and therefore diagnostic property. The absorption spectrum shows faint lines due to the "didymium" rare earths in the yellow at about 5850 Å.

Diagnostic Features
The high **S.G.** and **R.I.**, adamantine luster, and blue fluorescence characterize scheelite. However, scheelite synthesized by the Czochralski method has the same properties and is difficult to distinguish from the natural mineral. Although the synthetic is less flawed, lacking included tiny crystals and natural growth lines, it may show curved lines and bubbles.

Occurrence
Scheelite is a high-temperature mineral and its major occurrences are as contact metamorphic deposits in limestones adjacent to granite intrusions. It is also found as a vein mineral and in pegmatites. Scheelite is found in many places but at only a few are crystals sufficiently transparent for cutting. Fine crystals have come from Cornwall, England; Sonora, Mexico; and Austria and Switzerland. In the United States gem material has been found in California and Arizona.

Name
In honor of K. W. Scheele (1742–1786), the discoverer of tungsten.

Serpentine
Serpentine occurs in two distinct habits, a fibrous variety known as *chrysotile* and a platy variety known as *antigorite*. Chrysotile is important industrially as the chief source of asbestos. Antigorite occurs in massive form and is often used as a jade substitute, especially the hard, translucent variety known as *bowenite*. The data given here are for antigorite.

Crystallography
Monoclinic. Crystals unknown. Massive.

Physical Properties
H 2–5, usually 4; bowenite may have **H** 5–5½. **S.G.** 2.6–2.8. *Luster* greasy, waxlike. Translucent.

Color yellowish-green (bowenite), green, bluish-green, often variegated showing mottling in lighter and darker shades of green. *Williamsite* is a green variety containing black inclusions of chromite. *Verd antique* is an ornamental stone consisting of serpentine mixed with white marble.

Optical Properties
Essentially isotropic. **R.I.** 1.55–1.56.

Chemical Composition
A hydrous magnesium silicate, $Mg_3Si_2O_5(OH)_4$. Ferrous iron and nickel may be present in small amounts.

Luminescence and Absorption Spectra
Serpentine is nonfluorescent. Absorption spectra are unimportant.

Diagnostic Features
Bowenite and williamsite may closely resemble jadeite, nephrite, and chrysoprase. Lower hardness will distinguish from chrysoprase, and lower **R.I.** and **S.G.** from jadeite and nephrite.

Occurrence
Serpentine is a common mineral and widely distributed, usually as an alteration product of some magnesium silicate, especially olivine, pyroxene, and amphibole. Frequently associated with magnesite, chromite, and magnetite. Occurs in large rock masses called serpentinite.

Bowenite used for carving comes from China and Afghanistan; also from New Zealand; Cornwall, England; and Austria. In the United States bowenite is found in Pennsylvania and Rhode Island, and williamsite comes from Pennsylvania near the Maryland border. The classic locality for verd antique is near Larissa, Italy.

Name

The name serpentine refers to the green serpentlike cloudings of the massive variety.

Sillimanite

Sillimanite is a rock-forming mineral that occasionally forms transparent crystals suitable for cutting. Such crystals are not common and thus sillimanite gemstones are rare.

Crystallography

Orthorhombic. Occurs in long slender crystals often in parallel groups. May be in single fibers or in bundles or matted aggregates, and for this reason is called *fibrolite*.

Physical Properties

Cleavage perfect parallel to the side pinacoid. **H** 6–7. **S.G.** 3.23–3.27. *Luster* vitreous. Transparent to translucent. A chatoyant effect is present in some fibrous material.

Color most commonly colorless but gems may be blue, bluish-, brownish-, or grayish-green, or yellow.

Optical Properties

Biaxial (+); $\alpha = 1.657$, $\beta = 1.658$, $\gamma = 1.677$; biref. 0.020; $2V = 20°$. Pleochroism in deep-colored stones: α pale green to yellow, β green, γ blue. Disp. 0.015.

Chemical Composition

Aluminum silicate, Al_2SiO_5, with little variation from this composition. Polymorphous with andalusite and kyanite.

Diagnostic Features

Sillimanite may resemble euclase which has similar **R.I.** but can be distinguished by its slightly higher **S.G.**

Occurrence

Sillimanite is present in the highest grade of metamorphic rocks. Gem material is rarely recovered from the rocks in which it formed but usually as waterworn pebbles, as from its two major sources, Moguk, Burma and the gem gravels of Sri Lanka.

Name

In honor of Benjamin Silliman (1779–1864), professor of chemistry at Yale University.

Sinhalite

For many years brown stones in numerous gem collections were thought to be brown peridot because of similar refractive indices. In 1952 examination of one of them by X-ray diffraction revealed that it was a new mineral species.

Crystallography

Orthorhombic. Prismatic crystals rare; usually found as rolled pebbles in stream gravels.

Physical Properties

H 6½. **S.G.** 3.47–3.49. *Luster* vitreous. Transparent to translucent.

Color pale yellow-brown to greenish-brown and black (see Plate IX).

Optical Properties

Biaxial (−). $\alpha = 1.668$, $\beta = 1.699$, $\gamma = 1.707$; biref. 0.039; $2V = 50°$; disp. 0.018. Pleochroism distinct, pale brown, greenish-brown, and dark brown.

Chemical Composition

Magnesium aluminum borate, $MgAlBO_4$.

Luminescence and Absorption Spectra

Nonfluorescent. Absorption spectrum, which is similar to peridot, shows bands at 4930, 4750, 4630, and 4520 Å.

Diagnostic Features

Resembles brown chrysoberyl and brown zircon. In properties it is close to peridot but is distinguished by its higher specific gravity and lower $2V$.

Occurrence

Found in the gem gravels of Sri Lanka, and rarely at Mogok, Burma. Also in Warren County, New York, but not in gem quality.

Name
From *Sinhala,* the Sanskrit name for Ceylon.

Smithsonite
Pure smithsonite is a white to gray mineral but because of the presence of impurity elements it is found in attractive colors and used as an ornamental stone and gem material. Although it may be yellow or pink, the translucent bluish-green variety is best known and is cut in cabochon stones. Smithsonite is a carbonate mineral and like all carbonates its low hardness places restrictions on its use.

Crystallography
Hexagonal-rhombohedral. Rarely in small rhombohedral or scalenohedral crystals. Usually in banded botryoidal or stalactitic masses.

Physical Properties
Cleavage perfect rhombohedral but not seen in the common massive material. **H** 4–4½. **S.G.** 4.30–4.45. *Luster* vitreous, may be silky on a fractured surface. Translucent.

Color usually white or gray but the rarer gem material may be green, blue, yellow, or pink.

Optical Properties
Uniaxial (−); $\omega = 1.850$, $\epsilon = 1.623$; biref. 0.227; disp. 0.030. In the smithsonite usually encountered in gemstones, the **R.I.** is difficult or impossible to determine and the dispersion is not apparent.

Chemical Composition
Zinc carbonate, $ZnCO_3$. Small amounts of other metals substitute for zinc giving rise to the following color varieties: copper, green to bluish-green; cobalt, pink; cadmium, yellow. Smithsonite, as do most other carbonate minerals, effervesces in cold hydrochloric acid.

Luminescence
Some greenish-yellow smithsonite fluoresces a brilliant orange-red under ultraviolet light but most specimens of the mineral are nonfluorescent.

Diagnostic Features
Characterized by its low hardness and effervescence in hydrochloric acid. These tests should be performed on the back of a cabochon stone. Also the **S.G.** is higher than that of other gems with which it might be confused.

Occurrence
Smithsonite is a secondary mineral having formed by the alteration of primary zinc minerals, particularly sphalerite. It is found as a minor ore mineral of zinc at many localities but occurrences of ornamental material are restricted. The most notable localities for the translucent greenish-blue smithsonite are Laurium, Greece and in the United States at Kelley mine, Magdalena, New Mexico. It is also found at Tsumeb, South-West Africa and Santander, Spain. Sardinia is noted for yellow stalactites with concentric banding.

Name
In honor of James Smithson (1754–1829), the founder of the Smithsonian Institution, Washington, D.C.

Sodalite
Sodalite is a feldspathoid. That is, it is a member of the mineral group called feldspathoids that are composed of essentially the same elements as the feldspars but with a smaller percentage of silica. Sodalite is present in the mineral aggregate, lapis lazuli (q.v.), as a minor constituent but it is also found in a pure blue massive form and used as a gem and ornamental stone (see Plate I).

Crystallography
Isometric. Crystals are rare, usually dodecahedrons. Commonly massive.

Physical Properties
Cleavage perfect dodecahedral (six equivalent directions). **H** 5½–6. **S.G.** 2.15–2.3. *Luster* vitreous. Transparent to translucent.

Color usually blue, also white, gray, pink, and green. It is the blue material that is used as a gem.

Optical Properties

Isotropic. **R.I.** = 1.48.

Chemical Composition

A sodium aluminum silicate with chlorine, $Na_4(AlSiO_4)_3Cl$.

Luminescence and Absorption Spectra

Blue sodalite does not fluoresce but hackmannite, a pink variety, fluoresces a brilliant red under ultraviolet light. Absorption spectra are unimportant.

Diagnostic Features

Sodalite resembles lapis lazuli but does not have the rich ultramarine blue of lapis and appears to have a purple hue when the two are compared side by side. Also, sodalite does not contain the yellow particles of pyrite found in lapis, and the specific gravity of lapis (2.75–2.95) is higher than that of sodalite.

Occurrence

Sodalite is a comparatively rare mineral associated with nepheline and other feldspathoid minerals in nepheline syenites and related rocks. The massive blue variety used as a gemstone is found at Litchfield, Maine; in Ontario, Quebec, and British Columbia, Canada; South-West Africa; and recently in Minas Gerais, Brazil.

Name

Named in allusion to its sodium content.

Sphalerite

Sphalerite, called *blende* or *zinc blende* by the English, is best known as the major ore mineral of zinc. Most commonly it is dark brown to black but transparent crystals of a golden-brown or reddish-brown are sometimes faceted as gemstones. Because of its low hardness and good cleavage, it is rarely worn in jewelry and cut stones are seen mostly in private collections and in museums (see Plate IX).

Crystallography

Isometric. Crystals may be in tetrahedrons and dodecahedrons but usually malformed and twinned. Usually in coarse to fine cleavable masses.

Physical Properties

Cleavage perfect dodecahedral (six directions). **H** 3½–4. **S.G.** 3.9–4.1. *Luster* resinous to adamantine. Transparent to nearly opaque.

Color. Colorless when pure and green, pale yellow, or red when nearly pure. Most often yellow, brown, or black.

Optical Properties

Isotropic. **R.I.** = 2.37, nearly as great as diamond; disp. 0.157, three times that of diamond.

Chemical Composition

Zinc sulfide, ZnS. Rarely pure; nearly always contains some iron so the formula can be written (Zn,Fe)S. The color darkens with increasing iron content. Manganese and cadmium may also be present in small amounts.

Luminescence and Absorption Spectra

In general sphalerite is nonfluorescent, although a yellow glow in ultraviolet light has been noted in black sphalerite from some localities. Some, but not all, transparent sphalerite shows three absorption bands in the red at 6900, 6670, and 6510 Å.

Diagnostic Features

Sphalerite is characterized by its high luster, **R.I.,** and dispersion. In appearance it may be confused with synthetic rutile, sphene, and zircon but can be distinguished by its single refraction.

Occurrence

Sphalerite as an ore mineral has a worldwide distribution but transparent material has come from only a few localities. Some of the most important are Přibram, Czechoslovakia; Binnental, Switzerland; Sonora, Mexico; Santander, Spain; and in the United States, Franklin, New Jersey.

Name

Sphalerite from the Greek word meaning treacherous; blende from the German word meaning deceiving. Both names were given for the same reason: although black, resembling the lead ore, galena, it yielded no lead.

Sphene (Titanite)

Sphene is a rare gem of exceptional brilliance and fire. Because it has inferior hardness, it is little used in jewelry but is highly prized by collectors.

Crystallography

Monoclinic. Often in crystals with a characteristic wedge shape resulting from a prominent, steeply inclined basal plane (Fig. 13.61).

Physical Properties

Cleavage distinct prismatic (two directions). **H** 5–5½. **S.G.** 3.50–3.54. *Luster* resinous to adamantine. Transparent to translucent.

Color yellow, brown, green, black.

Optical Properties

Biaxial (+); α = 1.882–1.918, β = 1.894–1.921, γ = 2.014–2.054; biref. 0.105; $2V$ = 23–50°; disp. 0.051. Pleochroism strong: α nearly colorless, β greenish-yellow, γ reddish-yellow; the exact colors depend on the color of the stone.

Chemical Composition

Calcium titanium silicate, $CaTiSiO_5$. Iron is usually present in small amounts. With increase in iron content, the stones become darker accompanied by a decrease in refractive index and birefringence.

Luminescence and Absorption Spectra

Nonfluorescent. Absorption spectra are unimportant.

Diagnostic Features

Sphene has refractive indices and birefringence higher than most gems and dispersion higher than diamond. It is characterized, therefore, by high dispersion and luster, although the dispersion may be masked by the color. Because of its high birefringence, doubling of the back facets is readily observed. It may be confused with sphalerite, diamond, and some garnets all of which are isotropic; and with synthetic rutile and zircon both of which have appreciably higher **S.G.**

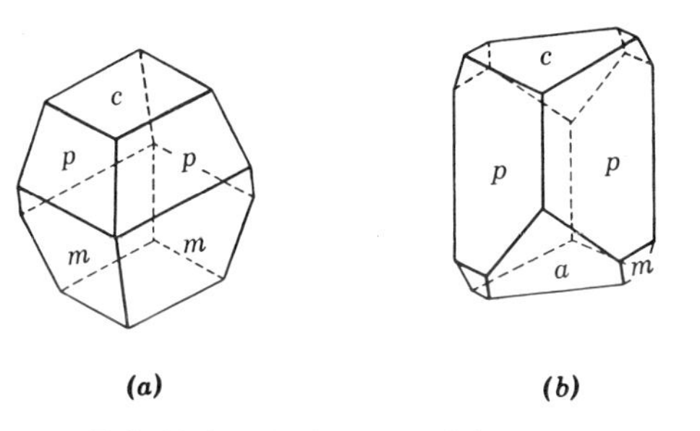

FIG. 13.61. Sphene crystals.

Occurrence

Sphene is a common accessory mineral in igneous rocks as small crystals. Crystals of larger size are found in metamorphic rocks and low-temperature veins. The finest gem sphene is the latter type and has come from veins in the Austrian and Swiss Alps. Gem material has also come from Renfrew County, Ontario, Canada; Madagascar; Mogok, Burma; Baja California, Mexico; and Brewster, New York.

Name

From the Greek word meaning wedge, in allusion to the characteristic shape of the crystals. Sphene is sometimes called *titanite,* so named because of the titanium content.

Spodumene

Spodumene is of interest to the economic geologist as the principal source of the lightest of all metals, lithium. It is of interest to the gemologist because of the transparent crystals that are cut into lovely pink, yellow, and green gems (see Plate VI).

Crystallography

Monoclinic. Ordinary spodumene occurs in rough crystals which in places reach gigantic proportions. A single crystal from the Etta mine, South Dakota measured 47 ft long and weighed 90 tons. Gem crystals are rarely more than 8 in. in length; they are commonly tabular, elongated on the c axis, and striated parallel to that direction. Many crystals show deep etching, indicating partial solution after formation.

Physical Properties

Cleavage prismatic perfect (two directions). The tendency to break along the excellent cleavage makes faceting difficult. **H** 6½–7. **S.G.** 3.18 ± 0.01. *Luster* vitreous. Transparent.

Color colorless, lilac-pink, yellow, yellowish-green, and green to emerald green. The lilac-pink type most commonly encountered in gems is called *kunzite;* the emerald-green variety is called *hiddenite.* No varietal names are used to designate the yellow and yellow-green; they go under the mineral name, spodumene. An originally deep-colored spodumene gem, particularly kunzite, may fade with time and become essentially colorless.

Optical Properties

Biaxial (+); α = 1.660, β = 1.666, γ = 1.676; biref. 0.016; $2V$ = 58°. Pleochroism: strong absorption in deep-colored stones with $\alpha > \beta > \gamma$. For kunzite α deep violet, β violet, γ nearly colorless. For hiddenite α emerald green, β blue-green, γ pale yellow-green. Disp. 0.017.

Stones are usually cut with the table nearly perpendicular to the c crystal axis. For with this orientation light moving perpendicular to the table has the greatest absorption producing the deepest color.

Chemical Composition

Lithium aluminum silicate, $LiAlSi_2O_6$, with little variation from this composition. However, small amounts of chromophores may be present; chromium in hiddenite, manganese in kunzite, and iron in yellow spodumene.

Luminescence and Absorption Spectra

Kunzite is the only spodumene to show significant luminescence in ultraviolet light. Under both long- and shortwave u.v. it has a peach-colored to orange glow. Characteristic absorption spectra are lacking in kunzite. The presence of iron in yellow spodumene is evidenced by two bands in the blue, one at 4375, the other at 4330 Å. Hiddenite shows bands resulting from chromium; a doublet in the deep red at 6905 and 5860 Å and a broad band at about 6200 Å.

Diagnostic Features

Kunzite is characterized by its lilac-pink color. Spodumene of other color may resemble citrine quartz, beryl, or topaz, but all these gems have a lower **R.I.** Quartz and beryl also have lower **S.G.** but the **S.G.** of topaz (3.5) is higher than that of spodumene. The pleochroism of spodumene is more pronounced than in the other gems.

Pink glass and synthetic pink spinel have been manufactured to simulate kunzite, but both are isotropic and thus nonpleochroic.

Occurrence

Spodumene both as an industrial mineral and as gem crystals is found almost exclusively in pegmatites where it is associated with other lithium-bearing minerals. It was not until the 1870s in Minas Gerais, Brazil that gem spodumene made its appearance as transparent yellow crystals. Shortly thereafter (1879) emerald-green spodumene crystals, thought at the time of their discovery to be diopside, were found associated with emerald at Stony Point, North Carolina. (Later the name of the town was changed to Hiddenite.) The first lilac-colored spodumene, kunzite, was found at Pala, California early in the nineteenth century and is still recovered from pegmatites in the area. Brazil today is the principal producer of yellow and yellow-green spodumene and is also a major source of kunzite. All these varieties have been found in Madagascar. Emerald-green hiddenite is the rarest spodumene. The rich color of the original crystals has not been duplicated at other localities, although paler green stones from Brazil are sold as hiddenite.

Spodumene, unlike many gemstones, is not recovered as a placer mineral. This is because of its good cleavage that causes it to break when tossed about in stream gravels.

Name

Spodumene from the Greek word meaning ash-colored; kunzite in honor of George F. Kunz, the American gemologist; hiddenite after W. E. Hidden, who discovered the green crystals.

Staurolite

Although transparent crystals of staurolite are sometimes cut as gemstones, the mineral attracts more interest as natural uncut twinned crystals. Those particularly valued, especially in Christian countries, are composed of two individuals crossing at about 90° (Fig. 13.62*a*). Locally, where they are found, they are known as *fairy stones* and are sold to be worn as charms. Because the demand frequently exceeds the supply, "fairy stones" are sometimes carved from a soft rock and dyed brown to simulate staurolite.

Crystallography

Monoclinic, pseudoorthorhombic. Elongated individuals are bounded by the side and basal pinacoids and second- and third-order prisms. Even more common are two types of cruciform twins composed of two interpenetrating individuals; one with the crystals crossing at nearly 90° (Fig. 13.62*a*), and the other with the crystals at about 60° (Fig. 13.62*b*).

Physical Properties

Fracture conchoidal. **H** 7–7½. **S.G.** 3.65–3.75. Both **H** and **S.G.** lowered by impurities. *Luster* resinous to vitreous when fresh, dull when altered or impure. Transparent to nearly opaque.

Color brown, red-brown, yellow-brown to brownish-black.

Optical Properties

Biaxial (+). $\alpha = 1.739–1.747$, $\beta = 1.745–1.753$, $\gamma = 1.752–1.761$; biref. 0.013; $2V = 85°$. Pleochroism: α colorless, β pale yellow, γ deep yellow to red. Disp. 0.023.

Chemical Composition

Iron aluminum silicate, $Fe_2Al_9O_6(SiO_4)_4(O,OH)_2$. Iron is usually replaced by some magnesium and rarely by zinc and cobalt.

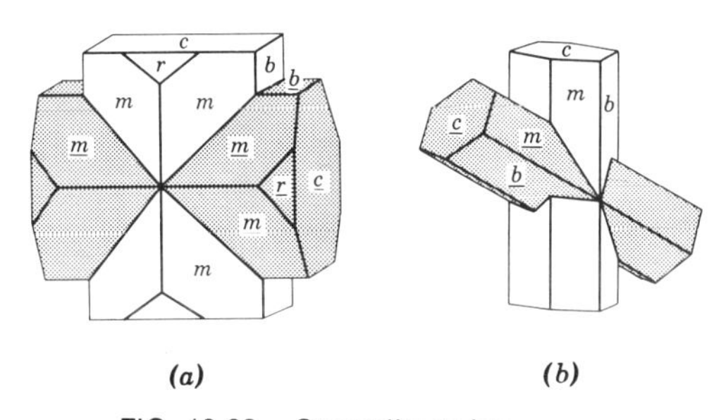

FIG. 13.62. Staurolite twins.

Fluorescence and Absorption Spectrum

Staurolite does not fluoresce under ultraviolet light. A strong absorption band is present at 4490 and a weaker band at 5780 Å.

Diagnostic Features

Staurolite twins as charms are distinguished from carved imitations by higher specific gravity and hardness. Cut stones are characterized by **R.I.** and pleochroism.

Occurrence

Staurolite is a regional metamorphic mineral associated with garnet and kyanite. Notable localities are St. Gotthard, Switzerland; Brittany, France; Brazil; U.S.S.R.; and in the United States in New Mexico and Montana. The states most noted for fairy stones, and hence their imitations, are Virginia, North Carolina, and Georgia.

Name

From the Greek meaning cross, in allusion to the cruciform twins.

Variscite

Variscite is a massive green to bluish-green mineral, resembling turquoise. Some has been used for cabochon gems but the most highly prized specimens are polished slices of nodules from near Fairfield, Utah.

Crystallography

Orthorhombic. Crystals very rare. Normally microcrystalline. Occurring in compact masses in nodules and veinlets.

Physical Properties

Fracture conchoidal. **H** 4–5. **S.G.** 2.4–2.6. *Luster* dull to waxy.

Color yellow-green to deep green in compact type. Pale green to almost white in earthy, porous type.

Optical Properties

Biaxial (−). $\alpha = 1.563$, $\beta = 1.588$, $\gamma = 1.594$; biref. 0.031; $2V = 50°$. **R.I.** near 1.58 for the massive, gemmy type.

Chemical Composition
Hydrous aluminum phosphate, $AlPO_4 \cdot 2H_2O$. Some ferric iron is usually present substituting for aluminum.

Luminescence and Absorption Spectra
Nonfluorescent. Absorption spectrum shows a strong line at 6880 Å and a weaker line at 6500 Å.

Diagnostic Features
Bears close resemblance to turquoise in texture but not in color, except in poorer turquoise colors. Variscite may be distinguished from turquoise by its lower **R.I.** and **S.G.**

Occurrence
Variscite is a secondary mineral found in near-surface (surface) deposits. The finest material comes from Clay Canyon, near Fairfield, Utah. Also found at other localities in Utah and Nevada.

Name
From *Variscia,* an ancient name for the Vogtland district of Germany, where the mineral was first identified.

Zoisite
Zoisite has long been used as the pink ornamental stone, *thulite,* and as one of the constituents of a chrome green mineral mixture somewhat resembling jade. However, in 1967 transparent crystals having an amethystine appearance were found in Tanzania. Since that time zoisite has become a popular gem to which the name *tanzanite* has been given. Except when otherwise noted, the following description is of these gem crystals.

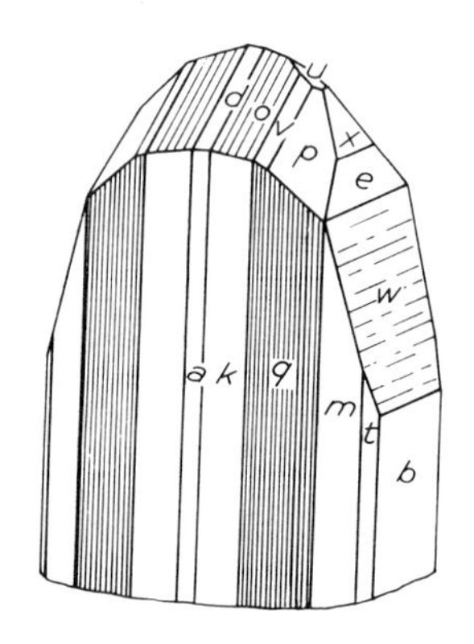

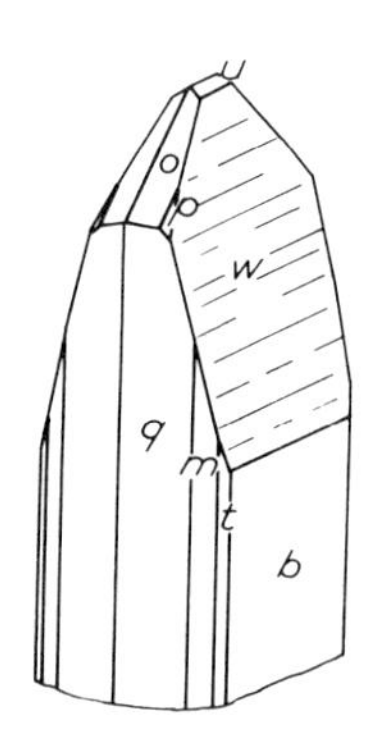

FIG. 13.63. Zoisite (tanzanite) crystals, Tanzania.

Crystallography
Orthorhombic. Well-formed crystals are commonly elongated on the *c* axis and frequently striated parallel to that direction (Fig. 13.63).

Physical Properties
*Cleavage** difficult parallel to the front and side pinacoids. Usually shows conchoidal fracture. **H** 6½. **S.G.** 3.36. *Luster* vitreous. Transparent.

Color of untreated crystals generally violet-blue. But because of extreme pleochroism, the color varies with the crystallographic direction in which it is viewed; parallel to *a* red-violet, *b* blue, *c* yellow or green. Most gem tanzanite is heat treated, a process which renders the stone, except when viewed along the *a* axis, a deep sapphire-blue.

The color of ordinary zoisite may be gray, brown, yellow, or green; the variety *thulite* is rose-red.

Optical Properties
Biaxial (+); $\alpha = 1.693$, $\beta = 1.694$, $\gamma = 1.702$; biref. 0.009; $2V = 53°$. Pleochroism: α red-violet, β blue, γ yellow-green (natural), γ blue (heat treated) (see Plate VI).

Chemical Composition
Hydrous calcium aluminum silicate, $Ca_2Al_3Si_3O_{12}(OH)$. Small amounts of vanadium present in Tanzanian crystals are believed to be responsible for the color. The pink color of thulite is ascribed to the presence of manganese.

Luminescence and Absorption Spectra
Zoisite does not fluoresce under ultraviolet light. The absorption spectrum has a broad band in the orange at 5950 Å and weaker bands in the green and blue at 5280 and 4550 Å, respectively.

* Although cleavage is difficult, the mineral may cleave when subjected to thermal shock or the vibrations of an ultrasonic cleaner.

Diagnostic Features
Tanzanite resembles blue sapphire in color but has both lower **R.I.** and **S.G.** as well as more marked pleochroism.

Occurrence
Zoisite is a mineral formed as a product of regional metamorphism. The crystals from Tanzania are found in metamorphosed limestone near the contacts with interlayered schists. Also in Tanzania a bright green granular zoisite is the host material for large tabular rubies.

Name
Zoisite for Baron von Zois who supplied the original specimens for description at the beginning of the nineteenth century. Thulite after Thule, an ancient name for Norway.

Glasses (Natural)
The natural glasses include volcanic glass, glass formed by lightning striking sand or rock, glass associated with meteor craters, and tektites. The natural glasses are not important gem materials, but their origins are interesting, and in some cases, controversial.

Glasses are amorphous. They usually have no exact chemical composition and are, therefore, variable in their physical and optical properties.

Obsidian
The most abundant and best known of the natural glasses is obsidian, a black or dark-colored volcanic glass of rhyolitic composition. Obsidian has been used since prehistoric times for making arrowheads, other sharp weapons and tools, jewelry, and art objects (see Fig. 5.3).

Physical Properties
Conchoidal fracture. **H** 5–5½. **S.G.** 2.33–2.50, variable depending on abundance of bubbles and inclusions. *Luster* vitreous. Transparent to opaque.

Color usually dark brown to black; also green, yellow, red. Black material with white patches is known as *flowered* or *snowflake* obsidian. *Flame obsidian* is dark brown to black material with golden reflections from oriented inclusions, similar to aventurine.

Optical Properties
Isotropic. **R.I.** 1.48–1.52.

Chemical Composition
Obsidian is a glass of rhyolite composition, rhyolite being the extrusive equivalent of granite. A typical example would have the following composition: SiO_2 76.8, Al_2O_3 12.1, Fe_2O_3 0.6, FeO 0.8, MgO 0.1, CaO 0.6, Na_2O 3.8, K_2O 4.9, H_2O 0.1%.

Inclusions
Obsidian always contains large numbers of minute inclusions called crystallites, which are usually more or less uniformly distributed, but may also be concentrated in layers. Round or elongated bubbles are also frequently present.

Diagnostic Features
Obsidian may be confused with other natural glasses, artificial glass, and sometimes black chalcedony. The vitreous luster of the fracture surface distinguishes it from the latter.

Occurrence
Obsidian has very widespread occurrence. Some of the best known localities are Iceland, and numerous places in Mexico. In the United States obsidian occurs in many localities in the western states; the best known of these is Obsidian Cliffs in Yellowstone National Park, Wyoming.

Name
Usage of the term obsidian goes back as far as Pliny, who described the rock from Ethiopia.

Tektites
Tektites are small, rounded, pitted bodies of silicate glass (Fig. 13.64), usually found in groups in widely separated areas of the earth's surface and bearing

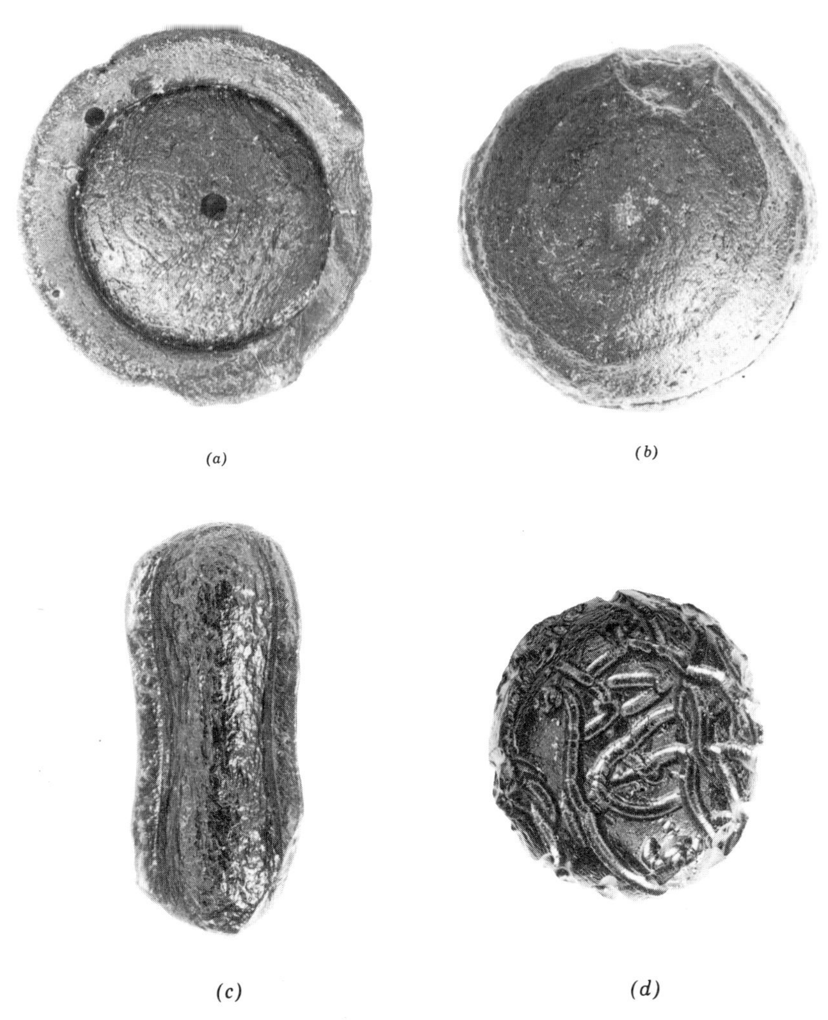

FIG. 13.64. Tektites. (*a*) Front view and (*b*) back view of a flanged tektite from Australia, 2.5X. (*c*) A flanged tektite from Australia, 2X. (*d*) Tektite from Indonesia, actual size.

no relation to the associated geologic formations. Their origin is not known, although they have been the subject of intensive scientific investigation for many years. They have relatively little value as gemstones, and because of their scientific value should be kept in their original form, not cut and polished.

Physical Properties

Conchoidal fracture. **H** 5½. **S.G.** 2.34–2.51, variable depending on composition and abundance of bubbles. *Luster* vitreous. Transparent to opaque.

Color green, greenish-brown, brown, black.

Optical Properties

Isotropic. **R.I.** 1.49–1.53.

Chemical Composition

A silicate glass of unusually high silica content (68–82%) and very low water content (average 0.005%). Their composition is unlike that of obsidian and more like that of shale.

Inclusions

Spherical or elongated bubbles are usually abundant, and there are no crystallites such as are seen in obsidian. Swirl striae are a prominent feature.

Diagnostic Features

Tektites may be confused with other natural glasses and artificial glass. In a cut stone positive identification may be difficult.

Occurrence

Tektites have various shapes (teardrop, dumbbell, button, flanged) strongly suggesting they were modeled by aerodynamic forces (Fig. 13.64). A widely held theory of their origin, that they are moonsplash, formed as gravity-escaping ejecta following lunar impacts, has been largely disproved by studies of lunar rocks. A second theory is that they are the product of large meteorite or comet impacts on terrestrial rocks.

The first noted occurrence of tektites was in 1787 in what is now Czechoslovakia, near the Moldau river, and were called *moldavites*. Other localities for tektites are in Indonesia (billitonites), Thailand, Vietnam, the Philippines, Australia (australites), and the Ivory Coast. In the United States they are found in Texas and Georgia.

Name

The term tektite was proposed in 1900 by Suess, who believed they were meteorites which at one time underwent melting; from the Greek *tektos*, meaning molten.

Other Natural Glasses

There are other occurrences of natural glasses, essentially fused silica, that are of scientific interest but of no gemological importance. Some of these follow:

Lechatelierite. A general term used for naturally fused silica formed either by lightning or meteorite impact. **R.I.** 1.462.

Fulgurites. Irregular, glassy, often tubular structures or crusts produced by the fusion of loose sand or compact rock by lightning.

Libyan Desert Glass. A nearly pure silica glass of unknown origin found as irregular lumps in the Libyan desert.

Impact Glass. A vesicular, glassy material produced by the fusion of rock by the heat generated from the impact of a large meteorite. It occurs in and around the resulting crater and is also known as *crater glass*.

Other Gems and Ornamental Stones

Included in this group are minerals that have been fashioned into gems by species collectors of gemstones but otherwise are of little importance. The list could be considerably expanded, for small gem quality portions of a large number of minerals are occasionally found and cut as gems.

The following minerals, with only brief descriptions, are considered in this group.

OTHER GEMS AND ORNAMENTAL STONES

Amblygonite	Cassiterite	Pollucite
Amphibole	Chlorastrolite	Prehnite
Pargasite	Datolite	Rhodizite
Smaragdite	Ekanite	Sepiolite
Tremolite-actinolite	Hambergite	Taaffeite
Apophyllite	Howlite	Talc
Anatase	Leucite	Thomsonite
Augelite	Painite	Willemite
Beryllonite	Petalite	Zincite

Amblygonite

Amblygonite is a lithium aluminum fluorophosphate, $LiAlPO_4(F,OH)$, occurring as a pegmatite mineral, usually in rough colorless to white crystals. Yellow crystals are less common but it is they that are usually cut as the rare gemstone. Properties: triclinic, two good and two poor cleavages, **H** 5½–6, **S.G.** 3.0–3.1, luster vitreous, transparent. Optics: biaxial (−), α = 1.59, β = 1.60, γ = 1.61 (all ±0.01). Large crystals have been found in Maine and California but most gem material comes from Brazil.

Amphibole

Amphibole is the name of a mineral group in which the individual members vary in chemistry and

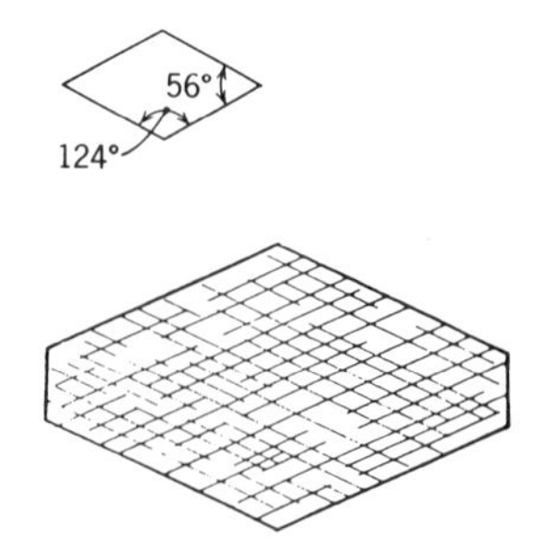

FIG. 13.65. Cleavages at 124° and 56° are characteristic of all amphiboles.

hence in some of their physical properties. They all have in common two good cleavages at 56° and 124° to each other (Fig. 13.65).

The *tremolite–actinolite* series has as its most important gem representative the felted aggregate, nephrite jade (see jade under *Important Gemstones*). However, two varieties of single tremolite crystals are occasionally cut as gems. One is transparent pink crystals called *hexagonite* from St. Lawrence County, New York. The other is in greenish crystals from Ontario, Canada that show a marked chatoyancy when cut cabochon. Properties: **H** 5½–6, **S.G.** 3.0, luster vitreous, transparent to translucent. Optics: biaxial (−), $\alpha = 1.602$, $\beta = 1.618$, $\gamma = 1.631$.

Smaragdite is a thin-foliated grass-green to emerald-green variety of amphibole with a chemical composition near that of actinolite (see nephrite jade under *Important Gemstones*). It has been used to simulate jadeite and has properties (**S.G.** 3.3 and **R.I.** about 1.65) close to that mineral.

Pargasite is a bluish-green amphibole from Pargas, Finland, only rarely seen as a gemstone. Properties: **H** 6, **S.G.** 3.1. Optics: biaxial (−), mean **R.I.** about 1.62; pleochroism blue, and yellow.

Apophyllite

Apophyllite is associated with zeolites in cavities in basaltic rocks notably in India, Mexico, Switzerland, and at several localities in the eastern United States. It is a hydrated potassium calcium silicate with fluorine, $KCa_4(Si_4O_{10})_2F \cdot 8H_2O$. The color, usually white or colorless, may be green, pink, or yellow. Properties: tetragonal, cleavage perfect basal, **H** 4½–5, **S.G.** 2.3–2.4, luster pearly parallel to the cleavage elsewhere vitreous, transparent to translucent. Optics: uniaxial (+), $\omega = 1.537$, $\epsilon = 1.535$.

Anatase

Anatase is titanium oxide, TiO_2, polymorphous with rutile and brookite. It has been found as a vein mineral at many localities, particularly in the Alps, but is only rarely cut as a gemstone. Properties: tetragonal; perfect cleavage parallel to the base and second-order dipyramid; **H** 5½–6; **S.G.** 3.9–4.0; luster adamantine; color various shades of brown also blue, green, and lilac; transparent. Uniaxial (−), $\omega = 2.561$, $\epsilon = 2.488$.

Augelite

Augelite is a basic aluminum phosphate, $Al_2(PO_4)(OH)_3$. It has been found at Potosi, Bolivia and in the White Mountains of California in colorless crystals that have been cut as gemstones, but for collectors only. Properties: monoclinic, two perfect and three imperfect cleavage directions, **H** 4½–5, **S.G.** 2.70, luster vitreous, transparent to translucent. Optics: biaxial (+), $\alpha = 1.574$, $\beta = 1.576$, $\gamma = 1.588$.

Beryllonite

Beryllonite is a sodium beryllium phosphate, $NaBePO_4$, found at Stoneham and Newry, Maine associated with phenakite and beryl. Faceted colorless stones are collectors' items. Properties: monoclinic (pseudoorthorhombic), one perfect and one good cleavage, **H** 5½–6, **S.G.** 2.81, luster vitreous to pearly. Optics: biaxial (−), $\alpha = 1.552$, $\beta = 1.558$, $\gamma = 1.561$.

Cassiterite

Cassiterite is tin oxide, SnO_2, the principal ore mineral of tin. Although pure tin oxide is white or colorless, the mineral is usually brown to black owing to the presence of iron; the more iron, the deeper the color. The occasional crystal with a low iron content may have a yellow to reddish color and be sufficiently transparent to be cut as a faceted stone. Properties: tetragonal, **H** 6–7, **S.G.** 7.0 ± 0.1, luster adamantine. Optics: uniaxial (+), ω = 2.00, ϵ = 2.10; disp. 0.062, unusually high.

Chlorastrolite

Chlorastrolite is a green to bluish-green fibrous but compact aggregate composed of several minerals, chiefly pumpellyite. Originally it was described as a mineral species but later was shown to be a mixture. In waterworn pebbles chlorastrolite is found on the shores of Lake Superior, particularly on Isle Royal in the lake, having weathered from the basic igneous rocks of the vicinity. The surface of cabochon stones have an attractive green and white mottled appearance. Properties: **H** 5–6, **S.G.** 3.1–3.3, **R.I.** about 1.7.

Datolite

Datolite is a calcium borosilicate, $CaBSiO_4(OH)$, found characteristically in cavities in basaltic rock as transparent crystals either colorless or with a greenish tinge. These crystals are sometimes faceted as collectors' items. Datolite also occurs as a microcrystalline aggregate, resembling unglazed porcelain, that is cut as cabochons. This latter type is usually white but may be yellowish, greenish, or reddish. Specimens of the microcrystalline type from the Lake Superior copper district containing finely divided native copper are highly prized (Fig. 13.66). Properties: monoclinic, **H** 5–5½, **S.G.** 2.8–3.0, crystals transparent with vitreous luster, microcrystalline opaque with dull luster. Optics: biaxial (−), α = 1.624, β = 1.652, γ = 1.668. **R.I.** for microcrystalline about 1.65.

FIG. 13.66. Massive datolite containing native copper (darkened areas). Lake Superior copper district, Michigan.

Ekanite

In 1961 a green gem mineral from the gravels of Sri Lanka (then called Ceylon) was described as a new species, *ekanite*. It is essentially a thorium calcium silicate, metamict and radioactive. Properties: **H** 6–6½, **S.G.** 3.28, isotropic with **R.I.** 1.60. Some stones show a four-rayed star.

Hambergite

Hambergite is a rare beryllium borate, $Be_2(BO_3)(OH)$, found in colorless crystals of gem quality at only one locality, in a pegmatite in Madagascar. It also has been reported from the gem gravels of Kashmir, India. Properties: orthorhombic, one perfect and one good cleavage, **H** 7½, **S.G.** 2.36, luster vitreous, transparent. Optics: biaxial (+), α = 1.554, β = 1.587, γ = 1.628, biref. 0.074, unusually high.

Howlite

Howlite is a hydrous calcium borosilicate, $Ca_2SiB_5O_9(OH)_5$. It occurs as compact nodules which break with a conchoidal fracture yielding a surface resembling unglazed porcelain. Its principal occurrence is in California, where it is associated with other borate minerals. The milky white mineral, sometimes with black veinlets, finds a minor use as an ornamental stone. Properties: monoclinic (?), **H** 3½, **S.G.** 2.53–2.59, color white but sometimes dyed blue. Biaxial with mean **R.I.** 1.59.

Leucite

Leucite, a potassium aluminum silicate, $KAlSi_2O_6$, is a colorless to white rock-forming mineral found in lava flows as well-formed trapezohedral crystals. The most important source is the lava of Mount Vesuvius, Italy. It is also found in Germany and in the United States at several localities in Wyoming and Montana. Leucite forms as isometric crystals at high temperatures and inverts to a tetragonal polymorph. Because of this the crystals show double refraction even though the external isometric crystal form is preserved. Properties: **H** 5½–6, **S.G.** 2.47, luster vitreous to dull, **R.I.** 1.51.

Painite

A single transparent, garnet-red crystal weighing 8½ carats from Mogok, Burma was described in 1957 as a new mineral, *painite*. Chemical analysis of a small portion of the crystal showed it to be a calcium aluminum borosilicate, essentially $Ca_4BAl_{20}Si_6O_{38}$. Until more of the mineral is found, it must be considered a "freak," but is worth noting because of its desirable gem properties. Furthermore, in the past painite, incorrectly identified, may have been cut as gemstones. Properties: hexagonal, **H** 8, **S.G.** 4.0. Optics: uniaxial (−), $\omega = 1.816$, $\epsilon = 1.788$; pleochroism: ω brownish-orange, ϵ ruby-red.

Petalite

Petalite is a colorless mineral only rarely cut as faceted stones. It occurs in pegmatites with spodumene and has a somewhat similar composition, $LiAlSi_4O_{10}$. Properties: monoclinic, one perfect and two imperfect cleavages, **H** 6–6½, **S.G.** 2.4, luster vitreous to pearly, transparent to translucent. Optics: biaxial (+); $\alpha = 1.505$, $\beta = 1.511$, $\gamma = 1.518$.

Pollucite

Pollucite is a white to colorless pegmatite mineral containing the rare element cesium as a major chemical component. It is a cesium aluminum silicate, $CsAlSi_2O_6 \cdot H_2O$, that is found sparingly at only a few localities. It has been found in Sweden, Rhodesia, and in the United States in South Dakota and Connecticut. But its major world occurrence is Newry, Maine where it was at one time mined for cesium. Properties: isometric, **H** 6½, **S.G.** 2.90, luster vitreous, transparent to translucent, **R.I.** 1.52.

Prehnite

Prehnite is a light green to yellow-green mineral only rarely cut as cabochons or carved as small ornaments. It is a hydrous calcium aluminum silicate, $Ca_2Al_2Si_3O_{10}(OH)_2$. The characteristic occurrence is in cavities in basaltic rocks as reniform groupings of subparallel platy crystals. The most notable localities in the United States are Farmington, Connecticut; Paterson and Bergen Hill, New Jersey; Westfield, Massachusetts; and the Lake Superior copper district, Michigan. Properties: **H** 6–6½, **S.G.** 2.8–2.95 (gem material usually about 2.90), luster vitreous to waxy, translucent, rarely transparent. Optics: biaxial (+), $\alpha = 1.616$, $\beta = 1.626$, $\gamma = 1.649$; a single **R.I.** of about 1.63 usually observed.

Rhodizite

Rhodizite is complex aluminum borate containing cesium, beryllium, and lithium, essentially $CsAl_4Be_4B_{11}O_{25}(OH)_4$. It is a rare mineral, usually rose-red but sometimes yellow or colorless, and has been found in pegmatites in the Ural Mountains and in Madagascar. Properties: isometric, poor octahedral cleavage, **H** 8½, **S.G.** 3.44, luster vitreous to adamantine, transparent to translucent, **R.I.** 1.694.

Sepiolite

Sepiolite is a hydrous magnesium silicate, $Mg_4Si_6O_{15}(OH)_2 \cdot 6H_2O$. It lacks all the normal attributes of gem material but is used for carving pipe bowls. Properties: orthorhombic but in microcrystalline aggregates, **H** 2–2½, **S.G.** 2, luster dull, opaque. Although the **S.G.** is 2, the microscopic particles of which it is composed are so loosely packed that it may have an apparent density of less than 1 and thus float on water. For this reason it is called *meerschaum* from the German word meaning sea foam. The mean **R.I.** is about 1.53 but the shadow edge of the refractometer is indistinct. The principal source of meerschaum has been Turkey but it is also mined in Greece, Spain, Morocco, and Kenya and in the United States in Pennsylvania, Utah, California, and New Mexico.

Taaffeite

A small (1.419 carats) mauve gemstone, supposedly spinel, was described in 1951 as a new mineral, *taaffeite*. Since then very few other stones have been identified as taaffeite. The origin of the stones is unknown but the type of cut suggests Sri Lanka. The mineral is a beryllium magnesium aluminum oxide, $BeMgAl_4O_8$, intermediate between spinel and chrysoberyl. Properties: hexagonal, **H** 8, **S.G.** 3.6. Optics: uniaxial (−), $\omega = 1.721$, $\epsilon = 1.717$.

Talc

Talc, a hydrous magnesium silicate, $Mg_3Si_4O_{10}(OH)_2$, is a soft platy mineral that lacks the normal qualifications of a gem material. However, the massive impure variety, *soapstone* or *steatite*, occurs abundantly and is used for carving ornamental objects. Because of its low hardness, it was easily worked with primitive tools and thus some of the oldest carvings, such as the scarabs of Ancient Egypt, are of soapstone. Modern carvings executed by the North American Eskimos can be purchased in many gift shops. Properties: monoclinic; color white to greenish-gray, often streaked; **H** 1 (has a greasy feel); **S.G.** 2.7–2.8; luster pearly to greasy; translucent to opaque. Optics: biaxial (−), $\alpha = 1.539$, β and $\gamma = 1.589$, an indistinct reading on the refractometer at about 1.55.

Much of the material that goes under the name of soapstone or steatite is undoubtedly basically *pyrophyllite*, $Al_2Si_4O_{10}(OH)_2$ for talc and pyrophyllite have essentially identical properties. To distinguish them one must resort to chemical or X-ray tests or a more precise measurement of refractive indices than given by the refractometer.

Thomsonite

Thomsonite is a hydrous sodium calcium aluminum silicate, $NaCa_2Al_5Si_5O_{20}\cdot 6H_2O$, a member of the zeolite mineral group. It characteristically occurs as slender fibrous crystals in cavities in lava flows and as such is found in many localities. More rarely it is in compact spherical forms resulting from closely packed fibers of varying color that radiate from several centers. It is this massive type that is used as a gem and when polished, white, red, yellow, and green "eyes" are seen on the rounded surface (Fig. 13.67). This material is found as pebbles on the shores of Lake Superior and particularly on the shores of Isle Royal in the lake. Properties: orthorhombic, **H** 5, **S.G.** 2.3–2.4, luster vitreous, translucent. Optics: biaxial (+), **R.I.** variable between 1.50 and 1.54.

Willemite

Willemite is found at several localities and at one, Franklin, New Jersey, it is a major ore mineral of zinc. It is a zinc silicate, Zn_2SiO_4, which when pure is colorless, but at Franklin the mineral may be yellowish-green, reddish, or brown. It is the rare transparent crystals from this locality that are cut as gems. Properties: hexagonal, good basal cleavage, **H** 5½, **S.G.** 3.9–4.2, luster vitreous to resinous, transparent to translucent, fluoresces green in ultraviolet light. Optics: uniaxial (+), $\omega = 1.691$, $\epsilon = 1.719$.

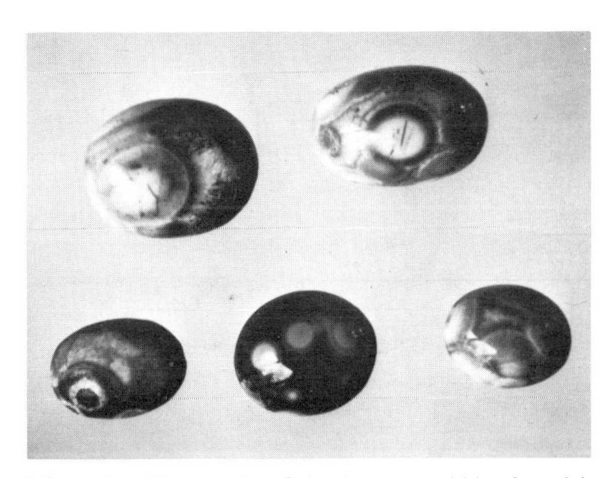

FIG. 13.67. Thomsonite. Cabochon cut pebbles from Isle Royal, Lake Superior.

Zincite

At the famous mines at Franklin and Sterling Hill, New Jersey, zincite, a red zinc oxide, ZnO, is an abundant mineral. But only rarely is it found as transparent material of sufficient quality to be cut as gemstones. Properties: hexagonal, perfect basal cleavage, **H** 4, **S.G.** 5.68, luster subadamantine, translucent rarely transparent. Optics: uniaxial (+), $\omega = 2.013$, $\epsilon = 2.029$.

Organic Gems

Included in this group are those substances used as gem materials which have an organic origin. The number is small but includes pearl which ranks high on the list of gems. Thus pearl is considered first in the following descriptions.

ORGANIC GEMS	
Pearl	Ivory
Amber	Jet
Coral	

Pearl

Pearl has long been considered to be among the most precious of gems, even though it is much softer than the standard demanded of other gem materials. Pearl may well have been the first gemstone used by ancient man since it requires no lapidary treatment to enhance its beauty.

The distinction between natural and cultural pearl is the most difficult identification problem faced by the gemologist. In most cases positive identification requires the use of X-ray equipment and can only be done in laboratories having this special equipment.

Natural Pearl

Pearl is formed when a mollusk deposits a substance called *nacre* around an irritant. Any mollusk can produce a pearl, but only those species that have a pearly, or nacreous (mother-of-pearl) lining can form a lustrous pearl suitable for use as a gem. Gem quality pearls are produced in salt water almost exclusively by the mollusk species *Pinctada,* previously known as either *Margaritifera* or *Meleagrina*.

Physical Properties

H 2½–4½. **S.G.** 2.68–2.75. Luster pearly. The luster of pearl, also known as *orient,* is an optical effect caused by thin overlapping plates of aragonite.

Color white, delicate shades of pink, yellow, blue, gray; also gunmetal, bronze, black.

Chemical Composition

Pearl is formed of 82–86% calcium carbonate in the form of aragonite (q.v. under *Less Important Gemstones*) 2–4% water, and 10–14% *conchiolin,* a hornlike substance. Nacre is formed when the spaces in a weblike deposit of conchiolin are filled with microscopic crystals of aragonite.

Structure

Natural pearl is formed when a mollusk deposits nacre around an irritant, which may be a parasitic organism or a grain of sand. The nacre is deposited in concentric layers of aragonite, building up in overlapping platelets (Fig. 13.68). The aragonite crystals are oriented with their long (c) axes perpendicular to the layers (Fig. 13.69a). The fact that natural pearl has this concentric and radiating structure is important in testing, as will be discussed later.

Spherical, or "free" pearls are formed entirely within the soft parts of the mollusk. A *blister pearl* (Fig. 13.70) is formed when the irritant gets between the shell and the mantle and the pearl forms as a bulge on the shell. Nacreless pearls form in many bivalve mollusks; they are usually purplish-brown in color and have no value. *Conch pearls* are light orange-red or pink concretions formed within a conch, and are without nacre. *Tridacna pearls,* found in the species *Tridacna gigas,* are characterized by lack or orient.

Cultured Pearl

Cultured blister pearls may have been made by the Chinese as early as the thirteenth century. In 1896 Kokichi Mikimoto patented a process for producing

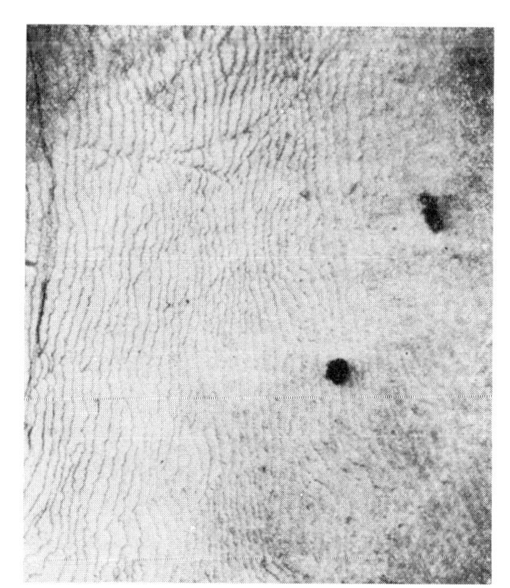

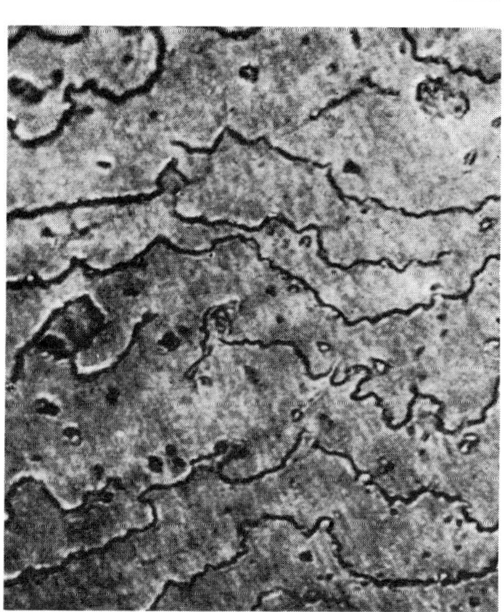

FIG. 13.68. The surface of a natural pearl as seen under a microscope, showing the overlapping plates of nacre. Courtesy of the Gemological Institute of America.

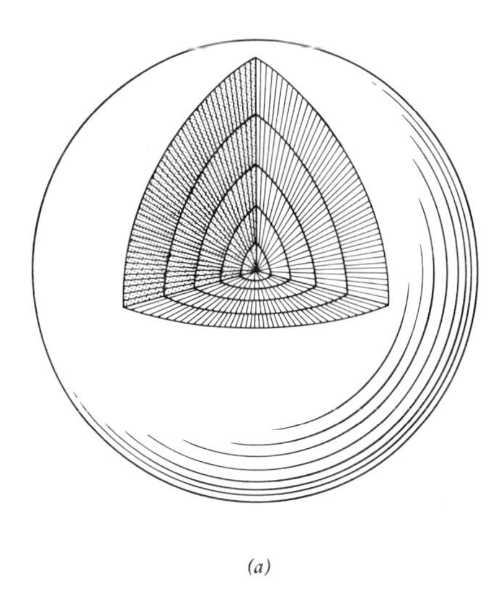

(a)

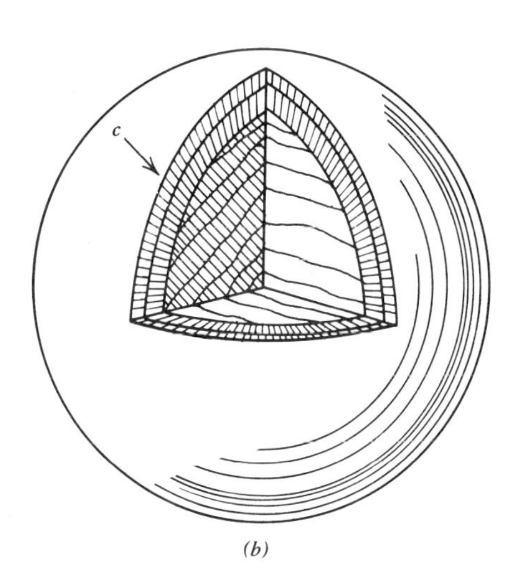

(b)

FIG. 13.69. (a) Schematic diagram of a natural pearl. The cutout section shows the nacre deposited in concentric spherical layers with the long (c) axes of the aragonite crystals perpendicular to the layers. (b) Diagrammatic section of a cultured pearl. The mother-of-pearl core is made up of layers of aragonite, essentially parallel, surrounded by concentric layers of nacre deposited by the mollusk. The left side of the cutout section is at right angles to the layers of mother-of-pearl. It can be seen that in only one direction, c in the diagram, are the c axes of the layers of aragonite in the core essentially parallel to the c axes of the aragonite crystals deposited by the oyster.

cultured blister pearls. Two other Japanese, Mise and Nishikawa, are credited with having independently developed a process for culturing whole pearls and in 1908 they signed a joint ownership agreement of the method. In 1916 Mikimoto was granted a patent for culturing pearls by a complex method no longer used, and all commercial production of cultured pearls has been by the Mise–Nishikawa

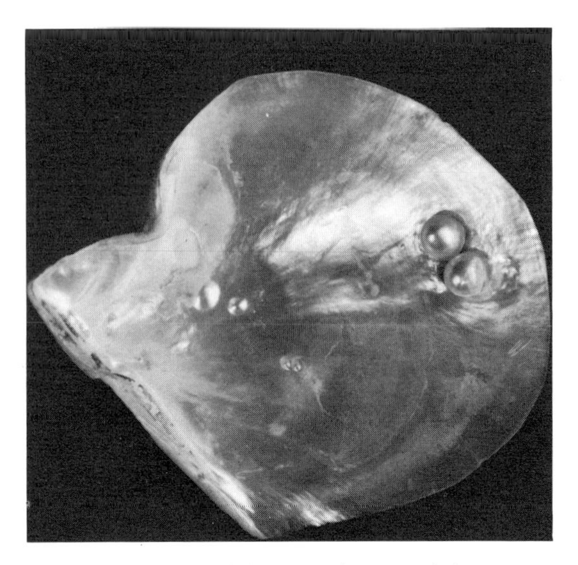

FIG. 13.70. Blister pearls, natural size.

method. Although Mikimoto did not develop the method, he was the first to produce cultured pearls on a large commercial scale, beginning in 1921.

Method

Saltwater pearls are cultured by inserting a mother-of-pearl bead and a piece of mantle tissue into an incision in the foot of the mollusk *Pinctada martensii* (Fig. 13.71). The oysters are then held in cages suspended under rafts, at a depth of 7–10 ft, for a period of 3–6 years, usually about 3½ years, by which time the pearl is about 1 mm larger than the inserted bead.

FIG. 13.71. Technician inserting a mother-of-pearl bead into an incision in the foot of the mollusk, one step in the process of culturing pearls.

The mother-of-pearl beads used as nucleii are mostly cut from the shells of a freshwater species found in the waters of the Mississippi river.

Freshwater pearls are cultured at Lake Biwa in Japan using a clam species *Hyriopsis schlegeli*. In this method a solid nucleus is not used. The mantle of the clam is cut in several places and a piece of mantle tissue inserted. Most of the pearls produced are baroque, that is, pearls with irregular shape.

Structure

The structure of a saltwater cultured pearl is shown in Figure 13.69*b*. The mother-of-pearl core is made up of parallel layers (essentially aragonite), surrounded by concentric layers of nacre deposited by the oyster.

Testing

The distinction between pearl imitations and either natural or cultured pearl is simple. However, since the surface material of a natural pearl and a cultured pearl are the same, the distinction between them is difficult, especially if the pearl is undrilled. A few simple tests, such as specific gravity, candling, and examination under magnification, can be used but these are not always conclusive. The endoscope, an instrument for examining drilled pearls, is now largely of historical interest. The same may be said of a method using X-ray diffraction patterns. The most reliable and most widely used testing method for natural/cultured pearls is a combination of X-radiography and X-ray fluorescence.

Specific Gravity

Because cultured pearls have a large core of mother-of-pearl there is generally a difference in specific gravity between natural and cultured pearls. In a solution with a specific gravity of 2.713 most natural saltwater pearls float and most cultured pearls sink.

Candling

In this method individual pearls are examined by holding them over an intense, shielded light source. In a cultured pearl the parallel layers of the mother-of-pearl bead may often be seen, whereas a natural pearl will show only decreasing illumination from edge to center. The method is not entirely reliable for an individual pearl, but can often give information about the identity of a strand, since strands are not normally mixed.

Examination of Drill Hole

A test that is simple to apply, but not always reliable, is to examine the drill hole of a pearl with hand lens or microscope. Since the nacreous layer of a cultured pearl is thin (1 mm or less), the boundary between this layer and the mother-of-pearl bead can often be seen. However, natural pearls may show a similar structure.

The Endoscope

This interesting instrument, now little used, employs a hollow needle with mirrors at 45° to its length, for examining the internal structure of a drilled pearl (Fig. 13.72). Before X-ray methods were developed it was considered to be the most reliable, although it had shortcomings, among which were the need to remove the pearls from a strand, and the inability to examine undrilled pearls.

Both single and double mirror methods are used, but only the latter is briefly described here. As shown in Figure 13.73, in a natural pearl, as the needle reaches the center, a beam of light is reflected from the first mirror, follows the concentric layers around to the second mirror, and is seen in the microscope as a bright flash. Because of the parallel-layer structure of the center of a cultured pearl, no such reaction is observed.

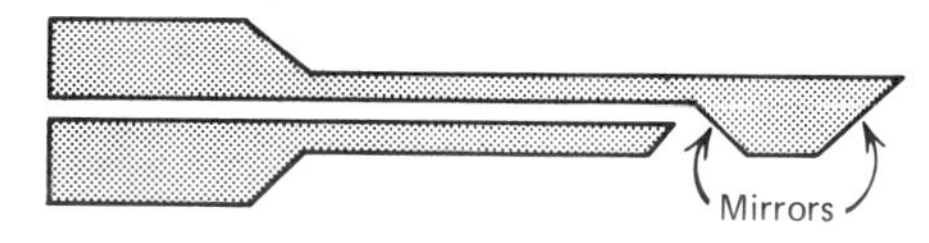

FIG. 13.72. Longitudinal section of an endoscope. The instrument is essentially a hollow needle through which light passes to be reflected by mirrors.

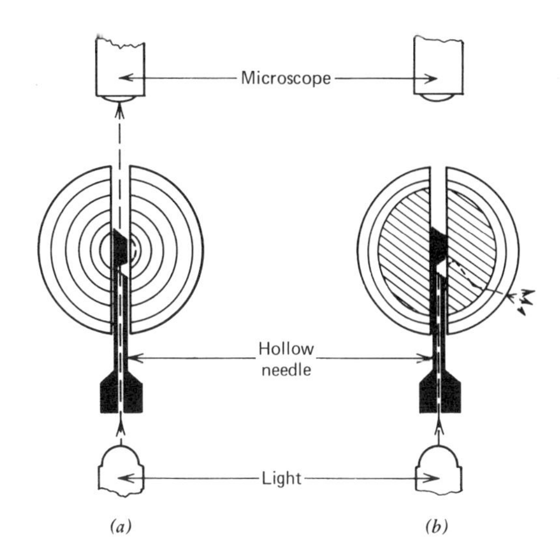

FIG. 13.73. Diagram showing how an endoscope is used to differentiate between a drilled natural pearl (*a*) and a drilled cultured pearl (*b*). Courtesy of the Gemological Institute of America.

X-Ray Diffraction

In this method a narrow beam of X-rays is passed through a pearl and the resulting diffraction pattern recorded on a piece of film (see Chapter 4). In a natural pearl all the aragonite crystals are arranged radially from the center, and regardless of the orientation of the pearl, the X-rays travel parallel to the length of the crystals. This direction is the pseudohexagonal, *c* axis of aragonite. The pseudohexagonal symmetry is revealed in the resultant X-ray diffraction pattern which looks like those shown in Figure 13.74. For a cultured pearl, for most positions, the X-rays are traveling more or less perpendicular to the length of the aragonite crystals, with a resultant pattern as in Figure 13.75. Note that for one direction of X-rays (*c*, Fig. 13.69*b*) natural and cultured pearls will give similar patterns. This method is reliable but cumbersome and slow and now little used.

X-Radiography

This method is similar to that used in medicine and dentistry. A single pearl, or an entire strand, is placed on a piece of photographic film and exposed to a broad beam of X-rays. Details in the photograph

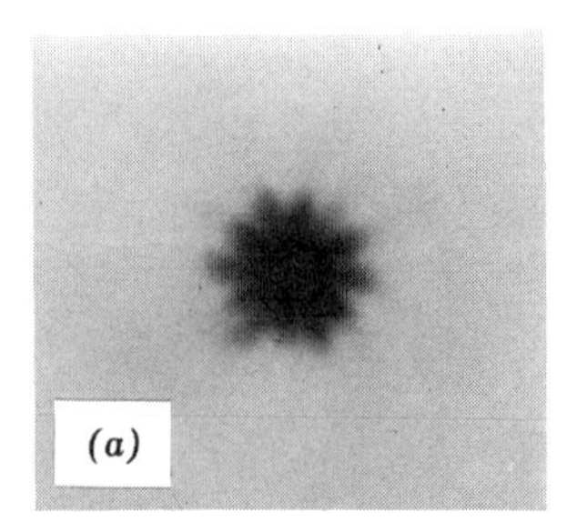

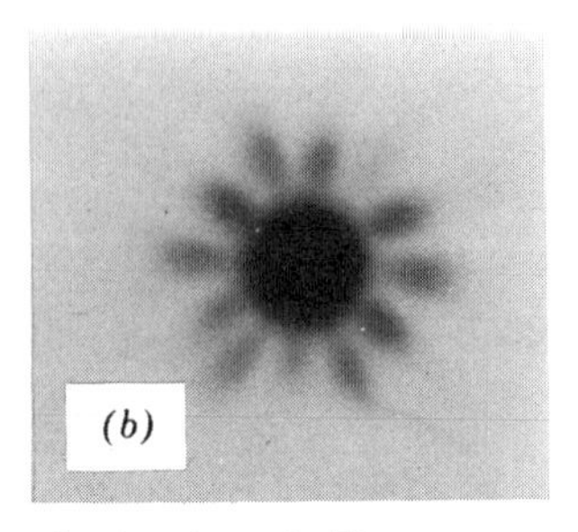

FIG. 13.74. X-ray diffraction patterns of natural pearls. The X-ray beam moved parallel to the pseudohexagonal *c* axes of aragonite crystals producing the 6-fold patterns. Courtesy of the Gemological Institute of America.

are the result of different degrees of transparency to X-rays of different components of the object. Cultured pearls (Fig. 13.76) show a dark line separating the core and nacre due to a conchiolin layer that is transparent to X-rays. Natural pearls (Fig. 13.77) may show several gradations in X-ray transparency but not the perfectly spherical core seen in cultured pearls.

X-Ray Fluorescence

Observation of the fluorescence under the same X-ray beam used in radiography is a useful supplementary test. All cultured pearls having a freshwater mother-of-pearl bead fluoresce under X-rays, whereas natural saltwater pearls do not.

Imitations

Imitation pearls are made by coating spheres with a lacquerlike substance containing in suspension something that when dry will produce a pearl-like appearance. The resultant surface is smooth, not platy as in Figure 13.68. This may be observed under the microscope, or felt by drawing the pearl across the cutting edge of the teeth. Spheres of three types are used, wax-filled, solid glass, and mother-of-pearl. A widely used material for coating the beads contains in suspension crystals of guinine, obtained from the membrane coating of fish scales, sometimes called *essence d'orient.*

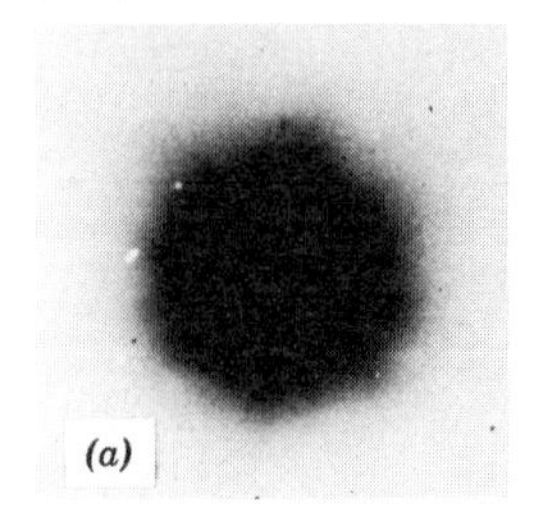

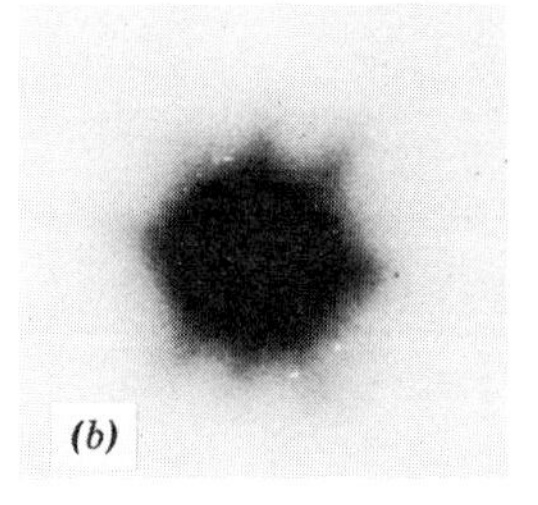

FIG. 13.75. X-ray diffraction patterns of cultured pearls. Courtesy of the Gemological Institute of America.

Occurrence

The finest natural pearls come from fisheries in the Persian Gulf, especially those off the islands of Bah-

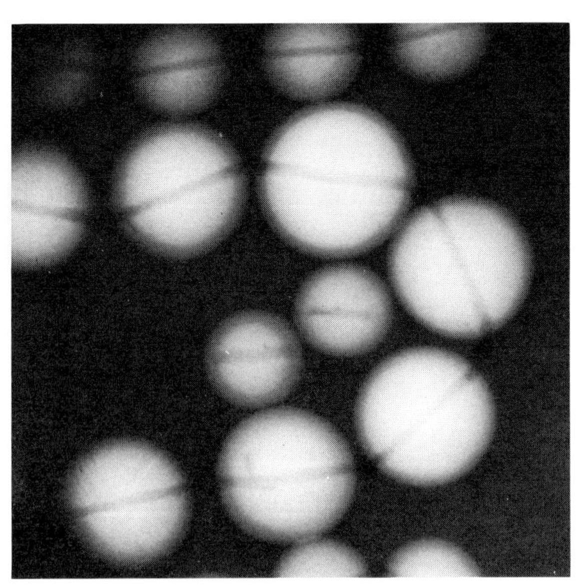

FIG. 13.76. X-radiograph of a strand of cultured pearls. Courtesy of the Gemological Institute of America.

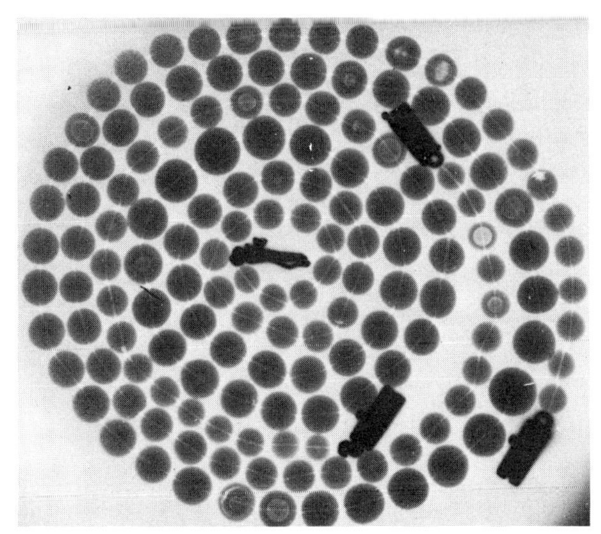

FIG. 13.77. X-radiograph of strands of natural pearls. Courtesy of the Gemological Institute of America.

rain. Other less important sources are in the Gulf of Mannar and in the Red Sea. Minor occurrences of natural pearl include waters adjacent to such widely scattered places as Australia, numerous islands in Micronesia and Polynesia, Venezuela, Mexico, Panama, and Colombia.

Fine freshwater natural pearls are found in rivers in Europe and North America, especially in Scotland, Wales, Ireland, France, and in the United States in the Mississippi River and its tributaries.

Most cultured saltwater pearls are produced in the Inland Sea of Japan, with lesser production from Australia and the Philippines.

Amber

The mineralogist does not consider amber (Fig. 13.78) to be a mineral for, aside from being a natural product, it satisfies none of the other criteria. Nevertheless, in gemology it occupies an important place as a gem and ornamental material (Fig. 13.79) that has been in use since at least the late Stone Age, about 6000 years ago.

Amber is a fossil resin, that is, a resin exuded from trees millions of years ago which has hardened through oxidation, loss of volatiles, and lengthy burial. Although found at several localities, most of the world's amber has come from the southern coast of the Baltic Sea and to many this is known as "true amber." It contains 3–8% succinic acid and has thus been given the mineral name *succinite*. The following description is for Baltic amber.

Crystallography

Amorphous. Occurs in masses weighing up to several pounds, usually irregular but sometimes having a teardrop shape.

Physical Properties

Brittle with conchoidal fracture. **H** 2½–3; can not be scratched by the fingernail but is easily scratched by a needle and less easily by a copper wire. With **S.G.** 1.05–1.09, it sinks in fresh water but is suspended or floats in a saturated salt solution. *Luster* greasy to resinous. Transparent to translucent, some nearly opaque.

FIG. 13.78. Baltic amber. (*a*) Unusually large piece. (*b*) Transparent nodular fragments. Courtesy of Judith W. Frondel, Harvard University.

FIG. 13.79. Chinese carving of Baltic amber. Courtesy of Judith W. Frondel, Harvard University.

Because amber is a mixture of several constituents, it does not have a definite melting point. It melts between 295 and 395°C, higher than most fossil resins, and gives a characteristic aromatic odor. If small fragments of clear amber are heated to a temperature of about 180°C, they soften and become welded together when subjected to high hydrostatic pressure. The resulting *pressed amber* or *ambroid* is then processed as a large single piece.

When rubbed, amber develops a negative electric charge permitting it to pick up tiny fragments of light materials. This fact was known to the Greeks and it is from *electron,* their word for amber, that our word *electricity* is derived. The development of the negative charge is not a distinguishing criterion, for most resins and many artificial substances made to resemble amber have this property.

Color yellow with shades of orange, brown, and more rarely reddish or greenish; milky white is "bone amber."

Optical Properties

Isotropic. **R.I.** 1.54. Some specimens show anomalous birefringence.

Chemical Composition

Amber is not a specific resin but a mixture of several, each with a different chemistry and physical properties. The bulk chemical analysis of Baltic amber gives carbon 79%, hydrogen 10.5%, oxygen 10.5%, expressed by the formula $C_{10}H_{16}O$. Amber burns in a match flame yielding a pleasant aromatic odor.

Luminescence

Some, but not all, amber fluoresces a blue-white under longwave u.v. Under shortwave u.v. the fluorescence is a dull yellow-green.

Diagnostic Features

Amber's low specific gravity, the lowest of any gem material, distinguishes it from the synthetic materials bakelite (**S.G.** 1.25) and celluloid (**S.G.** 1.38) which are made to simulate it. These synthetics will sink in the salt solution in which amber floats. *Copal,* recent fossil resin, has properties similar to those of amber and is more difficult to distinguish. However, true amber is less soluble in organic solvents than is copal, and a drop of ether placed on copal renders the surface sticky but on amber evaporates, leaving the original hard surface.

There are several plastics, particularly polystyrene, that when properly colored are used to imitate amber. Their specific gravities may be close to that of amber but they soften rapidly in organic solvents such as acetone or toluene. Also, if a heated needle is pressed into amber, one can smell the characteristic odor that is lacking when a plastic is so tested. Pressed amber can usually be distinguished by the presence of flattened bubbles and parallel bands or flow structure.

Inclusions

Of great interest, particularly to the entymologist, are the presence in amber of a variety of perfectly preserved insects (Fig. 13.80). The insects trapped in the viscous resin became overwhelmed by more resin exuded from the living tree. In addition to insects, amber may include pine needles and other plant remains. Unfortunately the presence of an in-

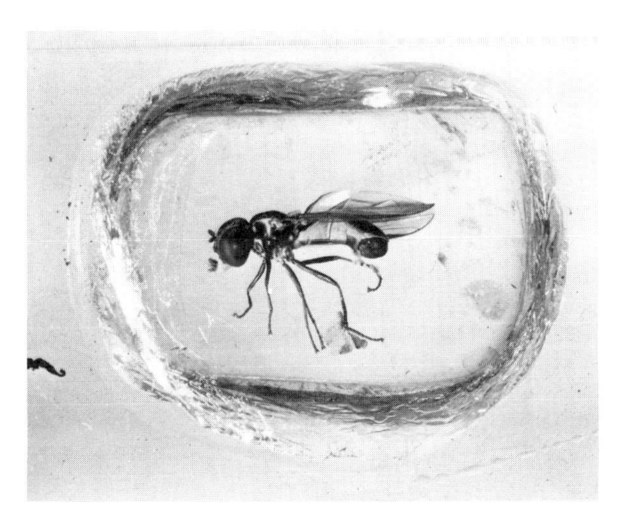

FIG. 13.80. A flower fly (*Pseudosphegina carpenteri*) in amber dating from the Oligocene Epoch nearly 40 million years ago. Courtesy of Frank M. Carpenter, Harvard University.

sect does not guarantee the authenticity of the material, for flies and other insects have been incorporated into manufactured materials.

Occurrence

Amber was early recovered from the shores of the Baltic Sea. Because of its low density, it sometimes traveled great distances to be tossed onto the beaches by wave action, and at low tide it was separated from seaweed in which it had become entangled. The amber-bearing deposits from which the sea amber has been washed are themselves worked extensively by open pit methods. The principal mining is near Kaliningrad, U.S.S.R. (formerly Königsberg, East Prussia).

Amber with properties differing only slightly from Baltic amber has been found in Rumania (*rumanite*), Sicily (called *simetite* from the Simeto River), and Burma (*bermite*).

Name

The English word amber comes from the Middle–Old Latin *ambrum,* which in turn came from the Arabic *anbar*.

Coral

Coral is essentially calcium carbonate (calcite) secreted from seawater by tiny marine animals, polyps, to form their skeletons. The coral polyps live in clear warm water, not as isolated individuals, but in colonies composed of myriad individuals with skeletons attached to one another. Corals made their appearance on earth about 500 million years ago, and since then many different forms have developed and flourished as shown by fossil remains in the rocks. It is not these ancient corals but only one type of the many present-day corals that is used as a gem. This type, known as precious coral, *Corallium rubrum,* is red or pink. The polyps that create it live in branching colonies that increase in size by "budding," that is, by the addition of new individuals to the established colony. Single branches of the bushlike colony may develop to a foot in length.

Although precipitated organically, coral is essentially fine-grained calcite and has properties similar to that mineral (see calcite under *Less Important Gemstones*).

Crystallography

Hexagonal-rhombohedral. In fine-grained aggregates.

Physical Properties

Cleavage of calcite not seen. **H** 3½–4, harder than pure calcite, probably because of impurities. **S.G.** 2.6–2.7 (calcite 2.72). Luster vitreous.

Color commonly white but in precious coral red or pink. The color is attributed to the presence of organic matter. Black coral is recovered in a few places. It has a lower **S.G.** (1.34±) than precious coral resulting from a high percentage of organic material.

Optical Properties

Essentially those of calcite: uniaxial (−); $\omega = 1.66$, $\epsilon = 1.49$, but difficult to determine.

Diagnostic Features

Effervescence in hydrochloric acid distinguishes coral from most imitations. Coral is sometimes used to imitate the pink conch pearl which effervesces in hydrochloric acid but less vigorously than coral. Furthermore, the conch pearl is essentially aragonite with higher **S.G.,** about 2.84.

Occurrence

Because coral grows in warm water, most species are confined to the equatorial zone between the latitudes 30°N and 25°S. Here are the major coral formations such as the Great Barrier Reef of Australia and the coral atolls of the South Pacific. Farther north, however, precious coral flourishes in the waters of the Mediterranean Sea and is recovered from the shallow water along its coast. The depth of water usually does not exceed 150 ft but some poorer grade coral has come from a depth of 900 ft. Coral is fished with nets principally along the coasts of Algeria, Tunisia, Italy, and Spain. Although these are the world's principal producing areas, precious coral is also fished in the waters of Japan and Malaysia.

Name

From the Latin word *corrallium.*

Ivory

Because it is relatively soft and easily worked, ivory has been a favored medium for carving since the earliest civilizations. The most elaborate carvings were executed by the Chinese during the seventeenth and eighteenth centuries. The Japanese also produced intricately carved pieces, frequently depicting a story.

Ivory is dentin, the material of which our teeth, and the teeth of all mammals, is largely composed. It is only when the tooth, as the tusk of an elephant, is sufficiently large that it is considered ivory, a material for carving.

Crystallography

Massive; cryptocrystalline in part.

Physical Properties

H 2½. **S.G.** 1.70–1.98; varies with origin. Walrus ivory has a higher density than elephant ivory.

Color white becoming somewhat yellow with age.

Optical Properties

Essentially isotropic. **R.I.** about 1.54.

Chemical Properties

Composed of both organic and inorganic material. The inorganic is similar to apatite (q.v. under *Less Important Gemstones*) with composition $Ca_5(PO_4)_3(F,OH)$. The organic material, chiefly collagen, makes up 30–35%.

Fluorescence

Under ultraviolet light, most ivory has a white to bluish-white fluorescence.

Diagnostic Features

Ivory imitations made from bone have a slightly higher density (2.0) and when examined under the microscope, show small dark lines not present in ivory. Plastic imitations have a lower hardness and specific gravity.

Occurrence

Elephant tusks from Africa and to a lesser extent from India are the principal sources of ivory. Small amounts come from the elongated canine teeth (tusks) of the walrus and from the large (some weigh several pounds) canine teeth of the hippopotamus. Some ivory has come from tusks of the extinct mammoth that lived during the glacial period. This has been called "fossil ivory" but it is not fossil in the true sense for the mammoths were preserved by being frozen in ice. Although much mammoth ivory is unusable, some of it is nearly indistinguishable from ivory of the modern elephant.

Name

Ivory comes from *ivurie,* the Old French word for the material.

Odontolite, also called *fossil turquoise* or *bone turquoise,* is fossil tooth or bone in which apatite has replaced the organic material. It consists of microcrystalline apatite colored blue by impregnation of the iron phosphate vivianite. Properties: **H** 5, **S.G.** 3.0, **R.I.** 1.60 ± 0.02. It can be distinguished from true turquoise by its greater specific gravity.

Jet
Jet, like coal, is a black fossil wood that was at first altered chemically by surface waters and later physically by pressure resulting from deep burial for a long period of time. Because it is easily worked and takes a good polish, it has been used for carvings, beads, pendants, and in mourning jewelry. The use of jet for articles of personal adornment, which antedates the arrival of the Romans in England, continued through the nineteenth century. However, it is rarely if ever used in modern jewelry.

Properties
Amorphous. Conchoidal fracture. **H** 2½. **S.G.** 1.33. *Luster* vitreous on fracture surfaces. Opaque. *Color* black. Isotropic. **R.I.** 1.66, but reading on refractometer indistinct.

Diagnostic Features
Jet is easily scratched by a needle, whereas agate, obsidian, or glass are not scratched. If the test material can be scratched, fumes are given off when a heated needle is pressed into it. Jet gives the odor of burning coal, which is quite different from the odor given off by plastics, bakelite, or hard rubber imitations.

Occurrence
The principal occurrence of jet is on the Yorkshire coast of England near the town of Whitby. The classical material is thus referred to as *Whitby jet.* Other minor occurrences of jet are in Spain, France, Germany, and in the United States in the Henry Mountains, Utah.

Name
The English word jet is derived from the Old French *jaiet,* which in turn comes from the Latin *gagāt.*

Synthetic Gems
Many gem minerals have been synthesized in the laboratory, often only in crystals too small to be cut, or not of gem quality. The following naturally occurring gems have been made in sizes large enough to cut and polish:

SYNTHETIC MINERALS

Beryl	Diamond
Emerald	Garnet
Chrysoberyl	Opal
Alexandrite	Quartz
Corundum	Rutile
Ruby, sapphire	Spinel
star ruby,	Turquoise
star sapphire	

Of the above, garnet (as YAG and other garnetlike substances) and rutile are made primarily in colorless form to imitate diamond, not as substitutes for the natural minerals.

In addition to the above, the following gem materials that have no natural counterparts are made: *strontium titanate, cubic zirconia, silicon carbide.*

Many other man-made crystals are used in technology and more are being developed in research laboratories. From time to time other synthetic crystals having suitable properties may appear as cut gems and the gemologist must be constantly alert to this possibility.

For the synthetic gem descriptions that follow, additional information may be found under the descriptions of naturally occurring gems, given earlier in this chapter.

Synthetic Emerald
Emerald was successfully synthesized as early as 1848 but no crystals of gem quality were produced until 1934. At that time chemists with I.G. Farbenindustrie in Germany produced a few small crystals which they named *Igmerald.* The first important commercial production was in 1940 by Carroll Chatham of San Francisco. Although he maintains secrecy about his process, features of Chatham emeralds indicate they are produced by the flux growth process. Single crystals of more than 1000 carats have been grown, and cut stones of more than 5 carats are available. Flux-grown crystals are also made by Gilson in Switzerland (see Plate XII). Synthetic emerald grown by a hydrothermal process was made by Lechtleitner in Austria in 1961. This

product consists of an overgrowth of emerald on prefaceted natural beryl. The Linde Company has also made hydrothermally grown synthetic emerald.

Most synthetic emeralds produced by the flux growth process have refractive indices and specific gravities substantially lower than the lowest values encountered for natural emerald. However, some Gilson flux-grown material is closer to natural emerald as is the Linde hydrothermal product.

PROPERTIES OF NATURAL AND SYNTHETIC EMERALD

	R.I.		
Type of Emerald	ω	ϵ	**S.G.**
Natural Colombian	1.583	1.577	2.71–2.72
Hydrothermal	1.573	1.568	2.66–2.70
Flux-grown (some Gilson)	1.579	1.571	2.65–2.68
Flux-grown (usual)	1.564	1.561	2.65–2.66

Most synthetic emeralds fluoresce a dull red under longwave u.v., whereas most natural emeralds are nonfluorescent. Liddicoat reports that for a time Gilson produced synthetic emerald to which a small amount of iron had been added to quench the fluorescence. Natural emeralds and flux-grown synthetics also differ in the transmission of shortwave u.v. Natural emeralds are opaque to wavelengths below 3000 Å. They will not transmit 2537 Å shortwave u.v., whereas flux-grown synthetics will transmit ultraviolet light of this wavelength.

Inclusions in flux-grown synthetic emeralds are very distinctive and are described as wisplike. This product may also show irregularly shaped flux inclusions, angular crystal inclusions (phenakite), and cubic crystals of platinum (from the crucibles in which the crystals are grown). See Plate III.12. The inclusions in Linde hydrothermally grown emeralds are different from those in flux-grown. Two-phase inclusions, described as cottony in appearance, and conical spaces extending from phenakite crystals are characteristic.

Synthetic Chrysoberyl (Alexandrite)

Synthetic alexandrite is being produced by the flux growth and Czochralski methods with properties very similar to natural material. The color change from red in artificial (incandescent) light to green in daylight is the same. Although synthetic stones have slightly lower **R.I.** and **S.G.** than the natural, the differences are too small for positive distinction and identification rests largely on the nature of the inclusions.

PROPERTIES OF NATURAL AND SYNTHETIC CHRYSOBERYL

	R.I.		
	α	γ	**S.G.**
Natural chrysoberyl	1.746	1.755	2.73
Synthetic alexandrite	1.742	1.751	2.71

Flux-grown alexandrite contains inclusions of the flux and wisplike internal features typical of all flux-grown synthetics. Also present are triangular or hexagonal plates, probably platinum. The material made by the Czochralski process shows curved growth lines and sometimes minute gas bubbles.

Synthetic Corundum (Ruby and Sapphire)

Synthetic corundum is produced in large quantities by the flame fusion (Verneuil) process. It is made colorless and, by the addition of small amounts of various metallic oxides, in a wide range of colors, including some that do not occur naturally. See Plate XII.

Synthetic corundum is also grown using flux growth techniques, by the Czochralski crystal-pulling method, and hydrothermally. Clusters of flux-grown synthetic ruby crystals have been used in jewelry in their uncut form.

In 1947 the Linde Company introduced synthetic red and blue star corundum. The star material is made by the flame fusion process by adding to the feed mix 0.1–0.3% titanium oxide. The boules are then reheated and held at high temperature (1100–1500°C) for a period of several hours to several days. This causes exsolution of the titanium oxide in the form of oriented rutile needles, just as in natural star corundum. The rutile needles are much smaller than those in natural material and can be resolved only by a magnification of at least 50×. Synthetic star corundum is also made in West Germany, Japan, and other countries.

The nature of the so-called "reconstructed" rubies, reportedly made by sintering together fragments of natural rubies, has been investigated by K. Nassau and R. Crowningshield.* They concluded that the reconstructed rubies were manufactured by a pre-Verneuil form of flame fusion and using purified alumina, not natural ruby, as feed material.

The physical and optical properties of synthetic corundum are the same as natural corundum: **H** 9; **S.G.** 4.02; uniaxial (−), ω = 1.770, ϵ = 1.762; biref. 0.008. Hence the nature of the inclusions is all important in distinguishing between synthetic and natural corundum.

Characteristic of synthetic corundum made by the flame fusion method are spherical gas bubbles (Plate III.2). The bubbles are usually small and appear as pinpoints of light in dark field illumination. They are sometimes elongated owing to movement of the material during cooling. Some stones may be entirely free of bubbles. Also characteristic are curved growth lines or striae. They are caused by flow of molten alumina across the curved top of the growing boule. Curved striae are not visible in all stones and are more readily seen in some colors than others.

Synthetic ruby produced by flux methods has wisplike inclusions similar to those characteristic of synthetic emerald, and inclusions of flux (Plate III.3).

Both natural and synthetic rubies fluoresce strongly under long- and shortwave u.v. because of the small amount of chromium present as the chromophore. Synthetic ruby fluoresces somewhat more strongly than its natural counterpart but the difference is not sufficient to be conclusive. Some synthetic sapphire of other colors fluoresces and some does not.

Cracks along facet junctions are commonly seen in synthetic corundum. These are caused by heat from too rapid polishing, and may be considered an indication, but not proof, of synthetic origin.

* K. Nassau and R. Crowningshield, 1969. The Synthesis of Ruby. *Lapidary Journal*, Vol. 23, pp. 313–314, 334–338.

Synthetic Diamond

Synthetic industrial grade diamond has been manufactured since 1955, and in 1970 the General Electric Company succeeded in making small (up to 1 carat) gem quality diamonds. However, so far as is known, no synthetic gem quality diamonds have appeared in the gem trade. For additional information see diamond under *Important Gemstones,* and *Diamond Synthesis* in Chapter 10).

Synthetic Garnet

The synthetic garnet used as a diamond simulant has the garnet crystal structure, but unlike natural garnet contains no silicon. The chemical formula of natural garnet may be written $A_3B_2Si_3O_{12}$ where, for example, in pyrope A is magnesium and B is aluminum. In the synthetic garnets not only the atoms in the A and B positions are replaced, but the silicon as well. For example:

YAG	(yttrium aluminum garnet)	$Y_3Al_2Al_3O_{12}$
YIG	(yttrium iron garnet)	$Y_3Fe_2Fe_3O_{12}$
GGG	(gadolinium gallium garnet)	$Gd_3Ga_2Ga_3O_{12}$

YAG and GGG are colorless and make excellent diamond imitations. YIG is black and is not used as a gem material.

YAG is isometric, with **H** 8¼, **S.G.** 4.55. Its optical properties are: **R.I.** = 1.833, biref. none, disp. 0.028. Under shortwave u.v. its fluorescence is none to weak orange, and under longwave u.v. none to moderate orange. In addition to colorless material, YAG has been produced in blue, green, red, and yellow colors (see Plate XII). It is grown by flux and Czochralski techniques, and also by flame fusion.

The diagnostic features of YAG are high specific gravity, spherical gas bubbles under high magnification, single refraction, and moderate dispersion (about 0.6 times that of diamond).

YAG is sold under a large number of trade names including Diamonair, Linde Simulated Diamond, Diamite, and Triamond.

GGG is isometric with **H** 6½, **S.G.** 7.05. Its optical properties are **R.I.** = 1.975, biref. none, disp. 0.038. Under shortwave u.v. it fluoresces strong orange to orange-red; it is nonfluorescent under longwave u.v. Usually GGG has a very light brown color. Because of its high refractive index and dispersion, very close to that of diamond, it makes an excellent diamond imitation.

The diagnostic features of GGG are very high specific gravity, strong fluorescence under shortwave u.v., high refractive index, and strong dispersion. Some material contains angular inclusions as triangular plates, and minute gas bubbles may sometimes be seen under high magnification.

Synthetic Opal

A very fine synthetic opal, both black and white, was introduced by Gilson in 1974. The properties of the synthetic and natural materials are compared below. Only in hardness is there a significant difference.

Opal	R.I.	S.G.	H
Synthetic	1.44	2.0–2.1	4½
Natural	1.43–1.47	2.0–2.2	5½–6½

Methods developed by the Gemological Institute of America to determine whether an opal is natural or synthetic are summarized as follows:

1. **Appearance in Transmitted Light.** In transmitted light, under magnification, synthetic black opals show a mosaic structure unlike anything seen in natural opal.

2. **Transparency to Ultraviolet Light.** Synthetic opal tested so far (1975) is much more transparent to ultraviolet light than natural opal. The test is made by placing the unknown and known natural and synthetic stones on photographic paper and exposing to longwave u.v. for 2–3 sec. If, on the development of the paper, the unknown shows it has transmitted more ultraviolet light than the natural stone, it points to a synthetic origin.

Synthetic Quartz

Synthetic colorless quartz, described in Chapter 10, is made for technological uses in large quantities by the hydrothermal method. This synthetic "rock crystal" has probably never appeared in quantity in the gem trade because it is at least as expensive as very abundant natural quartz. Synthetic quartz in blue, green, violet, and yellow colors as well as smoky, is now being made principally in the U.S.S.R. and exported in large quantities (millions of carats) (see Plate XI). The synthetic colored quartz is grown on a colorless seed plate, which may be seen in uncut crystals. Color banding parallel to the seed plate is present, and "bread crumb" inclusions are present near the seed plate. Otherwise the synthetic and natural material are identical. Some of the colors being produced, for example, deep blue and green, are colors in which natural quartz does not occur.

Synthetic Rutile

Synthetic rutile has been produced by the flame fusion process since 1947. Natural rutile is always very dark-colored and unsuited for gem use, but the synthetic material is transparent and very light yellow in color and is used as a diamond simulant. Less commonly it is seen in other colors including blue, golden-brown, brownish-red; and green.

Synthetic rutile, TiO_2, is tetragonal with **H** 6–6½, **S.G.** 4.26. Optical properties: uniaxial (+), $\omega = 2.616$, $\epsilon = 2.903$; biref. 0.287; disp. 0.330. Both birefringence and dispersion are exceptionally high. The former causes pronounced doubling of the back facets, so much so that it has a "fuzzy" appearance. The dispersion is seven and a half times that of diamond causing an unnatural degree of fire. It is nonfluorescent in ultraviolet light.

The diagnostic features of synthetic rutile are its tremendous birefringence and dispersion. Spherical gas bubbles may be seen under magnification.

This material has been sold under a large number of trade names, including Titania, Kenya Gem, Miridas, Titangem, and Kimberlite Gem.

Synthetic Spinel

Synthetic spinel, made by the flame fusion process, is used primarily not to simulate natural spinel, but to imitate gems of colors in which natural spinel does not occur. The most frequently seen of these are blue and colorless, to imitate blue sapphire, aquamarine and diamond. Other colors include green to imitate peridot, tourmaline, and emerald, and yellowish-green to imitate chrysoberyl. A sintered form of synthetic spinel colored by cobalt is made to imitate lapis lazuli, and a cloudy form is made that is a good imitation of moonstone.

The chemical composition of natural (gem) spinel is $MgAl_2O_4$, which may also be written $MgO \cdot Al_2O_3$; hence spinel contains magnesium oxide (MgO) and aluminum oxide (Al_2O_3) in equal proportions. Synthetic spinel cannot easily be made using the 1 : 1 ratio, and it has been found that the boules form best when the ratio of Al_2O_3 to MgO is 2.5 : 1. This causes an increase in refractive index and specific gravity over natural spinel. This excess of alumina also causes the anomalous double refraction ("crosshatched" effect) characteristic of synthetic spinel.

PROPERTIES OF SYNTHETIC AND NATURAL SPINEL

	Av. R.I.	Av. S.G.
Synthetic spinel	1.73	3.64
Natural spinel	1.72	3.60

Characteristic inclusions in synthetic spinel, which may be taken as proof of origin, are spherical gas bubbles. Curved growth lines, usually present in synthetic corundum, are absent in synthetic spinel.

Synthetic Turquoise

Synthetic turquoise was introduced into the gem market by Gilson in 1972. The material has a fine blue color closely resembling some natural turquoise (see Plate XII), but under high magnification (50× or more) gives the appearance of compacted minute spheres. This feature is not seen in natural turquoise.

Synthetic turquoise lacks the absorption spectrum lines at 4300 and 4200 Å that are characteristic of the natural material.

PROPERTIES OF SYNTHETIC AND NATURAL TURQUOISE

Turquoise	R.I.	S.G.	H
Synthetic	1.60	2.62–2.67	5–6
Natural	1.61–1.63	2.6–2.8	5–6

Strontium Titanate

Strontium titanate has been produced since 1955 by the flame fusion process.

Strontium titanate, $SrTiO_3$, is isometric, with **H** 5–6 and **S.G.** 5.13. Its optical properties are **R.I.** 2.409 (diamond **R.I.** 2.42), disp. 0.190. It is nonfluorescent under ultraviolet light. Spherical gas bubbles may usually be seen under magnification.

The diagnostic features of strontium titanate are high **S.G.**, high dispersion (about four times that of diamond), and no doubling of the back facets. Because of its low hardness polishing scratches and rounded facet junctions are always present.

Strontium titanate is sold under many trade names, including Fabulite, Diagem, and Wellington.

Cubic Zirconia

Faceted gemstones of a new diamond simulant, cubic zirconia, have recently been described by K. Nassau.*

The properties of this material are **H** 7½–8½, **S.G.** 5.6–6.0, **R.I.** 2.15–2.18, biref. none, disp. 0.060. The composition is zirconium oxide, ZrO_2, plus a small amount of calcium. One specimen examined by Dr. Nassau had what appeared to be small bubbles, but were revealed under higher magnification as minute negative crystals.

Cubic zirconia has a combination of hardness, specific gravity, refractive index, and dispersion that make it an outstanding diamond imitation. The material examined by Dr. Nassau is said to have been made by a flux technique by the Djevahirdjian Co. of Monthey, Switzerland, under the trade name Djevalite. It is also reportedly being made in the U.S.S.R. under the name Phianite. In the United States it is being made by the skull melting process

* K. Nassau, 1977. Cubic Zirconia, the Latest Diamond Imitation. *Lapidary Journal,* Vol. 31, pp. 899–904, 922–926.

by the Ceres Corporation, Waltham, Massachusetts under the direction of Mr. J. F. Wenckus.

Silicon Carbide

Silicon carbide, SiC, the well-known abrasive sold under trade names such as Carborundum and Norbide, has occasionally been cut as a gem. The material is hexagonal, has **H** 9½, **S.G.** 3.17. Its optical properties are uniaxial (+), $\omega = 2.65$, $\epsilon = 2.69$; biref. 0.043; disp. about twice that of diamond. The color is bluish green similar to that of irradiated diamond. It is seldom seen and only in collections of rare and unusual gems.

References and Suggested Reading

Bauer, Max, 1896. *Edelsteinkunde*. Chr. Herm. Tauchnitz, Leipzig. Translated from the German by L. J. Spencer, 1904, as *Precious Stones*. C. Griffin & Co., Ltd., London. Reprinted, 1969. Charles E. Tuttle Co., Inc., Rutland, Vermont.

Schlossmacher, K., 1954. *Edelsteinen und Perlen*. Stuttgart.

Sinkankas, J., 1959. *Gemstones of North America*, Vol I; 1976, Vol. II. Van Nostrand Reinhold Co., New York.

Smith, G. F. H., 1912. *Gemstones*. Methuen & Co., Ltd., London. 14th ed. revised by F. C. Phillips, 1972. Chapman and Hall, London.

Webster, R., 1975. *Gems*, 3rd ed. Butterworth & Co., Ltd., London.

14 DETERMINATIVE TABLES

Introduction

The procedure to be followed in the identification of gem materials depends on whether the material is in rough form or cut and polished, and in the latter case depends to some extent on whether the gemstone is mounted or unmounted.

Identification of uncut material is made easier because some properties are more readily observed or determined than if the unknown is a cut stone. These may include crystal system (if in crystals), cleavage, fracture, luster, and hardness. In addition, information about chemical composition can be obtained and the highly definitive X-ray powder diffraction method can be easily and safely used.

Although most gemologists are infrequently called upon to identify rough gem materials, they should develop familiarity with the appearance of the gem minerals and have available mineralogy textbooks containing determinative tables designed especially for mineral identification.

The principal difference in the testing methods used in the identification of rough and cut gems is that any test may be used for rough, whereas for cut gems the tests must be nondestructive and in no way mar or damage the gem.

An experienced gemologist, on the basis of long familiarity with characteristics of color and luster, can often make accurate sight identifications of gemstones. However, with rare and unusual gemstones appearing in increasing number and with greater use in jewelry of synthetics and imitations, comfirmatory tests should always be made. Sometimes only one or two simple tests are necessary. Other times the entire range of instruments in a well equipped gemological laboratory are needed for positive identification. In all cases the chances of error are diminished if a systematic proceedure is followed.

Identification Procedure

In gem identification one normally proceeds progressively from easily determined properties to those whose determination is more difficult and time consuming. Obviously, then, the first step is to look at the stone and note its color and luster. Of the several measurable properties refractive index is not only the easiest made but is the most informative; it should thus be the first measurement. In the following table of gem properties, gems are arranged according to increasing refractive index. By looking up the measured value in the table, one can usually narrow the possibilities to a small number, sometimes one or two. Next to be determined should be single or double refraction and, if doubly refractive, optic sign, degree of birefringence, and pleochroism. Determination of specific gravity, although less easily made, is necessary in some cases, particularly if the refractive index is above the range of the refractometer. Accordingly a second table is given in which gems are arranged in order of increasing specific gravity. Observation of the nature of inclusions becomes mandatory if the species is determined to be one that is available as a synthetic, or to confirm that the unknown is glass or other imitation. Hardness should be used only as a last resort, and then with great care. In certain especially difficult cases use of spectroscopic and X-ray diffraction methods may be required for positive identification.

GEM PROPERTIES WITH GEMS LISTED IN ORDER OF INCREASING REFRACTIVE INDEX

No.	Gem	R.I.[a]	Biref.	Opt.[b]	S.G.	Color[c]
		< 1.50				
1	Fluorite	1.43	—	I	3.18	All colors
2	Opal	1.45 ± 0.01	—	I	2.10 ± 0.10	All colors
3	Obsidian	1.48 to 1.52	—	I	2.33–2.50	RBl
4	Silica glass	1.46	—	I	2.21	C
5	Moldavite	1.48	—	I	2.40 ± 0.04	G
6	Glass	1.48 to 1.70	—	I	2.3–4.5	All colors
7	Calcite	1.486–1.658	0.172	U−	2.72	CRYGBBrGrBl
8	Coral	1.49–1.66	0.17	U−	2.65 ± 0.05	PRBl
9	Sodalite	1.48	—	I	2.23 ± 0.07	CPGB
10	Celluoid	1.495 to 1.520	—	I	1.35 to 1.80	
		1.50–1.59				
11	Calcite	1.486–1.658	0.172	U−	2.72	CRYGBBRGrBl
12	Celluloid	1.495 to 1.520	—	I	1.35 to 1.80	
13	Lazurite	1.50	—	I	2.43 ± 0.03	B
14	Lapis lazuli	1.50	—	Agg	2.7 ± 0.2	B
15	Chrysocolla	1.50	—	I	2.2 ± 0.2	GBBrBl
16	Thomsonite	1.50–1.54	0.010	B+	2.35 ± 0.05	RYG
17	Petalite	1.505–1.518	0.013	B+	2.42 ± 0.03	C
18	Leucite	1.51	—	Agg	2.47	CW
19	Pollucite	1.52	—	I	2.90	CW

[a] For doubly refracting gems the highest and lowest refractive indices are given. Only when indices differ by more than ±0.003 from those listed is the variation given. For uniaxial crystals the variable index is underlined, for example, Calcite 1.486–1.658. For isotropic gems having a range of refractive index, the range is given from low *to* high as: glass 1.48 to 1.70. Several gems with high birefringence are listed in each section in which a single measurement of one of the indices might fall as: painite (1.788–1.816) is in sections **1.70–1.79 and 1.80–1.99.**

[b] I, isotropic; U, uniaxial + or −; B, biaxial + or −; Agg, fine-grained aggregate.

[c] Colors are indicated by letters as follows: C, colorless, W, white, P, pink; R, red; O, orange; Y, yellow; G, green; B, blue; V, violet; Br, brown; Gr, gray; Bl, black.

GEM PROPERTIES WITH GEMS LISTED IN ORDER OF INCREASING REFRACTIVE INDEX (*Continued*)

No.	Gem	R.I.	Biref.	Opt	S.G.	Color
20	Sanadine	1.522–1.528 ± 0.004	0.006	B−	2.59 ± 0.03	C
21	Orthoclase	1.518–1.526 ± 0.003	0.008	B−	2.56 ± 0.02	CYG
22	Gypsum	1.52–1.53	0.01	B+	2.32	CWYBrGr
23	Microcline	1.522–1.530 ± 0.003	0.008	B−	2.56	CWRYG
24	Aragonite	1.530–1.686	0.156	B−	2.95	CWY
25	Apophyllite	1.535–1.537	0.002	U+	2.35 ± 0.05	CWPYG
26	Chalcedony	1.537 ± 0.003	—	Agg	2.60 ± 0.02	All colors
27	Talc	1.539–1.589	0.050	B−	2.75 ± 0.05	WGGr
28	Iolite	1.54–1.55 ± 0.02	0.01	B±	2.61 ± 0.04	BV
29	Scapolite	1.540–1.568	0.02	U−	2.65 ± 0.10	PYGBGr
30	Amber	1.54 ± 0.02	—	I	1.07 ± 0.04	ROYBr
31	Ivory	1.54	—	I	1.84 ± 0.14	WY
32	Jasper	1.54 approx.	—	Agg	2.6 to 2.9	RYGBr
33	Albite	1.530–1.540	0.011	B+	2.63 ± 0.01	CWPB
34	Oligoclase	1.538–1.547 ± 0.009	0.009	B+	2.66 ± 0.02	CW
35	Quartz	1.544–1.553	0.009	U+	2.65	All colors
36	Serpentine	1.55–1.56	—	B	2.7 ± 0.1	YG
37	Pyrophyllite	1.55–1.60 ± 0.05	0.05	B−	2.8 ± 0.1	WYBGr
38	Beryllonite	1.552–1.561	0.009	B−	2.81	C
39	Hambergite	1.554–1.628 ± 0.003	0.074	B+	2.36	C
40	Labradorite	1.560–1.568 ± 0.005	0.008	B+	2.72 ± 0.02	CYGBGr
	Emerald (synthetic)					
41	Flux	1.561–1.565	0.004	U−	2.66 ± 0.01	G
42	Hydrothermal	1.568–1.573	0.005	U−	2.68 ± 0.02	G
43	Emerald (natural)	1.575–1.580 ± 0.003	0.006	U−	2.72 ± 0.03	G
44	Odontolite	1.57 to 1.63	—	Agg	3.0	BGr
45	Augelite	1.574–1.588	0.014	B+	2.7	C
46	Beryl	1.578–1.586 ± 0.008	0.007	U−	2.67–2.82	CPYGB
47	Rhodochrosite	1.579–1.816	0.237	U−	3.60 ± 0.1	PRBr
48	Variscite	1.58	—	Agg	2.50 ± 0.10	WYG
49	Howlite	1.59	—	Agg	2.58	W
50	Amblygonite	1.59–1.61 ± 0.01	0.02	B−	3.05 ± 0.05	CWY
		1.60–1.69				
51	Calcite	1.486–1.658	0.172	U−	2.72	CRYGBBrGrBl
52	Coral	1.49–1.66	0.17	U−	2.65 ± 0.05	PRBl
53	Rhodochrosite	1.579–1.816	0.237	U−	3.60 ± 0.10	PRBr
54	Ekanite	1.60	—	I	3.28	G
55	Brazilianite	1.602–1.621	0.019	B+	2.98	Y
56	Nephrite (trem.-actin.)	1.614–1.630 ± 0.014	0.025	B−	2.96 ± 0.06	WYGBBrBl
57	Prehnite	1.616–1.649	0.033	B+	2.87 ± 0.07	WYG
58	Lazulite	1.612–1.643	0.031	B−	3.05 ± 0.05	CGB
59	Pargasite	1.613–1.635	0.022	B−	3.1	Br
60	Topaz	1.618–1.627 ± 0.01	0.008	B+	3.50 ± 0.10	CPROYGBVBr
61	Turquoise	1.62 ± 0.01	—	Agg	2.7 ± 0.1	GB
62	Smithsonite	1.623–1.850	0.227	U−	4.37 ± 0.07	WPYGBGr
63	Datolite	1.624–1.668	0.044	B−	2.90 ± 0.01	WRYG
64	Tourmaline	1.625–1.646 ± 0.01	0.025	U−	3.12 ± 0.12	All colors
65	Danburite	1.630–1.636	0.006	B±	3.00	CPY
66	Andalusite	1.632–1.643	0.010	B−	3.18 ± 0.02	RYGBr
67	Jadeite	1.64–1.67	0.012	Agg	3.4 ± 0.1	All colors
68	Apatite	1.642–1.646	0.003	U−	3.17 ± 0.02	CYGBVBr

GEM PROPERTIES WITH GEMS LISTED IN ORDER OF INCREASING REFRACTIVE INDEX (*Continued*)

No.	Gem	R.I.	Biref.	Opt	S.G.	Color
69	Smaragdite	1.65 approx.	—	B−	3.3	G
70	Dioptase	1.651–1.703 ± 0.007	0.053	U+	3.3	G
71	Euclase	1.652–1.671	0.019	B+	3.08 ± 0.03	GB
72	Phenakite	1.654–1.670	0.016	U+	2.90 ± 0.10	CPRYBr
73	Peridot	1.654–1.690 ± 0.02	0.036	B±	3.34 ± 0.04	YG
74	Malachite	1.655–1.909	0.254	B−	3.96 ± 0.06	G
75	Sillimanite	1.657–1.677	0.020	B+	3.25 ± 0.02	CYGBBr
76	Enstatite	1.658–1.666	0.01	B+	3.35 ± 0.15	YGBrGr
77	Spodumene	1.660–1.676 ± 0.005	0.006	B+	3.18 ± 0.01	CPYGV
78	Jet	1.66	—	–	1.35	Bl
79	Kornerupine	1.666–1.679	0.013	B−	3.36 ± 0.10	CYGBr
80	Sinhalite	1.668–1.707	0.039	B−	3.48 ± 0.01	YGBrBl
81	Diopside	1.67–1.70 ± 0.01	0.03	B+	3.30 ± 0.05	YGBV
82	Axinite	1.675–1.685	0.01	B−	3.31 ± 0.04	YBBr
83	Willemite	1.691–1.719	0.028	U+	4.05 ± 0.15	RYGBr
84	Zoisite	1.693–1.702	0.009	B+	3.36	CRYGBGr
		1.70–1.79				
85	Rhodochrosite	1.579–1.816	0.237	U−	3.60 ± 0.10	PRBr
86	Smithsonite	1.623–1.850	0.227	U−	4.37 ± 0.07	WPYGBGr
87	Sinhalite	1.668–1.707	0.039	B−	3.48 ± 0.01	YGBrBl
88	Willemite	1.691–1.719	0.028	U+	4.05 ± 0.15	RYGBr
89	Zoisite	1.693–1.702	0.009	B+	3.36	CRYGBGr
90	Rhodizite	1.694	—	I	3.44	CRY
91	Chlorastrolite	1.7	Agg	I	3.2 ± 0.1	WG
92	Idocrase	1.701–1.729 ± 0.023	0.004	U−	3.35 ± 0.10	RYBBr
93	Spinel (nat)	1.71 to 1.72	—	I	3.65 ± 0.07	All colors
94	Kyanite	1.712–1.728	0.016	B−	3.60 ± 0.05	CGBGr
95	Spinel (syn)	1.72 to 1.75	—	I	3.64 ± 0.04	CROYGBV
96	Azurite	1.730–1.838	0.108	B+	3.77	B
97	Taaffeite	1.717–1.721	0.004	U−	3.6	V
98	Rhodonite	1.724–1.739 ± 0.008	0.013	B+	3.55 ± 0.15	PRBr
99	Epidote	1.733–1.784 ± 0.018	0.03	B−	3.37 ± 0.07	GBl
100	Pyrope	1.740 ± 0.010	—	I	3.72 ± 0.07	RVBr
101	Grossular	1.742 ± 0.007	—	I	3.68 ± 0.07	ROYGVBr
102	Staurolite	1.743–1.757 ± 0.004	0.013	B+	3.70 ± 0.05	RYBrBl
103	Chrysoberyl	1.745–1.755	0.010	B+	3.73 ± 0.02	RYGBr
104	Benitoite	1.757–1.804	0.047	U+	3.64	CB
105	Rhodolite	1.76 ± 0.01	—	I	3.84 ± 0.10	PR
106	Corundum	1.762–1.770 ± 0.004	0.008	U−	4.02	CPRYGVGrBr
107	Painite	1.788–1.816	0.028	U−	4.0	R
108	Almandine	1.79 ± 0.03	—	I	4.05 ± 0.12	RV
		1.80 to 1.99				
109	Rhodochrosite	1.579–1.816	0.237	U−	3.60 ± 0.1	PRBr
110	Smithsonite	1.623–1.850	0.227	U−	4.37 ± 0.07	WPYGBGr
111	Painite	1.788–1.816	0.028	U−	4.0	R
112	Spessartine	1.80 ± 0.01	—	I	4.18 ± 0.03	OYBr
113	Zircon (low)	1.80 ± 0.02	—	I	4.0 ± 0.07	G
114	YAG	1.833	—	I	4.55	CRYGB
115	Andradite	1.875 ± 0.02	—	I	3.84 ± 0.03	YGBl
116	Sphene	1.90–2.02 ± 0.015	0.11	B+	3.52 ± 0.02	YGBBr

GEM PROPERTIES WITH GEMS LISTED IN ORDER OF INCREASING REFRACTIVE INDEX (*Continued*)

No.	Gem	R.I.	Biref.	Opt	S.G.	Color
117	Zircon	1.92–1.98	0.06	U+	4.70 ± 0.03	CPROYGBBr
118	Scheelite	1.920–1.934	0.014	U+	6.0 ± 0.1	OYGBBr
		> **2.00**				
119	Sphene	1.90–2.02 ± 0.015	0.11	B+	3.52 ± 0.02	YGBBr
120	Cassiterite	2.00–2.10	0.10	U+	7.00 ± 0.10	CYBrBl
121	Zincite	2.013–2.029	0.016	U+	5.68	R
122	Zirconia (cubic)	2.15	—	I	5.9 ± 0.1	C
123	Sphalerite	2.37	—	I	4.0 ± 0.1	CRYGBrBl
124	Strontium titanate	2.409	—	I	5.13 ± 0.02	C
125	GGG	1.975	—	I	7.05	C
126	Diamond	2.417	—	I	3.52	CPYGBBl
127	Anatase	2.488–2.561	0.073	U−	3.95 ± 0.05	GBVBr
128	Rutile (syn)	2.616–2.903	0.287	U+	4.26	CY
129	Silicon carbide	2.65–2.69	0.04	U+	3.17	GB

SPECIFIC GRAVITY OF GEMS[a]

No.[b]	Name	S.G.
		< 2.3
30	Amber	1.05–1.09
	Plastics	1.05–1.55
78	Jet	1.35
10	Celluloid	1.35–1.80
31	Ivory	1.70–1.98
	Sepiolite	2.0
2	Opal	2.0–2.2
15	Chrysocolla	2.0–2.4
9	Sodalite	2.15–2.30
4	Silica glass	2.21
		2.3–2.49
25	Apophyllite	2.3–2.4
16	Thomsonite	2.3–2.4
6	Glass	2.3–4.5
22	Gypsum	2.32
39	Hambergite	2.36
5	Moldavite	2.36–2.44
17	Petalite	2.4
13	Lazurite	2.4–2.45
48	Variscite	2.4–2.6
3	Obsidian	2.33–2.50
18	Leucite	2.47
		2.5–2.69
14	Lapis lazuli	2.5–2.0
29	Scapolite	2.55–2.74
26	Chalcedony	2.58–2.62
21	Orthoclase	2.56
23	Microcline	2.56–2.58
20	Sanidine	2.56–2.62
49	Howlite	2.53–2.59
28	Iolite	2.58–2.66
8	Coral	2.6–2.7
36	Serpentine	2.6–2.8
61	Turquoise	2.6–2.8
32	Jasper	2.6–2.9
33	Albite	2.62–2.64
34	Oligoclase	2.64–2.68
35	Quartz	2.65
	Emerald (syn)	
41	Flux-grown	2.65–2.67
42	Hydrothermal	2.66–2.70
46	Beryl	2.67–2.82
	Pearl	2.68–2.75
43	Emerald (nat)	2.69–2.75
		2.7–2.99
6	Glass	2.3–4.5
29	Scapolite	2.55–2.74
36	Serpentine	2.6–2.8
61	Turquoise	2.6–2.8
32	Jasper	2.6–2.9
46	Beryl	2.67–2.82
	Pearl	2.68–2.75
43	Emerald (nat)	2.69–2.75
37	Pyrophyllite	2.70
40	Labradorite	2.70–2.74
27	Talc	2.7–2.8
11	Calcite	2.72
72	Phenakite	2.79–3.00
63	Datolite	2.8–3.0
57	Prehnite	2.8–2.95
38	Beryllonite	2.81
19	Pollucite	2.90
56	Nephrite	2.90–3.02
24	Aragonite	2.95
55	Brazilianite	2.98
		3.0–3.29
6	Glass	2.3–4.5
56	Nephrite	2.9–3.02
56	Tremolite	3.0
65	Danburite	3.0
44	Odontolite	3.0
50	Amblygonite	3.0–3.1
58	Lazulite	3.0–3.1
64	Tourmaline	3.0–3.24
99	Epidote	3.0–3.45
71	Euclase	3.05–3.10
59	Pargasite	3.1
91	Chlorastrolite	3.1–3.3
68	Apatite	3.15–3.20
66	Andalusite	3.16–3.20
129	Silicon carbide	3.17
77	Spodumene	3.17–3.19
1	Fluorite	3.18
76	Enstatite	3.2–3.5

[a] Arranged according to increasing specific gravity.
[b] The number refers to the gem number in the preceding table of gem properties, where more data are given.

SPECIFIC GRAVITY OF GEMS[a] (*Continued*)

No.[b]	Name	S.G.
75	Sillimanite	3.23–3.27
81	Diopside	3.25–3.35
82	Axinite	3.27–3.35
79	Kornerupine	3.27–3.45
54	Ekanite	3.28
		3.3–3.59
6	Glass	2.3–4.5
99	Epidote	3.0–3.45
81	Diopside	3.25–3.35
70	Dioptase	3.3
69	Smaragdite	3.3
67	Jadeite	3.3–3.5
73	Peridot	3.3–3.8
92	Idocrase	3.35–3.45
84	Zoisite	3.36
90	Rhodizite	3.44
60	Topaz	3.4–3.6
80	Sinhalite	3.47–3.49
47	Rhodochrosite	3.5–3.7
116	Sphene	3.5–3.54
126	Diamond	3.52
94	Kyanite	3.55–3.66
93	Spinel (nat)	3.57–3.99
		3.6–3.99
6	Glass	2.3–4.5
94	Kyanite	3.55–3.66
93	Spinel (nat)	3.57–3.72
97	Taaffeite	3.6
98	Rhodonite	3.6–3.7
95	Spinel (syn)	3.60–3.68
101	Grossular	3.61–3.75
104	Benitoite	3.64
102	Staurolite	3.65–3.75
100	Pyrope	3.65–3.80
103	Chrysoberyl	3.71–3.75
105	Rhodolite	3.74–3.94
96	Azurite	3.77
115	Andradite	3.81–3.87
127	Anatase	3.9–4.0
123	Sphalerite	3.9–4.1
83	Willemite	3.9–4.2
74	Malachite	3.90–4.03
113	Zircon (low)	3.93–4.07
		4.0–4.99
6	Glass	2.3–4.5
123	Sphalerite	3.9–4.1
88	Willemite	3.9–4.2
74	Malachite	3.9–4.03
113	Zircon (low)	3.93–4.07
107	Painite	4.0
108	Almandine	4.00–4.25
106	Corundum	4.02
112	Spessartine	4.12–4.18
128	Rutile	4.24–4.28
62	Smithsonite	4.30–4.45
114	YAG	4.55
117	Zircon	4.67–4.73
		> 5.0
	Pyrite	5.02
124	Strontium titanate	5.11–5.15
	Hematite	5.26
121	Zincite	5.68
122	Zirconia (cubic)	5.8–6.0
118	Scheelite	5.9–6.1
120	Cassiterite	6.9–7.1
125	GGG	7.05

APPENDIX

Interference Figures and Their Use

In the discussion of optical properties in Chapter 6, it was pointed out that from the movement of the shadow edges on the refractometer, a gemstone can be determined as uniaxial or biaxial and whether it is optically (+) or (−). The results are sometimes ambiguous and this information is more reliably determined using an interference figure. The appearance of interference figures is illustrated in Figures 4 and 5.

Interference figures are usually obtained on small crystals or crystal slices using a high-power polarizing (petrographic) microscope. The microscope, normally an orthoscope (i.e., with parallel light rays passing upward from the substage illuminator), is converted to a conoscope. This is accomplished by swinging a substage condensing lens into position so that the light passing through the specimen is strongly converging. With the high-power objective necessary, the working distance is usually too short to accommodate even a small gemstone. Moreover, the optic axis of the stone must be parallel, or nearly parallel, to the optical axis of the microscope. In the general case, then, it is necessary to rotate the stone to achieve the desired orientation. Devices have been manufactured to obtain interference figures of gems using the petrographic microscope but they are costly and time consuming to use.

Under certain conditions interference figures can be seen without special equipment, except that the stone or crystal must be viewed between crossed polars as in a polariscope. They are seen in crystal spheres, the sphere itself acting as the condensing lens. Interference figures also appear on the rounded top of a cabochon stone cut with the flat surface perpendicular to the optic axis. One ingenious method is to place a drop of a viscous liquid, for example, honey, on a faceted stone. The liquid drop acts as a condensing lens and an interference figure appears on its surface provided light moves through the stone parallel to an optic axis—a requirement usually difficult to fulfill. It is sometimes possible to observe an interference figure in a faceted stone by holding a 10× hand lens beneath the analyzer of the polariscope and observing it from about 18 in.

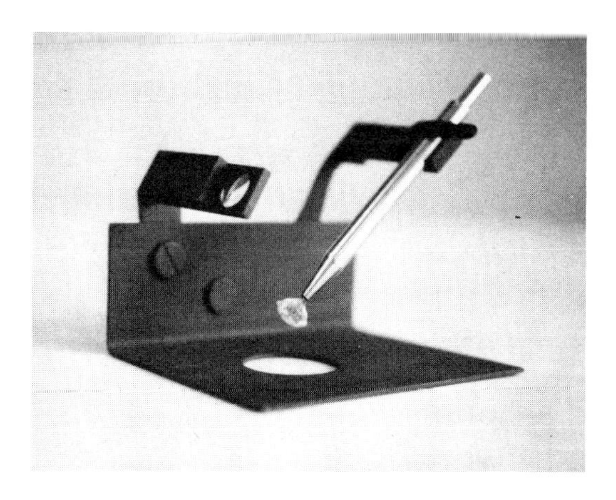

FIG. 1. The crystal orienter with gemstone held in three-prong tweezers.

Interference Figures Using a Low-Power Microscope

In 1975 an inexpensive device called a *crystal orienter* (Fig. 1) was described in *Gems and Gemology** by means of which an interference figure can be obtained on transparent gemstones ranging from 5 to 20 mm in diameter. One of its chief attributes is that it can be used with a low-power binocular microscope with a large working distance.

The Microscope

Several manufacturers make suitable microscopes (see Chapter 7). The principal requisite is that there be a means of substage illumination. The model illustrated (Fig. 2) is a Bausch and Lomb with a zoom lens giving a continuous variation in magnification from 7× to 30×. The instrument must further be converted to a polarizing microscope by equipping it with two polarizing filters, one beneath the stage and the other below the objectives. This is easily done by placing beneath the glass stage a sheet of Polaroid (the polarizer) and a smaller Polaroid sheet (the analyzer) beneath the objectives. In the conventional positions the polarizer is turned to permit light to vibrate in a north–south direction and the analyzer is adjusted to pass light vibrating in an east–west direction. Under these conditions no light reaches the eye and the microscope becomes a polariscope. Thus an anisotropic gemstone shows four positions of darkness when rotated 360° on the stage.

The Crystal Orienter

In Figure 1 is shown the crystal orienter with a gemstone held in three-prong tweezers, "Tricepts." These tweezers are more desirable than other types, for the thin wire prongs obstruct little of the light passing through the stone. The instrument is constructed so that the gemstone can be rotated about two mutually perpendicular axes. An arm, rotating about an axis parallel to the microscope stage, carries a support to which the tweezers are held in a semicylindrical groove by a spring clip. The long dimension of the tweezers is the second axis, and about it the stone can be rotated 360°. By adjusting

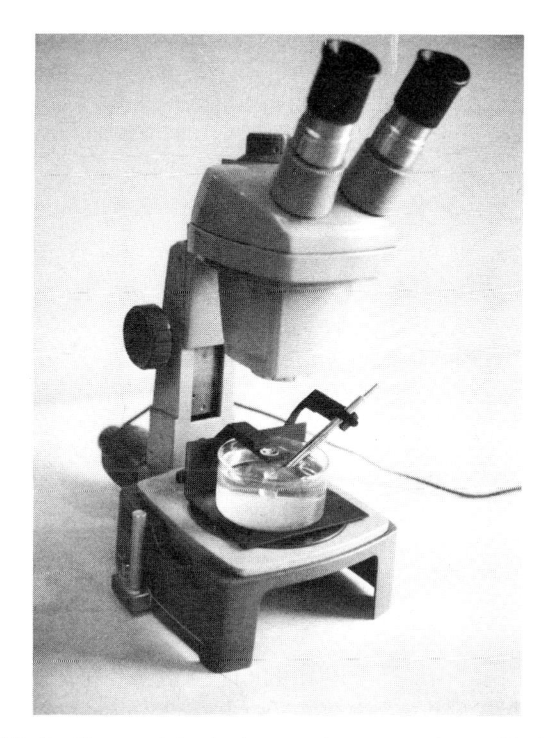

FIG. 2. The crystal orienter on microscope stage with the gemstone immersed in liquid and the accessory lens swung into position to observe interference figure.

* C. S. Hurlbut, 1975. A Device for Obtaining Interference Figures for Gemstones. *Gems and Gemology*, Vol. 15, pp. 66–71. The description and use of the crystal orienter given here are in part taken from this article and are reprinted with permission of the publisher.

the position at which the tweezers are supported, one can place the stone so that the two axes of rotation intersect at its center. This point of intersection is directly above a hole in the base of the instrument that permits light to pass through from the substage illuminator. When in use, a vessel containing a liquid of relatively high refractive index is placed on the base so that the stone can be examined while immersed in the liquid (Fig. 2). Immersion minimizes the refraction of light from facets, permitting one to see into the stone, and is essential in obtaining an interference figure.

Ideally the immersion liquid and the gemstone should have the same refractive index. However, this is not necessary and any of several liquids can be used. It is desirable to use a single liquid for, unless great care is used, contamination results in changing from one liquid to another. It has been found that monobromonaphthalene (**R.I.** 1.66) produces good results with stones having refractive indices of 1.45–1.95.

As we have seen, converging polarized light is necessary to obtain an interference figure. In using the crystal orienter, the effect is produced by placing a sheet of ground glass between the immersion vessel and the microscope stage. The glass scatters the light so it moves through the stone in all directions and not as parallel rays.

Orientation Procedure

Uniaxial crystals have one optic axis, that is, one direction in which light passes through them as through an isotropic substance; there is no birefringence. Biaxial crystals have two such directions. To obtain what is called a centered optic axis figure (either uniaxial or biaxial), the crystal must be rotated so that an optic axis is parallel to the direction in which light passes through the microscope. This desired position can be obtained quickly and easily using the crystal orienter.

When an anisotropic gemstone held in the orienter is viewed through the microscope, the probability is that it will appear uniformly bright against a dark field.* A search for an optic axis is made as follows: rotate the stone slowly 360° about the tweezer axis with the tweezers as nearly horizontal as the immersion vessel permits. If this rotation does not show proximity of an optic axis (how this is noted is described later), rotate upward at intervals of about 10° on the horizontal axis, each time rotating 360° about the tweezer axis. If an optic axis or the proximity to one has not thus been located, the stone must be reoriented in the tweezers and the above procedure repeated. If possible, the tweezer axis should make on reorientation an angle of about 45° with its initial position. The first and second positions should not be at right angles.

Since the rotation possible on the horizontal axis is only slightly more than 45°, there is a 50–50 chance that the first position in the tweezers will yield an interference figure. This is true with a randomly cut stone. However, many minerals produce the most pleasing gems when the table is either parallel to or perpendicular to the optic axis. Therefore, the chances of obtaining an interference figure with the first setting are considerably increased if the tweezer axis makes an angle of about 45° with the table.

Polarized light entering an anisotropic crystal is separated into two rays which interfere when they reach the analyzer, and certain wavelengths are eliminated. The thicker the crystal, the higher its birefringence, and the greater the angular deviation of the light path from an optic axis, the more wavelengths are eliminated but the whiter the crystal appears. The absolute birefringence of a gemstone is, of course, constant no matter what the orientation. However, since the stone is not of uniform thickness, the distance light travels through it varies greatly from edge to center. In orienting a stone one should observe the thinnest edges. Here may be seen interference colors, color bands, parallel to the edge as the stone is turned. The colors indicate that an optic axis has been brought more nearly parallel to the light path. Continued turning on either axis that results in broadening of the color bands is indication that an optic axis is approaching parallelism with the light path. Narrowing of the bands indicates a departure from parallelism. In Figure 3*a* color bands can be seen at the edge of a brilliant cut beryl; Figure 3*b* shows a broadening of the

* These conditions offer an excellent opportunity to examine the interior of the stone for imperfections, inclusions, etc.

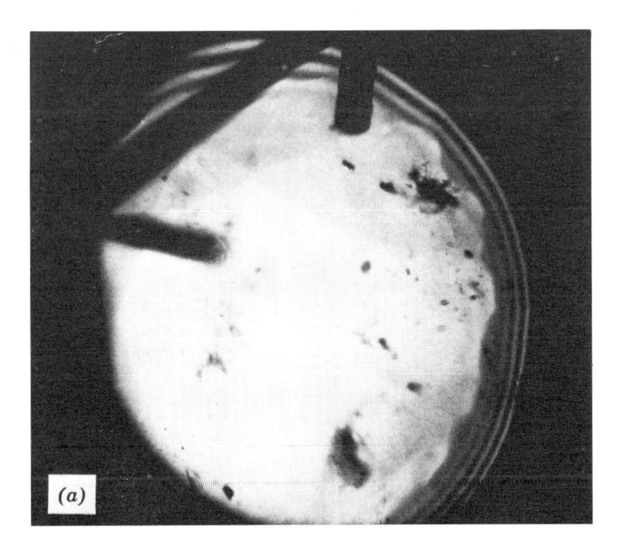

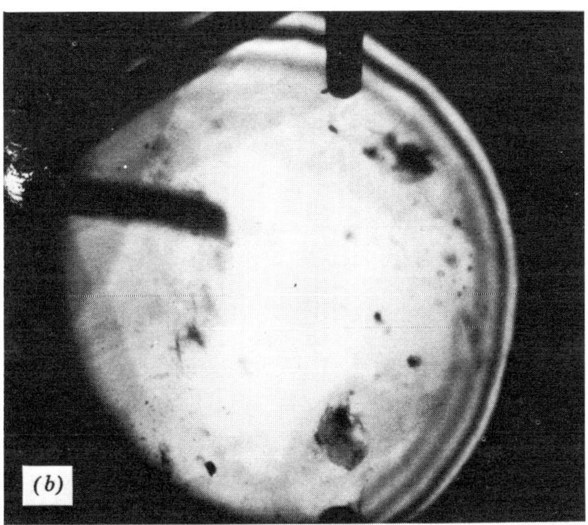

FIG. 3. A brilliant cut beryl held in three-prong tweezers and observed through the microscope between crossed polars. (*a*) The stone has been brought into position so that color bands appear on the thin edges. (*b*) Broadening of the color bands after the stone has been rotated 15°.

bands after the stone has been rotated 15°. When it is impossible to further broaden the bands by rotation, an optic axis is nearly vertical. This is the position desired for obtaining an interference figure.

Gems of low birefringence are easier to orient than those of high birefringence. Apatite (biref. 0.003) shows interference color bands when the optic axis is as much as 30–40° from its desired position, whereas in zircon (biref. 0.06) the axis must be within 10–15° of the vertical before color bands appear.

The Interference Figure

To observe the interference figure a small accessory lens on the crystal orienter is swung into position above the stone. The figure appears on the upper part of the lens and the microscope tube should be raised to bring it into sharper focus. If it is not centered, a slight rotation on one or both of the axes of the orienter will bring it to a centered position. Since with a binocular microscope the stone is viewed from two different angles, it is impossible to obtain a perfectly centered interference figure simultaniously from both halves of the optical system. One should, therefore, make the final observations using only one ocular.

A uniaxial optic axis interference figure (Figs. 4*a*,*b*) is composed of concentric circles of interference colors superimposed on a black cross. The biaxial optic axis interference figure (Fig. 6*b*) is made up of curved bands of interference colors crossed by a single dark brush which may be curved or straight. Figure 5 is an acute bisectrix figure showing the emergence of the two optic axes. If such a figure is obtained, the gem should be rotated to bring one optic axis in the center of the field.

The Optic Sign

Coupled with other observations the determination of the uniaxial or biaxial character of a gemstone may be sufficient to characterize it, and further data may be unnecessary. However, using an optic axis interference figure it is possible to go one step more and determine the optic sign. This can be obtained using a quartz wedge, an accessory available from any manufacturer of polarizing microscopes.

In a uniaxial interference figure the bars of the black cross will be N–S and E–W, parallel to the vibration directions of the polars. To determine the sign, slowly pass the quartz wedge over the accessory lens at an angle of 45° to the bars of the black cross (Fig. 6*a*). It is necessary that the thin edge of the wedge (always at the opposite end from the handle) be the leading edge in making the obser-

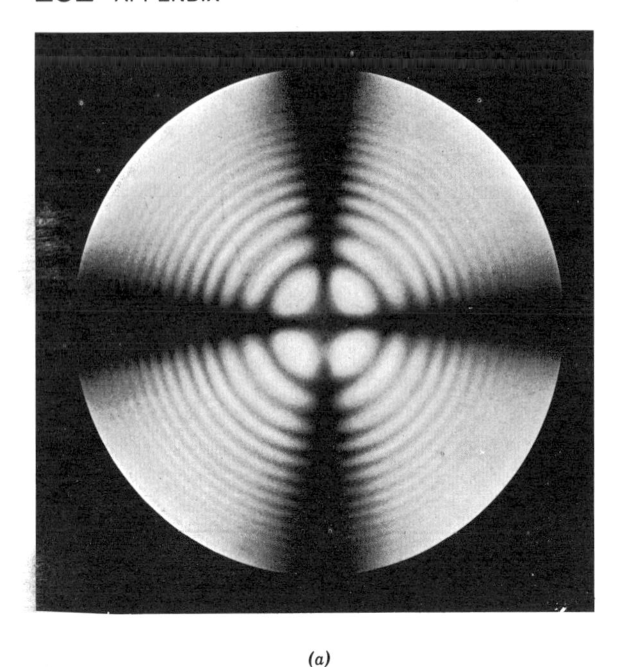

(a)

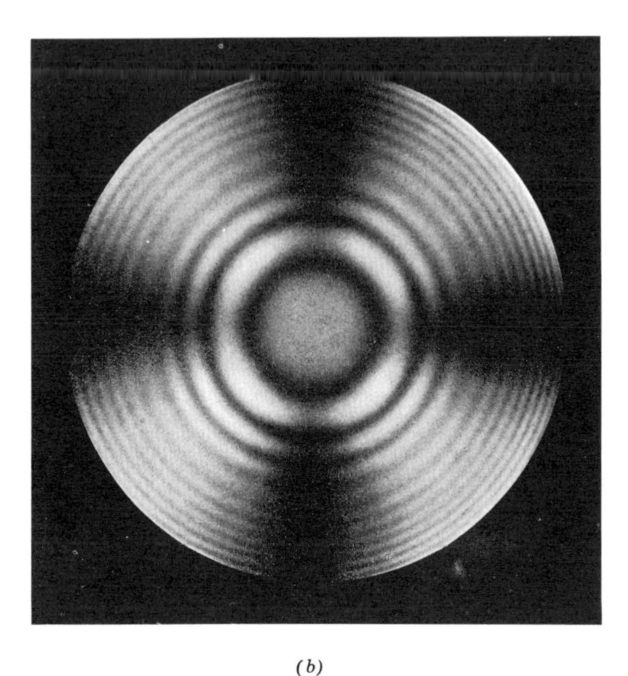

(b)

FIG. 4. Uniaxial optic axis interference figures. (*a*) Normal figure with black cross superimposed on circular color bands, the type that would be given by corundum, beryl, and most other uniaxial crystals. (*b*) The figure given by quartz. Note the center of the black cross is missing. This is the result of the rotary polarization of light and the figure is characteristic for the mineral.

vations. This causes the circles of interference colors to move. In two opposite quadrants they move toward the center of the figure and in the other two quadrants they move away from the center. If the movement toward the center is in the quadrants at right angles to the long dimension of the quartz wedge, the crystal is positive. If the movement of the bands is toward the center in the quadrants parallel to the long dimension of the wedge, the crystal is negative.

Determination of the optic sign of a biaxial crystal is not as straightforward as that of a uniaxial crystal. Before it can be accomplished, the single dark band of the biaxial optic axis figure must have its long dimension essentially at 45° to the vibration directions of the polars. That is, it must be in a NW–SE or NE–SW position (see Fig. 6*b*). This is also the position of the maximum curvature of the bar. On initial observation of the figure, the dark band may be in this position. If it is not the crystal (i.e., the whole crystal orienter) must be turned, perhaps as much as 45°, to accomplish it. In this orientation, known as the 45° position, the bar in most instances

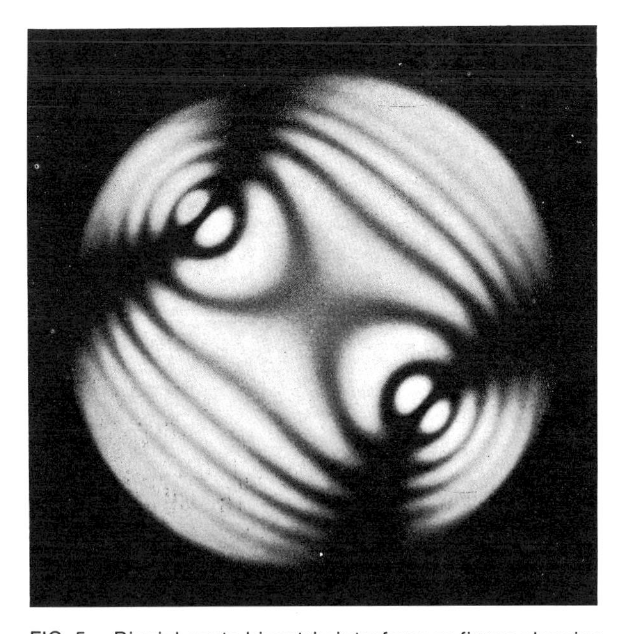

FIG. 5. Biaxial acute bisectrix interference figure showing the emergence of the two optic axes located at the center of the "eyes." If the crystal were rotated about a NE-SW line, a single optic axis figure would be seen in the center of the field as an "eye" crossed by a dark brush.

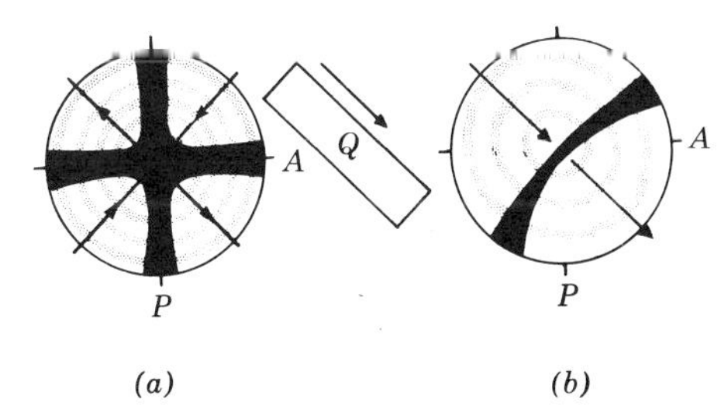

FIG. 6. Diagrams of optic axes interference figures. (*a*) Uniaxial. (*b*) Biaxial. The arrows indicate the movement of interference colors (stippled bands) in positive crystals when the quartz wedge (*Q*) is passed over the interference figures in the direction indicated. In negative crystals the color bands would move in the opposite directions. *A* is the vibration direction of the analyzer, *P* the vibration direction of the polarizer.

will be curved. The optic sign is then determined by observing the movement of the color bands as the quartz wedge is passed over the accessory lens. The wedge must be moved so that its thin edge approaches the convex side of the black bar. If the color bands move toward the convex side and outward from the concave side, the crystal is positive (Fig. 6*b*). In a negative crystal the color bands move in the opposite directions.

Some biaxial crystals yield optic axis interference figures in which the black bar remains essentially straight in all positions as the crystal orienter is turned. In these crystals the optic angle (2*V*), the angle between the two optic axes is 90°, or nearly 90°, and it is impossible to determine the optic sign. However, knowing that the optic angle is essentially 90° is in itself useful information.

Notable Diamonds

The largest diamond ever found was the Cullinan, weight 3106 carats, from South Africa. A mass of carbonado from Brazil was nearly as large, 3078 carats.* The *1971 International Diamond Annual* states that 2527 rough diamonds of 100 carats or more were found in South Africa between 1870 and 1970.

The largest polished diamond is the Cullinan I, weighing 530.20 carats, followed in size by the 317.40 carat Cullinan II. Both these incomparable gems are among the British crown jewels.

The following lists of notable rough and polished diamonds are taken from *Notable Diamonds of the World,* compiled by N. W. Ayer and Son, Inc., New York. Only rough diamonds of more than 400 carats are given here. In the list of polished diamonds are given all stones of more than 150 carats, plus a selection of other notable or historic stones of smaller size. For more comprehensive lists the reader is referred to the publication mentioned above.

* G. F. Herbert-Smith, in *Gemstones* (10th ed., 1945, Pitman Publishing Corporation, London), gives the weight of this stone as 3148 metric carats, larger than the Cullinan.

Birthstones of the Month

Birthstones, or gems suitable to the birth month, have been used since about 1562. The choice of gems is related to the twelve stones of the breastplate of the high priest (Exodus, Chapter 28) and the twelve foundation stones (Revelation, Chapter 21). The list has varied from time to time. The one now used by the Retail Jewelers Association and the American Gem Society is given below. It is the same as the official list issued by the National Association of Goldsmiths of Great Britain and Ireland, except in the latter no alternate stones are given.

January
  Garnet
February
  Amethyst
March
  Aquamarine or bloodstone
April
  Diamond
May
  Emerald
June
  Pearl, moonstone, or alexandrite
July
  Ruby
August
  Peridot or sardonyx
September
  Sapphire
October
  Opal or tourmaline
November
  Topaz or citrine
December
  Turquoise or zircon

ROUGH DIAMONDS

Carat Weight	Color	Name	Discovery Date	Discovery Country
3106	White	Cullinan	1905	South Africa
995.20	White	Excelsior	1893	South Africa
968.80	White	Star of Sierra Leone	1972	Sierra Leone
787.50	White	Great Mogul	1650	India
770.00	White	Woyie River	1945	Sierra Leone
726.60	White	Presidente Vargas	1938	Brazil
726.60	White	Jonker	1934	South Africa
650.80	White	Reitz	1895	South Africa
620.14	White	Unnamed	Unk.	South Africa
609.25	White	Baumgold Rough	1922	South Africa
601.25	Brownish	Lesotho	1967	Lesotho
600.00	White	Goyas	1906	Brazil
600.00	White	Unnamed (Jagersfontein)	1884	South Africa
593.50	White	Unnamed (Preimer)	1919	South Africa
572.25	White	Unnamed (Jagersfontein)	1955	South Africa
565.75	White	Unnamed Jagersfontein)	1912	South Africa
537.00	White	Unnamed (Jagersfontein)	Unk.	South Africa
532.00	Unknown	Unnamed (Woyie River)	1943	Sierra Leone
527.00	White	Lesotho B	1965	Lesotho
523.74	White	Unnamed (Premier)	1907	South Africa
514.00	White	Unnamed (Premier)	1911	South Africa
511.25	Yellowish	Venter	1951	South Africa
507.00	White	Unnamed (Jagersfontein)	1914	South Africa
503.50	White	Unnamed (Kimberley)	1896	South Africa
490.00	Yellow	Baumgold II	1941	South Africa
487.25	White	(Unnamed (Premier)	1905	South Africa
469.00	White	Victoria 1884	1884	South Africa
458.75	White	Unnamed (Premier)	1907	South Africa
458.00	White	Unnamed (Premier)	1913	South Africa
455.00	Brown	Darcy Vargas	1939	Brazil
444.00	White	Unnamed (Premier)	1926	South Africa
442.25	White	Unnamed (Dutoitspan)	1917	South Africa
440.00	White	Nizam	1835	India
435.00	White	Light of Peace	1969	Sierra Leone
430.50	White	Unnamed (Jagersfontein)	1913	South Africa
428.50	Yellowish	Victoria 1880	1880	South Africa
428.50	Yellowish	De Beers	1888	South Africa
427.50	White	Unnamed (De Beers)	1913	South Africa
426.50	White	Ice Queen	1954	South Africa
419.00	White	Unnamed (Jagersfontein)	1913	South Africa
416.25	Brown	Berglen	1924	South Africa
412.50	Unknown	Broderick	1928	South Africa
410.00	White	Pitt	1701	India
409.00	White	Unnamed (Premier)	1913	South Africa
407.68	White	President Dutra	1949	Brazil
407.50	White	Unnamed (Jagersfontein)	1926	South Africa
400.65	Unknown	Coromandel IV	1948	Brazil
400.00	White	Unnamed (De Beers)	1891	South Africa

POLISHED DIAMONDS

Carat Weight	Color	Name	Country of Origin	Last Reported Owner	Date of Report
530.20	White	Cullinan I	S. Africa	British crown jewels, London	1978
317.40	White	Cullinan II	S. Africa	British crown jewels, London	1978
280.00	White	Great Mogul	India	Lost after sack of Delhi in 1739	
277.00	White	Nizam	India	Nizam of Hyderabad	1934
250.00	Pink	Great Table	India	Cut into Darya-i-Nur and Nur-ul-Ain	1966
250.00	White	Indien	India	Unknown; listed by Duke of Brunswick in 1869	
245.35	White	Jubilee	S. Africa	Paul-Louis Weiller, Paris	1971
234.50	Yellow	De Beers	S. Africa	Unknown; sold to Indian prince about 1890	
228.50	Yellow	Victoria 1880	S. Africa	Unknown; sold to Indian prince about 1890	
205.00	Canary	Red Cross	S. Africa	Unknown; sold in London in 1918	
202.00	Black	Black Star of Africa	S. Africa	Unknown; exhibited in Tokyo in 1971	
189.60	White	Orloff	India	Russian Diamond Fund, Kremlin	1968
185.00	Pink	Darya-i-Nur (Iran)	India	Iranian crown jewels, Teheran	1968
184.50	White	Victoria 1884	S. Africa	Unknown; reported sold to Nizam of Hyderabad	1900
183.00	Yellowish	Moon	S. Africa	Unknown; sold at Southby's in 1942	
152.16	Cape	Iranian I	S. Africa	Iranian crown jewels, Teheran	1968
150.00	White	Light of Peace	S. Leone	Zale Corporation, Dallas	1971
150.00	White	Darya-i-Nur (Dacca)	India	Unknown; offered for sale in Dacca in 1959	
140.50	White	Regent	India	Louvre Museum, Paris	1978
137.27	Yellow	Florentine	India	Stolen from Austrian royal family about 1920	
128.51	Yellow	Tiffany	S. Africa	Tiffany and Co., New York	1978
127.02	White	Portuguese	Brazil	Smithsonian Institution, Washington, D.C.	1978
125.65	White	Jonker	S. Africa	Queen of Nepal	1959
108.83	White	Koh-i-noor	India	British crown jewels, London	1978
104.15	Bronze	Great Chrysanthemum	S. Africa	Julius Cohen, New York	1971
94.80	White	Star of the East	India	Unknown since exile of King Farouk in 1952	
94.40	White	Cullinan III	S. Africa	British crown jewels, London	1978
88.70	White	Shah	India	Russian Diamond Fund, Kremlin	1968
84.00	White	Spoonmaker's	India	Topkapi Museum, Istanbul	1966
71.73	White	Lesotho	Lesotho	Private owner; sold by Harry Winston	1971
70.20	White	Idol's Eye	India	Harry Levinson, Chicago	1971
69.42	White	Cartier	S. Africa	Elizabeth Taylor	1970
67.89	Champagne	Transvaal	S. Africa	Smithsonian Institution, Washington, D.C.	1978
60.00	Pink	Nur-ul-Ain	India	Iranian crown jewels, Teheran	1969

POLISHED DIAMONDS (*Continued*)

Carat Weight	Color	Name	Country of Origin	Last Reported Owner	Date of Report
55.00	White	Sancy	India	Astor family in England	1966
45.52	Blue	Hope	India	Smithsonian Institution, Washington, D.C.	1978
43.38	White	Nassak	India	Edward J. Hand, Greenwich, Conn.	
41.00	Green	Dresden Green	India	Dresden Historical Museum	1971
35.32	Blue	Wittelsbach	India	Private owner; sold by I. Konkommer	1966
31.00	Blue	Eugenie Blue	India	Smithsonian Institution, Washington, D.C.	1978
23.60	Pink	Willamson	Tanzania	Queen Elizabeth II	1971
12.42	White	Uncle Sam	U.S.A.	Private owner; sold by Sidney De Young	1971
11.15	White	Dewey	U.S.A.	Unknown since death of John Morrissey in 1898	
8.27	White	Star of Arkansas	U.S.A.	Private owner	1968

INDEX

GEM INDEX

In this index the gem name, page reference, and chemical composition are followed by commonly sought information. (1) Crystal system (Xl Sys.). (2) Optical category, that is, isotropic (I), uniaxial (U), or biaxial (B). For uniaxial and biaxial the U or B is followed by the optic sign (+) or (−). Below the optic sign the highest and the lowest refractive indices (Av. **R.I.**) are given, not the extreme range. (3) The average birefringence (Av. Biref.). (4) Range of specific gravity (**S.G.**). (5) Range of hardness (**H**). (6) Color. The colors are indicated by letters as follows: C, colorless; W, white; P, pink; R, red; O, organge; Y, yellow; G, green; B, blue; V, violet; Br, brown; Gr, gray; Bl, black. (7) Remarks. A few words are given for each entry which, for the sake of brevity, may be abbreviated; for example, var = variety, cl = cleavage.

Name, Page, Composition	Xl Sys.	Op. Sign, Av. R.I.	Av. Biref.	S.G.	H	Color	Remarks
Achroite, 148							Colorless tourmaline
Actinolite, 202 $Ca_2(Mg,Fe)_5Si_8O_{22}(OH)_2$	Mon	B(−) 1.615 1.641	0.026	3.1–3.3	5–6	GBl	The amphibole of nephrite
Adularia, 166 $KAlSi_3O_8$	Mon	B(−) 1.518 1.526	0.008	2.56	6	C	A transparent K feldspar
Agalmatolite, 143					2–2½	YGBrGr	Massive mica
Agate, 153 SiO_2				2.6	7		Concentric layers of chalcedony
Alabaster, 180				2.32	2	WYBrGr	Massive gypsum
Albite, 168 $NaAlSi_3O_8$	Tri	B(+) 1.525 1.536	0.011	2.62–2.64	6	CWPB	Na end member of plagioclase series
Alexandrite, 143							Var of chrysoberyl
Almandine, 158 $Fe_3Al_2(SiO_4)_3$	Iso	I 1.79		3.93–4.17	7	RVBr	A garnet
Amazonite, 166						G	Microcline
Amber, 211 Organic resins	Amor	I 1.54		1.05–1.09	2½–3	ROYGBr	Fossil resin
Amblygonite, 201 $LiAlFPO_4$	Tri	B(−) 1.59 1.61	0.02	3.0–3.1	6	CWY	2 good cleavages
Amethyst, 151						V	Violet quartz
Amphibole, 201							A mineral group
Anatase, 202 TiO_2	Tet	U(−) 2.488 2.561	0.073	3.9–4.0	5½–6	GBVBr	A rare gem
Andalusite, 169 Al_2SiO_5	Orth	B(−) 1.632 1.643	0.010	3.16–3.20	7½	RYGBr	Strong pleochroism
Andradite, 159 $Ca_3Fe_2(SiO_4)_3$	Iso	I 1.875	—	3.81–3.87	6½	YGBl	High dispersion; a garnet
Anhydrite, 181 $CaSO_4$	Orth	B(+) 1.570 1.614	0.034	2.89–2.98	3–3½	WBVGr	Usually massive
Apatite, 170 $Ca_5(PO_4)_3F$	Hex	U(−) 1.642 1.646	0.003	3.15–3.20	5	CYGBVBr	Low birefringence
Apophyllite, 202 $KCa_4(Si_4O_{10})_2F \cdot 8H_2O$	Tet	U(+) 1.535 1.537	0.003	2.3–2.4	4½–5	CWPYG	One good cleavage
Aquamarine, 134							Blue-green beryl
Aragonite, 171 $CaCO_3$	Orth	B(−) 1.530 1.686	0.156	2.95	3½–4	CY	Effervesces in HCl
Augelite, 202 $Al_2(PO_4)(OH)_3$	Mon	B(+) 1.574 1.588	0.014	2.70	4½–5	C	A rare gem
Australite, 201							Australian tektite
Aventurine, 153, 168							Oligoclase, quartz
Axinite, 171 CaFeMnAl borosilicate	Tri	B(−) 1.675 1.685	0.010	3.27–3.35	7	YBBr	Strong pleochroism in brown, yellow, green
Azurite, 186 $Cu_3(CO_3)_2(OH)_2$	Mon	B(+) 1.730 1.838	0.108	3.77	3½–4	B	Associated with malachite
Ballas, 125							Diamond aggregate
Ballas ruby, 163						R	See spinel
Benitoite, 172 $BaTiSi_3O_9$	Hex	U(+) 1.757 1.804	0.047	3.64	6½	CB	High disp., 0.044
Beryl, 134 $Be_3Al_2Si_6O_{18}$	Hex	U(−) 1.578 1.586	0.007	2.67–2.82	7½–8	CPYGB	1 poor cleavage
Beryllonite, 202 $NaBePO_4$	Mon	B(−) 1.552 1.561	0.009	2.81	5½	C	A rare gem

Name, Page, Composition	Xl Sys.	Op. Sign, Av. R.I.	Av. Biref.	S.G.	H	Color	Remarks
Billetonite, 201	. . .	. . .	. . .	. . .	. . .	. . .	A tektite
Bloodstone, 154	. . .	. . .	. . .	. . .	. . .	RG	Microgranular quartz; green with red spots
Bort, 125	. . .	. . .	. . .	. . .	. . .	. . .	Nongem diamond
Bowenite, 143, 191	. . .	. . .	. . .	. . .	. . .	YG	Translucent serpentine; jade substitute
Brazilianite, 172 $NaAl_3(PO_4)_2(OH)_4$	Mon	B(+) 1.602 1.621	0.019	2.98	$5\frac{1}{2}$	CWY	One good cleavage
Bronzite, 177 $(Mg, Fe)SiO_3$	Orth	B(−) 1.679 1.690	0.010	3.3–3.4	$5\frac{1}{2}$–6	YBr	A pyroxene; 2 cls at nearly rt angles
Calcite, 173 $CaCO_3$	Hex	U(−) 1.486 1.658	0.172	2.72	3	CWPRO YGBGr	Effervesces in HCl; 3 directions of cl
Californite, 182	. . .	. . .	. . .	. . .	. . .	G	Massive green idocrase
Carbonado, 125	. . .	. . .	. . .	. . .	. . .	GrBl	Industrial diamond
Carbuncle, 158							Garnet
Carnelian, 153	. . .	. . .	. . .	. . .	. . .	. . .	Orange-red chalcedony
Cassiterite, 202 SnO_2	Tet	U(+) 2.0 2.1	0.010	6.9–7.1	6–7	CYBrBl	Adamantine luster; high disp.
Cat's eye, 143	. . .	. . .	. . .	. . .	. . .	YGBr	Chrysoberyl; chatoyant
Chalcedony, 153 SiO_2	Hex	Agg 1.537		2.58–2.62	7	All colors	Microcrystalline quartz
Chert, 154	. . .	. . .	. . .	. . .	. . .	WGr	Microgranular quartz
Chiastolite, 169	. . .	. . .	. . .	. . .	. . .	. . .	See andalusite
Chlorastrolite, 203 Mineral mixture		1.7	. .	3.1–3.3	5–6	WG	Mottled
Chlorite, 143	. . .	. . .	. . .	. . .	. . .	G	Mica-like mineral; see jade
Chloromelanite, 142	. . .	. . .	. . .	. . .	. . .	GB	Var of jadeite
Chrysoberyl, 143 $BeAl_2O_4$	Orth	B(+) 1.745 1.755	0.010	2.71–2.75	$8\frac{1}{2}$	RYGBr	Vars: cat's eye, alexandrite
Chrysocolla, 174 $CuSiO_3 \cdot nH_2O$	Amor	I 1.50	—	2.0–2.4	2–4	GBBrBl	Usually mixed with quartz
Chrysoprase, 153	. . .	. . .	. . .	. . .	. . .	. . .	Green chalcedony
Citrine, 152	. . .	. . .	. . .	. . .	. . .	YBr	Var. quartz
Copal, 212	. . .	. . .	. . .	. . .	. . .	. . .	See amber
Coral, 213 Essentially $CaCO_3$	Hex	U(−) 1.49 1.66	0.17	2.6–2.7	$3\frac{1}{2}$–4	PRBl	Effervesces in HCl
Cordierite, 182 $(Mg, Fe)_2Al_4Si_5O_{18} \cdot nH_2O$	Orth	B(−) 1.545 1.550	0.01	2.58–2.66	7–$7\frac{1}{2}$	YGBV	Strong pleochroism in yellow, blue, violet
Corundum, 137 Al_2O_3	Hex	U(−) 1.762 1.770	0.008	4.02	9	CPRYG VGr	Vars: sapphire and ruby
Cymophane	. . .	. . .	. . .	. . .	. . .	. . .	See chrysoberyl
Danburite, 175 $CaB_2Si_2O_8$	Orth	B(+)(−) 1.630 1.636	0.006	3.0	7	CPY	A rare gem
Datolite, 203 $CaB(SiO_4)(OH)$	Mon	B(−) 1.624 1.668	0.044	2.8–3.0	5–$5\frac{1}{2}$	WYGR	A rare collector's gem

Name, Page, Composition	Xl Sys.	Op. Sign, Av. R.I.	Av. Biref.	S.G.	H	Color	Remarks
Demantoid, 159						G	Green andradite
Diamond, 125 C	Iso	I 2.417	—	3.52	10	CPYG BBrBl	High disp. and **R.I.**
Dichroite, 182							Iolite synonym
Diopside, 175 $CaMgSi_2O_6$	Mon	B(+) 1.67 1.70	0.03	3.2–3.3	5–6	YGBV	A pyroxene with 2 cls at 87° and 93°
Dioptase, 176 $CuSiO_3 \cdot H_2O$	Hex	U(+) 1.651 1.703	0.053	3.3	5	G	High disp. 0.036
Dravite, 148						BrBl	Na, Mg tourmaline
Ekanite, 203 $(Th,U)(Ca,Fe,Pb)Si_8O_{20}$	Amor	I 1.60	—	3.28	6–6½	G	May show 4-rayed star
Elbaite, 148						CPRYGB	Na, Li, Al tourmaline
Emerald, natural, 134						G	Deep green beryl
Emerald, synthetic, 215							
Enstatite, 177 $MgSiO_3$	Orth	B(+) 1.658 1.666	0.010	3.2–3.5	5½	YGBrGr	A pyroxene; 2 cls at 87° and 93°
Epidote, 178 $Ca(Al, Fe)Al_2Si_3O_{12}(OH)$	Mon	B(−) 1.733 1.784	0.03	3.30–3.45	6–7	YGBl	2 good cls at 65° and 115°
Essonite, 158						OYBr	Var of grossular
Euclase, 178 $BeAlSiO_4(OH)$	Mon	B(+) 1.652 1.671	0.019	3.05–3.10	7½	GB	1 perfect cl
Fairy stone, 197							Staurolite twin
Fayalite, 162							Iron-rich olivine
Feldspar, 165							A mineral group
Fibrolite, 192							See sillimanite
Flint, 154						BrBl	Microgranular quartz
Fluorite, 179 CaF_2	Iso	I 1.433	—	3.18	4	All colors	Good octahedral cleavage
Forsterite, 162						YGBr	Magnesium-rich olivine
Fuchsite, 143						G	Chrome mica; see jade
Gahnite, 163 $(Zn, Fe)Al_2O_4$	Iso	I 1.80	—	4.4–4.5	7½–8	G	Zinc spinel
Garnet, 156 $A_3B_2(SiO_4)_3$	Iso	I 1.74–1.88	—	3.5–4.3	6½–	WPROY GBrBl	A mineral group
GGG, 217 $Gd_3Ga_5O_{12}$	Iso	I 2.030	—	7.05	6½	C	Synthetic; high disp. 0.038
Glass, 103, 199 Various comp.	Amor	I 1.48–1.7+	—	2.2–4.5	5–6	All colors	Data for glass; imitation gems
Goldstone, 104						RBr	Glass imitation of sunstone
Goshenite, 134						C	Colorless beryl
Grossular, 158 $Ca_3Al_2(SiO_4)_3$	Iso	I 1.742	—	3.61–3.75	7	ROYG VBr	A garnet
Gypsum, 180 $CaSO_4 \cdot 2H_2O$	Mon	B(+) 1.520 1.530	0.010	2.32	2	CWP RBr	Vars: satin spar and alabaster
Hambergite, 203 $Be_2BO_3(OH)$	Orth	B(+) 1.554 1.628	0.074	2.36	7½	C	A rare gem
Heliodor, 134						Y	Golden beryl
Heliotrope, 154						RG	Bloodstone synonym
Hematite, 181 Fe_2O_3	Hex			5.26	5½–6½	RBl	Opaque, red streak

Name, Page, Composition	Xl Sys.	Op. Sign, Av. R.I.	Av. Biref.	S.G.	H	Color	Remarks
Hessonite, 158	. . .	. . .	. . .	. . .	. .	OYBr	Var of grossular
Hexagonite, 202	. . .	. . .	. . .	. . .	. .	PV	See tremolite
Hiddenite, 196	. . .	. . .	. . .	. . .	. .	G	Green spodumene
Hyalite, 140	. . .	. . .	. . .	. . .	. .	C	Colorless opal
Hypersthene, 177	. . .	. . .	. . .	. . .	. .	YBr	See enstatite, bronzite
Howlite, 203 $Ca_2B_5SiO_9(OH)_5$	Mon	Agg 1.59	—	2.58	3½	W	Fluoresces yellow
Iceland spar, 173	. . .	. . .	. . .	. . .	. .	C	Optical calcite
Idocrase, 182 Ca, Mg, Al silicate	Tet	U(−) 1.725 1.729	0.004	3.35–3.45	6	RYGBBr	Massive green var: californite
Indicolite, 148	. . .	. . .	. . .	. . .	. .	B	Blue tourmaline
Iolite, 182 $(Mg, Fe)_2Al_4Si_5O_{18} \cdot nH_2O$	Orth	B(−) 1.545 1.550	0.010	2.58–2.66	7–7½	YGBV	Pleochroism strong
Ivory, 214 Organic + inorganic	—	I 1.54	—	1.70–1.98	2½	WY	Cryptocrystalline in part
Jade, 141	. . .	. . .	. . .	. . .	. .	. . .	See jadeite, nephrite
Jadeite, 141 $NaAlSi_2O_6$	Mon	Agg 1.66	—	3.3–3.5	6½–7	All colors	Most valued type of jade
Jasper, 154 SiO_2	Hex	Agg 1.54	—	2.6–2.9	7	RG	Granular microcrystalline quartz
Jet, 215 Organic	Amor	I 1.66	—	1.35	2½	Bl	Heated gives burning coal odor
Kornerupine, 183 Mg, Al borosilicate	Orth	B(−) 1.666 1.679	0.013	3.27–3.45	6½	CYGBr	A rare gem
Kyanite, 184 Al_2SiO_5	Tri	B(−) 1.712 1.728	0.016	3.55–3.66	5–7	CGBGr	1 good cleavage
Kunzite, 196	. . .	. . .	. . .	. . .	. .	PRV	See spodumene
Labradorite, 168 $Ab_{50}An_{50}$–$Ab_{30}An_{70}$	Tri	B(+) 1.560 1.568	0.008	2.71–2.74	6	YGBVGr	A plagioclase feldspar
Lapis lazuli, 184 Mineral mixture	—	Agg 1.50	—	2.5–2.9	5–5½	GB	Usually contains pyrite
Larvikite, 166	. . .	. . .	. . .	. . .	. .	BGr	Feldspar rock
Lazulite, 185 $(Mg, Fe)Al_2(PO_4)(OH)_2$	Mon	B(−) 1.612 1.643	0.013	3.0–3.1	5½	CGB	Resembles azure-blue lazurite
Lazurite, 185 Na, Ca, Al silicate	Iso	I 1.50	—	2.40–2.45	5–5½	B	Major blue mineral in lapis lazuli
Lechatelierite, 201 SiO_2	Amor	I 1.46	—	2.2	6–7	C	Natural silica glass
Leucite, 203 $KAlSi_2O_6$	Tet	U(+) 1.508 1.509	0.001	2.47	5½–6	CW	Pseudoisometric
Liddicoatite, 148	. . .	. . .	. . .	. . .	. .	CRYGBBr	Ca, Al, Li tourmaline
Malachite, 186 $Cu_2CO_3(OH)_2$	Mon	B(−) 1.655 1.909	0.254	3.90–4.03	3½	G	Effervesces in HCl
Marcasite, 187	Orth	—	—	4.89	6–6½	Y	See pyrite
Matura "diamond," 161	. . .	. . .	. . .	. . .	. .	. . .	See zircon
Meershaum, 204	. . .	. . .	. . .	. . .	. .	. . .	See sepiolite
Mexican onyx, 174	. . .	. . .	. . .	. . .	. .	. . .	Banded calcite
Microcline, 165 $KAlSi_3O_8$	Tri	B(−) 1.522 1.530	0.008	2.56	6	CWRG	A feldspar; amazonite a green var
Moldavite, 201	Amor	. . .	. . .	. . .	. .	. . .	A natural glass
Moonstone, 166, 168	. . .	. . .	. . .	. . .	. .	CW	Orthoclase, albite

Name, Page, Composition	Xl Sys.	Op. Sign, Av. R.I.	Av. Biref.	S.G.	H	Color	Remarks
Morganite, 134	. . .	. . .	. . .	. . .	. . .	PR	See beryl
Nephrite, 142 $Ca(Mg, Fe)_5Si_8O_{22}(OH)_2$	Mon	B(−) agg about 1.62	—	2.90–3.02	6–6½	YGB	The amphibole var of jade
Obsidian, 199 Natural glass	Amor	I 1.48–1.52	—	2.33–2.50	5–5½	RYGBr Bl	Usually black; conchoidal fracture
Odontolite, 214 Largely apatite	—	Agg 1.57–1.63	—	3.01	5	B	Fossil bone replaced; microcryst
Oligoclase, 168 $Ab_{80}An_{20}–Ab_{60}An_{40}$	Tri	B(+) 1.538 1.547	0.009	2.64–2.68	6	CW	A plagioclase feldspar
Olivine, 161 $(Mg, Fe)_2SiO_4$	Orth	B(+)(−) 1.654 1.690	0.036	3.30–3.38	6½–7	YGBr	Data for gem olivine, peridot
Onyx, 154	. . .	. . .	. . .	. . .	. . .	. . .	Layered chalcedony
Onyx, marble, 174	. . .	. . .	. . .	. . .	. . .	. . .	Banded calcite
Opal, 139 $SiO_2{\cdot}nH_2O$	Amor	I 1.45	—	2.0–2.2	5½–6½	All colors	Internal reflections
Orthoclase, 165 $KAlSi_3O_8$	Mon	B(−) 1.518 1.526	0.008	2.56	6	WCRYG	A feldspar
Painite, 204 $Ca_4BAl_{20}SiO_{38}$	Hex	U(−) 1.788 1.816	0.028	4.0	8	R	Very rare gem; pleochroic
Pargasite, 202 Na, K, Ca, Mg, Fe, Al, Si	Mon	B(−) 1.613 1.635	0.022	3.1	6	GB	Pleochroic; an amphibole
Paste, 103	. . .	. . .	. . .	. . .	. . .	. . .	Glass
Pearl, 206 $CaCO_3$ + organic	—	Agg —	—	2.68–2.75	2½–4½	WPYB GrBl	Effervesces in HCl
Peridot, 161 $(Mg, Fe)_2SiO_4$	Orth	B(+)(−) 1.654 1.690	0.036	3.30–3.38	6½–7	YGBr	2 poor cleavages; gem olivine
Perthite, 166	. . .	. . .	. . .	. . .	. . .	. . .	Feldspar intergrowth
Peristerite, 168	. . .	. . .	. . .	. . .	. . .	. . .	Irisdescent albite
Petalite, 204 $LiAlSi_4O_{10}$	Mon	B(+) 1.505 1.518	0.013	2.4	6–6½	CW	One good cleavage
Phenakite, 187 Be_2SiO_4	Hex	U(+) 1.654 1.670	0.016	2.79–3.00	7½–8	CPRY Br	A rare gem
Plasma, 154	. . .	. . .	. . .	. . .	. . .	G	Fine-grained quartz
Pollucite, 204 $CsAlSi_2O_6{\cdot}H_2O$	Iso	I 1.520	—	2.90	6½	CW	Under u.v. fluoresces orange
Prase, 154	. . .	. . .	. . .	. . .	. . .	G	Green chalcedony
Prehnite, 204 $Ca_2Al_2Si_3O_{10}(OH)_2$	Orth	B(+) 1.616 1.649	0.013	2.8–2.95	6–6½	WYG	Waxy luster
Pseudophite, 143	. . .	. . .	. . .	. . .	. . .	G	Jade substitute
Pumpellyite, 203	. . .	. . .	. . .	. . .	. . .	. . .	Feldspar intergrowth
Pyrite, 187 FeS_2	Iso	. . .	. . .	5.02	6–6½	Y	Opaque; cut stones sold as marcasite
Pyrope, 157 $Mg_3Al_2(SiO_4)_3$	Iso	I 1.740	—	3.65–3.80	7	RVBr	A garnet
Pyrophyllite, 205 $Al_2Si_4O_{10}(OH)_2$	Mon	B(−) 1.55 1.60	0.05	2.65–2.9	1–2	WYBGr	Like talc
Quartz, 149 SiO_2	Hex	U(+) 1.544 1.553	0.009	2.65	7	All colors	Conchoidal fracture
Rhodochrosite, 188 $MnCO_3$	Hex	U(−) 1.579 1.816	0.237	3.5–3.7	3½–4	PRBr	3 cl directions; massive, banded
Rhodolite, 157 $(Mg, Fe)_3Al_2(SiO_4)_3$	Iso	I 1.76	—	3.74–3.94	7	PRV	A garnet

Name, Page, Composition	Xl Sys.	Op. Sign, Av. R.I.	Av. Biref.	S.G.	H	Color	Remarks
Rhodonite, 189 $MnSiO_3$	Tri	B(+) 1.724 1.739	0.013	3.4–3.7	6	PRBr	Black veinlets usually present
Rhodizite, 204 $CsAl_4Be_4B_{11}O_{25}(OH)_4$	Iso	I 1.694	—	3.44	$8\frac{1}{2}$	CRY	A rare gem
Rock crystal, 151						C	Colorless quartz
Rubellite, 148						PR	Tourmaline
Rubicelle, 163						R	See spinel
Ruby, 138						R	Red corundum
Rutile, synthetic, 218 TiO_2	Tet	U(+) 2.616 2.903	0.287	4.24–4.28	$6–6\frac{1}{2}$	CY	High dispersion
Sanidine, 165 $KAlSi_3O_8$	Mon	B(−) 1.522 1.528	0.006	2.56–2.62	7	CW	A feldspar
Saphir d'eau, 182							See iolite
Sapphire, 138							Corundum of all colors except red
Sard, 153						Br-R	Chalcedony
Sardonyx, 154						WRBrBl	White chalcedony layered with sard
Satin spar, 173, 180						W	Fibrous gypsum or calcite
Saussurite, 143 Mineral mixture	—	Agg 1.50–1.70	—	3.0	$6\frac{1}{2}$	G	Jade substitute; albite + zoisite
Scapolite, 189 Na, Ca, Al silicate	Tet	U(−) 1.540 1.568	0.02	2.55–2.74	5–6	PYGBGr	Good cleavage in 4 directions
Scheelite, 190 $CaWO_4$	Tet	U(+) 1.920 1.934	0.014	5.9–6.1	$4\frac{1}{2}–5$	OYGBr	Fluoresces blue
Schorl, 148						Bl	Black tourmaline
Selenite, 181						C	Colorless gypsum
Sepiolite, 204 $Mg_4Si_6O_{15}(OH)_2 + H_2O$	Orth	Agg 1.53	—	2.0	$2–2\frac{1}{2}$	WY	Microcrystalline
Serpentine, 191 $Mg_3Si_2O_5(OH)_4$	Mon	Agg 1.555	—	2.6–2.8	2–5	YGB	Jade substitute
Silicon carbide, 220 SiC	Hex	U(+) 2.65 2.69	0.043	3.17	$9\frac{1}{2}$	GB	Rarely seen as a gem
Sillimanite, 192 Al_2SiO_5	Orth	B(+) 1.657 1.677	0.020	3.23–3.27	6–7	CYGBBr	Common mineral; rare as cut gem
Sinhalite, 192 $MgAlBO_4$	Orth	B(−) 1.668 1.707	0.039	3.47–3.49	$6\frac{1}{2}$	YGBrBl	Resembles peridot
Smaragdite, 202 Near actinolite	Mon	B(−) about 1.65	—	3.3	5–6	G	An amphibole; jade substitute
Smithsonite, 193 $ZnCO_3$	Hex	U(−) 1.623 1.850	0.227	4.30–4.45	$4–4\frac{1}{2}$	WPYG	Usually massive; ornamental mineral
Soapstone, 205							See talc
Sodalite, 193 $Na_4(AlSiO_4)_3Cl$	Iso	I 1.48	—	2.15–2.30	$5\frac{1}{2}–6$	WPGBGr	Usually blue
Spessartine, 158 $Mn_3Al_2(SiO_4)_3$	Iso	I 1.80	—	4.15–4.21	7	OYBr	A garnet
Sphalerite, 194 ZnS	Iso	I 2.37	—	3.9–4.1	$3\frac{1}{2}–4$	CRYG BrBl	Resinous luster; high disp.
Sphene, 195 $CaTiSiO_5$	Mon	B(+) 1.919 2.035	0.105	3.50–3.54	$5–5\frac{1}{2}$	YGBBr	High disp., 0.051
Spinel, 163 $MgAl_2O_4$	Iso	I 1.72	—	3.57–3.72	8	All colors	Gems most commonly red

Name, Page, Composition	Xl Sys.	Op. Sign, Av. R.I.	Av. Biref.	S.G.	H	Color	Remarks
Spodumene, 195 $LiAlSi_2O_6$	Mon	B(+) 1.660 1.676	0.006	3.17–3.19	6½–7	CPYGV	2 good cls
Staurolite, 196 $Fe_2Al_9O_6(SiO_4)_4(O, OH)_2$	Mon	B(+) 1.743 1.757	0.013	3.65–3.75	7–7½	RYBrBl	Cruciform twins; pseudoorthorhombic
Steatite, 205	. . .	. . .	. . .	. . .	. . .	. . .	See talc
Stichtite $Mg_6Cr_2CO_3(OH)_{16}\cdot H_2O$	Hex	B(−) 1.52 1.55	0.030	2.15–2.22	2½	PRV	A rare gem
Strontium titanate, 219 $SrTiO_3$	Iso	I 2.409	—	5.13	5–6	C	Synthetic; high disp., 0.109
Succinite, 211	. . .	. . .	. . .	. . .	. . .	. . .	Baltic amber
Sunstone, 168	. . .	. . .	. . .	. . .	. . .	. . .	Oligoclase with red spangles
Swiss lapis, 185	. . .	. . .	. . .	. . .	. . .	B	Agate dyed blue
Taaffeite, 205 $BeMgAl_4O_8$	Hex	U(−) 1.717 1.721	0.004	3.6	8	V	A rare gem
Talc, 205 $Mg_3Si_4O_{10}(OH)_2$	Mon	B(−) 1.539 1.589	0.050	2.7–2.8	1	WGGr	Used for carvings
Tanzanite, 198	. . .	. . .	. . .	. . .	. . .	B	Gem zoisite
Tektite, 199	. . .	. . .	. . .	. . .	. . .	. . .	A natural glass
Thomsonite, 205 $NaCa_2Al_5Si_5O_{20}\cdot 6H_2O$	Orth	B(+) 1.52 1.53	0.01	2.3–2.4	5	WRYG	Shows "eyes" on cabochon stones
Thulite, 198	. . .	. . .	. . .	. . .	. . .	PR	Massive zoisite
Tiger's eye, 152	. . .	. . .	. . .	. . .	. . .	. . .	Chatoyant quartz
Titanite, 195	. . .	. . .	. . .	. . .	. . .	. . .	See sphene
Topaz, 145 $Al_2SiO_4(F, OH)_2$	Orth	B(+) 1.618 1.627	0.008	3.4–3.6	8	CPROY	One good cleavage
Topazolite, 159	. . .	. . .	. . .	. . .	. . .	Y	Andradite garnet
Tourmaline, 146 Complex borosilicate	Hex	U(−) 1.625 1.640	0.025	3.0–3.2	7–7½	All colors	Colored stones pleochroic
Tremolite, 202 $Ca_2Mg_5Si_8O_{22}(OH)_2$	Mon	B(−) 1.620 1.631	0.031	3.0	5½–6	W	An amphibole; 2 good cls
Turquoise, 155 $CuAl_6(PO_4)_4(OH)_8\cdot 4H_2O$	Tri	Agg 1.62	—	2.6–2.8	5–6	GB	Waxy luster; may be crossed by veinlets
Uvarovite, 159 $Ca_3Cr_2(SiO_4)_3$	Iso	I 1.87	—	3.7–3.9	7½	G	Crystals too small to cut as gems
Unakite, 178	. . .	. . .	. . .	. . .	. . .	. . .	Epidote rock
Uvite, 148	. . .	. . .	. . .	. . .	. . .	Br	Ca, Mg tourmaline
Variscite, 197 $AlPO_4\cdot 2H_2O$	Orth	Agg 1.58	—	2.4–2.6	4–5	WYG	Resembles turquoise
Verd antique, 191	. . .	. . .	. . .	. . .	. . .	YG	Serpentine rock
Verdelite, 148	. . .	. . .	. . .	. . .	. . .	G	Green tourmaline
Verdite, 143 Mineral mixture	—	Agg 1.58	—	2.9	6½	G	Green rock; jade substitute
Vesuvianite, 182	. . .	. . .	. . .	. . .	. . .	. . .	See idocrase
Violan, 175	. . .	. . .	. . .	. . .	. . .	BV	Blue-violet diopside
Willemite, 205 Zn_2SiO_4	Hex	U(+) 1.691 1.719	0.028	3.9–4.2	5½	RYGBr	Fluoresces yellow-green
Williamsite, 191	. . .	. . .	. . .	. . .	. . .	G	Translucent serpentine
YAG, 217 $Y_3Al_5O_{12}$	Iso	I 1.833	—	4.55	8	C	Synthetic; disp. 0.028

Name, Page, Composition	Xl Sys.	Op. Sign, Av. R.I.	Av. Biref.	S.G.	H	Color	Remarks
Zincite, 206 ZnO	Hex	U(+) 2.013 2.029	0.016	5.68	4	R	Rare as a gem
Zircon, 160 $ZrSiO_4$	Tet	U(+) 1.92 1.98	0.06	4.67–4.73	7–7½	CPROY GBBr	High biref.; diamond substitute
Zircon, low, 160 $ZrSiO_4$	Amor	I 1.80	—	3.93–4.07	6–6½	OG	A metamict mineral
Zirconia, cubic, 219 ZrO_2	Iso	I 2.15	—	5.6–6.0	8½	C	Synthetic; high disp. 0.06
Zoisite, 198 $Ca_2Al_3Si_3O_{12}(OH)$	Orth	B(+) 1.693 1.702	0.009	3.36	6½	CRYG	Gems usually blue; thulite var pink